Lecture Notes in Physics

Edited by H. Araki, Kyoto, J. Ehlers, München, K. Hepp, Zürich
R. Kippenhahn, München, H. A. Weidenmüller, Heidelberg
J. Wess, Karlsruhe and J. Zittartz, Köln
Managing Editor: W. Beiglböck

293

Th. Dorfmüller R. Pecora (Eds.)

Rotational Dynamics of Small and Macromolecules

Proceedings of a Workshop
Held at the Zentrum für interdisziplinäre Forschung
Universität Bielefeld, Bielefeld, FRG, April 21–23, 1986

Springer-Verlag
Berlin Heidelberg GmbH

Editors

Th. Dorfmüller
Universität Bielefeld, Fakultät für Chemie
Postfach 8640, D-4800 Bielefeld 1, FRG

R. Pecora
University of Stanford, Department of Chemistry
Stanford, CA 94305, USA

ISBN 978-3-662-13612-6   ISBN 978-3-540-48079-2 (eBook)
DOI 10.1007/978-3-540-48079-2

© Springer-Verlag Berlin Heidelberg 1987
Originally published by Springer-Verlag Berlin Heidelberg New York in 1987
Softcover reprint of the hardcover 1st edition 1987

2153/3140-543210

Preface

Under the auspices of the project "Complex Liquids" of the "Zentrum für interdisziplinäre Forschung" of the University of Bielefeld a workshop on the "Rotational Motion of Small and Macromolecules in Complex Liquids" was held in the period 3 - 5 April 1986. This volume is based on the lectures held at this workshop.

We usually think of the liquid state as being a *condensed disordered* state of matter displaying a *large mobility* as compared to solids. By mobility we mean the macroscopic flow properties as well as molecular mobility, since both are to some extent interconnected. Any definition of the liquid state is far from being unambiguous even for simple molecular liquids, and becomes even more complicated if we consider molecules which are highly anisotropic in shape, for example rod-like molecules, or molecules which interact via anisotropic forces, such as the case where hydrogen bonding occurs, or finally, molecules which display high internal mobility like bulk polymers. Liquids whose properties are significantly influenced by these factors are usualy termed "complex liquids".

In complex, as well as in simple liquids, a high molecular mobility is typical for the liquid state. Actually this statement should be made more specific since in a given liquid we may have degrees of freedom whose dynamic time-ranges differ by several orders of magnitude. Thus in macromolecular systems some of the intermolecular degrees of freedom may lie in the picosecond to nanosecond time-range whereas others, involving for example large-scale cooperative motions, may lie in the time-range of seconds. Also liquid crystals display rather fast rotational motions around the long molecular axis, while rotations of this axis will be cooperative and much slower. Rotational motions, which have time scales differing by orders of magnitude, generally also have a quite different character. In fact rotations can be classified as inertially determined rotations, or as rotations whose rate is determined by random interactions with the environment. The latter case represents a specific type of randomly fluctuating constraint, controlling the rate of molecular reorientation, and which can be quantified by the introduction of an appropriate friction constant thus connecting the microscopic to the mathematically more tractable macroscopic description. We also often describe rotational modes on the molecular level by different more specific types of constraints, like for example entanglements, or by the cooperative motion of several molecules. In view of this variety of rotational modes, the experimental techniques to study them, and the theoretical frames to describe them also display a large variety. Thus while inertia controlled rotation generally lies in the picosecond and subpicosecond time scale - depending on the molecular moment of inertia - viscosity controlled rotation may extend from the picosecond time scale down to macroscopic times, depending on the molecule, the particular degree of freedom considered, and the viscosity of the liquid. In order to study experimentally the whole spectrum of rotational motions we must use several techniques from those presently available. However, despite the fact that almost the entire time-range from the femtosecond to the macroscopic range are covered by appropriate experiments, the information given by all these techniques is usually of a different type, and usually precludes simple extrapolation from one to another. The same situation applies to theory, where the techniques used in dealing with different kinds of motion are quite different. Also molecular dynamics simulation faces similar problems, since a simulation which is commonly used in the picosecond time-range cannot simply be extended to a slower time scale unless adequate algorithms are developed to cope with the discrepancy between the different time scales and the finite capacity of computers.

In preparing this volume for publication we wish to express our gratitude to the ZiF of the University of Bielefeld for providing the financial support for the project "Complex Liquids". We are grateful to the staff of the ZiF, in particular to Ms. Hoffmann who expertly handled the organization of the meeting. We are also very pleased to acknowledge the expert assistance of Mrs. L. Jegerlehner in the preparation of this volume, and for her excellent typing of the manuscript. One of us (R.P) wishes to acknowledge the "Alexander von Humbold-Stiftung" for a "Senior U.S. Scientist Award".

Bielefeld, October 1987   Thomas Dorfmüller
                                                           Robert Pecora

Contents

ROTATIONAL CORRELATION TIMES FOR SMALL MOLECULES IN LIQUIDS

Daniel Kivelson
Department of Chemistry, University of California
Los Angeles, CA 90024 / USA

Introduction

This article is a summary of my talk given at the conference; it includes things I said, things I wanted to say, things that I should have said, and perhaps some things that I should not have said. This presentation has been modified as a result of discussions at the conference, particularly in response to comments by M. Fixman J. Freed, K. Spears and Th. Dorfmüller.

I will focus on the rotational motion of small molecules in liquids. This motion is often "slow" compared to all other relevant molecular motions, which is equivalent to saying that rotations are Markovian or diffusional. In this case the correlation function for a normalized spherical harmonic, $Y_{\ell m}(\cos\theta)$, relaxes exponentially with a decay time τ_ℓ. In broad terms we can identify two classes of exponential rotational relaxation in liquids: one which is diffusional in that the motion takes place via small random angular jumps, and the other in which the molecule rotates quite rapidly but not frequently because most of the time it is held in an oscillatory well which itself loosens up only infrequently [1]. This latter motion corresponds to random large angular jumps; although for this motion the slow decay mode does indeed appear to be exponential, the librational modes are important and give rise to an appreciable high frequency rotational spectrum. It is likely that the large angular jumps become increasingly important at high viscosity. Whether or not the motion is exponential, we can always define a correlation time [2]

$$\tau_\ell = \int_0^\infty dt \ <Y_{\ell m}(t)Y_{\ell m}(0)^*> \tag{1}$$

where $< \ >$ represents an equilibrium average. The study of the functional form of the relaxation process and its description in terms of dynamical processes is an interesting one, but one which we shall consider only perfunctorily.

One of the most striking features of molecular rotations in liquids is the fact that the rotational correlation time obeys a form of Walden's rule, i.e. it is linear in (η/T), where η is the coefficient of shear viscosity and T the temperature [2]. It is not hard to find deviations from this result, but so much success has been achieved with this relationship that it would be profligate to discard it as one of

Supported in part by the National Science Foundation through grant No. CHE-85-01019.

the underpinnings of a theory of molecular rotations in liquids. More precisely, one finds that [3]

$$\tau = A\eta/T + \tau^0 \tag{2}$$

where A and τ^0 are relatively independent of temperature, and possibly of pressure. (See Fig. 1.) There is still considerable uncertainty about the molecular origin of this relation [4], and I shall say a few words about this. There are also cases where the deviations from this expression are considerable, too great to be ignored, and it is these cases which, I believe, are particularly "interesting".

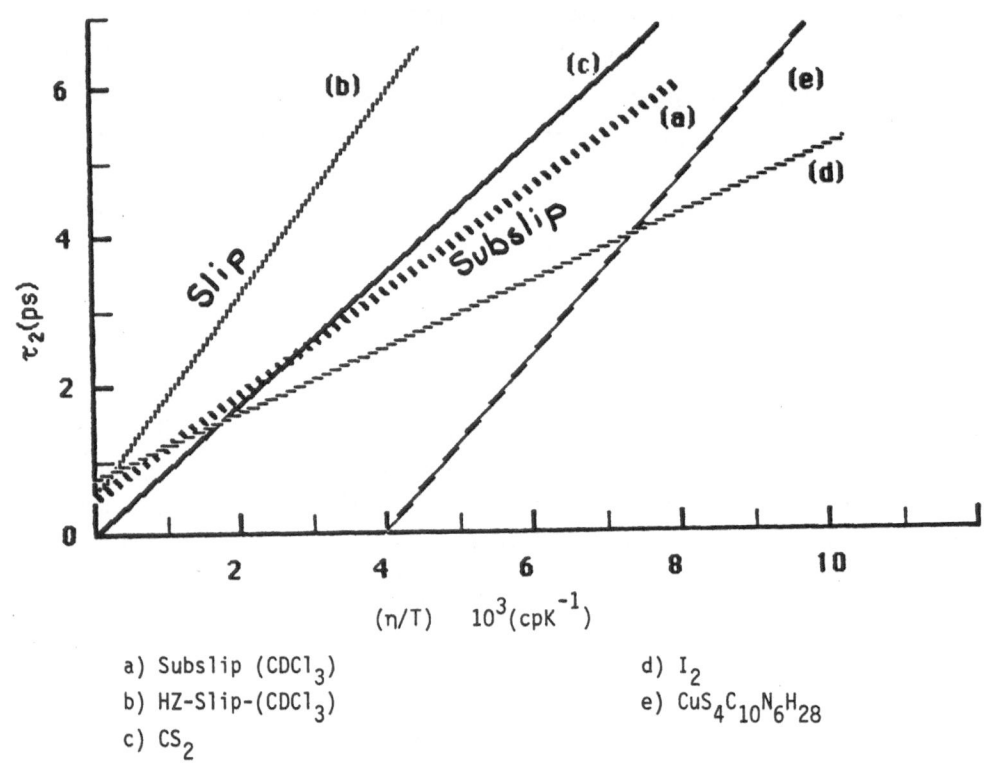

a) Subslip $(CDCl_3)$

b) HZ-Slip-$(CDCl_3)$

c) CS_2

d) I_2

e) $CuS_4C_{10}N_6H_{28}$

<u>Figure 1</u>: τ_2 vs. (η/T). Lines through experimental points, extrapolated to zero η or zero τ_2. See Eq. (2). Plots of experimental points can be found in Reference [2] and in the References given under (iv) below.

(a) $CDCl_3$. Subslip. (i)
(b) Hu-Zwanzig Slip for Spheroid Equivalent to $CDCl_3$. (τ^0 adjusted as in (a).)
(c) CS_2. $\tau^0 = 0$. (ii)
(d) I_2. $\tau^0 > 0$. (iii)
(e) $Cu\{S_2(CN[(CH_3)_2NH]_2\}_2$. $\tau^0 < 0$. (iv)

(i)   Neat. (Vary T). D.L. Vander Hart, J. Chem. Phys. <u>60</u>, 1858 (1974)
(ii) Neat. (Vary T). T.L. Cox, M.R. Battaglia and P.A. Madden, Mol. Phys. <u>38</u>, 1539 (1979). Also R.R. Vold, S.W. Sparks and R.L. Vold, J. Magn. Reson. <u>30</u>, 497 (1978).
(iii) Dilute in alkanes. (Const. T). M.R. Battaglia and P.A. Madden, Mol. Phys. <u>36</u>, 1601 (1978).
(iv) Dilute in Toluene. (Vary T). F.G. Herring and P.S. Phillips, J. Chem. Phys. <u>73</u>, 2603 (1980)

Before proceeding, it seems worthwhile to respond to a question posed by Marshall Fixman; why does one care much about molecular rotations? There is, of course, interest in molecular rotations as a phenomenon in its own right, but perhaps it is true that it has been over-examined and that the new insights developed over the past years have been rather meager. However, I believe that a good understanding of molecular rotations, and in particular of the relationship in Eq. (2), can be useful in identifying other molecular processes in liquids. Molecular rotations can serve as a rather sensitive probe of molecular dynamics and molecular structure, much more so than translational motion. One can, therefore, take the point of view that behavior such as that predicted by Eq. (2) is in some respects "universal" and, if I may risk a controversial statement, "uninteresting", but that deviations from Eq. (2) can yield "interesting" specific information about the system under study. With this as motivation, it is useful to understand Eq. (2) as thoroughly as possible, to understand some of the general principles which underlie it and some of the general phenomena that can lead to deviations; one can then subtract out the "uninteresting universal" aspects of the problem, and if there is an "appreciable" remainder we can associate it with specific "interesting" phenomena such as solvation, hydrogen-bonding, aggregation or other molecular rearrangement. I shall say more about this later.

The Parameter A in Eq. (2)

In hydrodynamic theories of molecular rotations, the τ^0 term in Eq. (2) is missing. The Stokes-Einstein-Debye theory [5] models a rotating molecule as a rotating sphere in a continuous, homogeneous fluid; it is a hydrodynamic theory and it yields $A = v[\ell(\ell+1)k_B]^{-1}$, where v is the molecular volume, k_B the Boltzmann constant, and ℓ the order of the rotational function being studied. Perrin [6] extended the theory to ellipsoidal bodies; the factor A in this theory is augmented by a geometrical factor which is one for a sphere and increases with increasing asphericity [7]. These theories assume stick boundary conditions at the surface of the "tagged" particle; this means that the first layer of fluid sticks to the surface, but this is not equivalent to bona fide solvation because the fluid molecules are infinitesimal in size. (See Fig. 2.) Hu and Zwanzig [8] extended the theory to slip boundary conditions; for a sphere there is then no interaction between the tagged particle and the fluid so that $\tau_\ell = 0$, but rotating non-spherical particles interact by pushing fluid aside. (See Fig. 3.) One can conveniently express A as [5,6]

$$A = v[\ell(\ell+1)]^{-1}fC \qquad (3)$$

where f is a shape factor associated with stick boundary conditions, and C is a factor depending upon shape and boundary conditions. $f \geq 1$, and $0 \leq C \leq 1$.

How should the formula above be used? One can model the molecular shape and size as accurately as possible, and one can then calculate A for both stick and slip

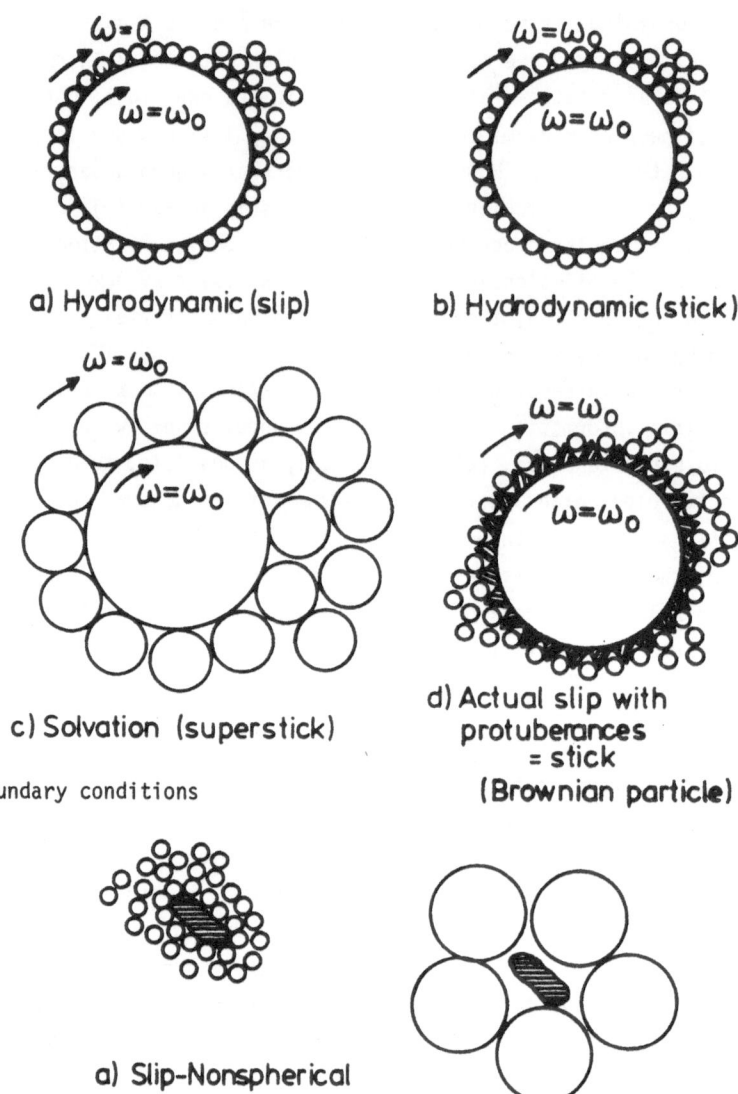

a) Hydrodynamic (slip)

b) Hydrodynamic (stick)

c) Solvation (superstick)

d) Actual slip with protuberances
= stick
(Brownian particle)

Figure 2: Boundary conditions

a) Slip-Nonspherical

b) Subslip-Nonspherical

Figure 3: a) Small solvent molecules = continuous medium; hydrodynamic slip.
b) Large solvent molecules; nearly free rotation.

boundary conditions [7,9]. Usually one is less ambitious and replaces the exact shape with an equivalent spheroid for which A can readily be obtained from the Perrin formulas for stick and from the Hu and Zwanzig tables for slip boundary conditions. Of course, for a spherical top molecule or for rotations under slip boundary conditions about a C_{3v} axis, this approach is invalid, and one must include the "knobs" or protuberances that jut out, thereby destroying the cylindrical symmetry. (See Fig. 4.) Formulas exist which enable us to obtain results easily for a variety of high symmetry shapes [7,9]. But it is not rewarding merely to state that the boundary conditions are slip, stick or inbetween. We seek a better picture.

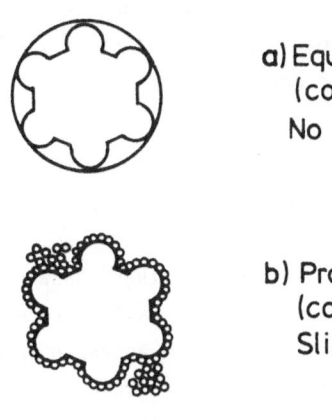

a) Equivalent Spheroid (continuous solvent) No slip resistance

b) Protuberances (continuous solvent) Slip resistance

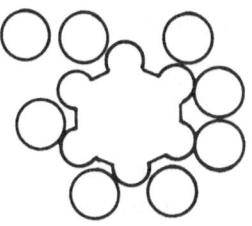

c) Protuberances (molecular solvent) Little resistance

Figure 4: Models for benzene. Slip boundary conditions.

It is found that in most cases, for small molecules A is considerably less than expected from stick boundary conditions, and rather close to that expected from slip boundary conditions. (See Fig. 1.) For large Brownian particles, however, stick boundary conditions seem to hold. One can rationalize these results by assuming slip boundary conditions and focusing on the relative size of the tagged molecule to the solvent molecules as a significant factor: solvent molecules flowing over smooth surfaces behave like a "slipping" fluid, but if the surface has cracks and protuberances, and if the solvent molecules are sufficiently small, then they may make much better contact with the surface than expected for a smooth surface. Therefore, a large Brownian sphere may appear very rough to small solvent molecules, and though slip boundary conditions may be appropriate, they are slip boundary conditions for a rough, rather than for a smooth sphere, a condition which is equivalent to stick boundary conditions for a smooth sphere. (See Fig. 2.) Thus if one ignores the details of molecular shape, one would expect slip boundary conditions to hold whenever solvent molecules are comparable to or larger than the tagged particle or the protuberances on the tagged particles, and one would expect stick conditions if the reverse were true. If this picture is correct, we could, in principle, get sensible results by calculating correlation times with slip boundary conditions and realistic shapes. In some cases, however, one might get poorer results with realistic than with smoothed surfaces because the hydrodynamic theories assume infinitesimal solvent molecules that can flow into any molecular crack, no matter how small, thereby exaggerating the contribution of a crack or protuberance to rotational friction [7]. (See Fig. 4.) Thus, it appears that though for rotational motion of benzene about its six-fold axis, it is essential to consider the hydrogen protuberances or else there would be no rotational relaxation, for rotation about its other principle axes better results are obtained if these protuberances are smoothed out [9].

In some cases one finds that the τ_ℓ's are larger than those predicted by the Stokes-Einstein-Debye (SED) theory with stick boundary conditions; we might label this "superstick". (See Fig. 2c.) On the other hand, for nonspherical molecules one sometimes finds that the τ_ℓ's are smaller than those predicted by SED theory with slip boundary conditions: we label this "subslip". (See Fig. 1, curves a and b.) In both cases this might be taken as evidence of a breakdown of hydrodynamical theory, a need to consider the discrete structure of the fluid, or equivalently, the influence of molecularity on the fluid [7]. Subslip conditions can be understood by envisaging a small nonspherical tagged particle rotating among large solvent molecules with large interstitial gaps; the small tagged molecule can rotate almost freely in these gaps, whereas hydrodynamic theory, which treats the fluid as a continuum, would picture the viscous solvent in intimate frictional contact with the tagged molecule [2]. (See Fig. 3b.) Plastic crystals represent the condition of subslip. Superstick conditions might be effected by bona fide solvation, which is equivalent to stick conditions but with solvent molecules of finite size; in this case the volume v of the tagged molecule might be appreciably enlarged, thereby increasing τ_ℓ appreciably. (See Fig. 2c.) But before we jump to this conclusion we have to make sure that this superstick condition isn't the result of a "universal" effect such as dielectric friction [10,11]. I discuss this below.

Theory for the A-Parameter

We have already discussed slip and stick models for rotations. We have examined conditions under which they might be expected to apply, when one might expect intermediate boundary conditions, and when subslip or superstick conditions might pertain. In many cases, if the temperature and pressure are changed, the coefficient A in Eq. (2) is almost invariant for a given solute-solvent system, but if one changes solvent the A parameter may change appreciably [2]. These changes might be partially ascribed to the size of the solvent molecules as discussed above, but it appears that they may also be dependent upon compressibility, dipole moment, free volume and a host of other properties [2].

Besides the various forms of the slip and stick models combined with specific molecular shapes, as described above, there are numerous other models which attempt to incorporate some of the molecular features we have listed. We mention a few models that have been used. The microviscosity model of Gierer and Wirtz [12] is a quasi-hydrodynamic model which takes account of the finite size of the "spherical" solvent molecules while <u>imposing the same boundary condition between "spherical" solute and solvent as between layers of solvent</u>. This model has been extended [2] to arbitrary boundary conditions between solute and solvent, including stick conditions which then lead to true solvation. Another model [2], called "the free space model", has had greater success than other models in rationalizing data from a wide range of systems, but agreement with experiment could hardly be called quantitative. In this model the tagged particle is envisaged as rotating in a free space which is statistically determined; the rotating particle exchanges energy and momentum with the bath, as described by hydrodynamics, but it couples to the bath (boundary conditions) by bumping into the "cavity walls". In this model the size of the local free space relative to that of the tagged molecule, and the resiliency of the "cavity walls", both affect the coupling, i.e. the parameter C in Eq. (2); these phenomena are expressed in terms of independently determined macroscopic quantities such as compressibility, viscosity, free volume and molar volume, and in terms of the volume (and shape) of the tagged molecule. This model has been quite successful in describing the large variations observed in going from solvent to solvent, changes ranging from subslip to stick, but for a given solvent, where C varies little, the model tends to introduce a greater change in A with changes in temperature and pressure than is observed. This has been clearly demonstrated by Zager and Freed [13] who have carried out a detailed and broad pressure-temperature study of a tagged molecule in a single solvent. In these experiments the variation in C is very slight, far less and far too subtle to be described adequately by the free space model

The theories discussed above might be called "universal" since they do not depend upon specific structural features but relate τ_ℓ to molecular size and independently measured macroscopic properties of the fluid. If we had a truly reliable "universal model" of this kind, we would be able to associate deviations from universal behavior with specific structural, dynamic and chemical effects, i.e. to the truly "interesting" phenomena. Thus a better theory is called for. Formal statistical mechanical theories have been developed, but as yet they do not yield actual values for the parameter C [14].

Can superstick have a "universal" origin? Solvation or the formation of "solvent-bergs" can be classified as a specific phenomenon [15]. (See Fig. 2c.) But dielectric friction is "universal" in that it is always present, regardless of the details of the system. Rotational dielectric friction arises from the polarization of the surrounding medium by the dipole on the tagged molecule. If the tagged molecule were stationary, the surrounding dipolar solvent would be polarized so as to minimize the free energy; if now the tagged dipole were to be rotated slightly, it would find itself in an orientation other than that yielding the free energy minimum, and consequently, it would be pushed back towards its original position. Of course, as the tagged molecule turns, the surrounding dipoles reorient to keep the free energy at a minimum, and it is the lag in response of the surrounding dipoles to the motion of the tagged dipole that determines the dielectric friction. This process is "universal" in that it is determined by the dielectric constant and relaxation of the medium, and the dipole moment and size of the tagged particle. Theories of dielectric friction exist, all of which contain serious approximations, but they allow reasonable estimates [11,12,16]. The conclusion is that dielectric friction is a small effect [10] unless the tagged molecule has a large dipole moment such as that of an ion with the charge asymmetrically placed far from the rotation center [17]; in this latter case it may be appreciable.

Concluding Statement About Calculations of A

With the help of simple "universal" theories we can try to calculate the parameter A in Eq. (2). If we have reasonable confidence in the theory, we compare the calculation with the observed value of A. If there are large discrepancies we can then attribute these to specific effects such as hydrogen-bonding, solvation, molecular aggregations, near neighbor rearrangements, local heating, etc. Since, as we have seen, the "universal" models can yield values of A that range from subslip to superslip, great caution must be excercised before "unusual" τ_ℓ values are associated with these specific effects. Jiri Jonas and co-workers and Jack Freed and co-workers have been emphasizing the need to carry out pressure dependent measurements in order to obtain constant density curves as a prelude for developing a truly adequate theory. But, at best, it is unlikely that the "universal" theories used to subtract out the "uninteresting" effects can be very accurate, and only if the discrepancies between theory and experiment are large can we confidently model the specific effects as the cause of the discrepancies. Such large, interesting discrepancies have been found and reported to this meeting by Professors K- Spears and Th. Dorfmüller.

Intercept τ^0

We now turn our attention to the intercept in Eq. (2). We note that it is the intercept for the viscosity extrapolated to zero, not the behavior at zero viscosity. What is its significance? And do we care? Since it cannot have a hydrodynamic origin and since our interpretation of A in Eq. (2) is, at least in part, based upon hydrodynamics, we run the risk of not understanding the framework of our analysis if we have no explanation for τ^0. Furthermore, in predicting rotational correlation times, one has to have an estimate of τ^0. And finally, it is possible that buried within τ^0 is interesting molecular information.

To date there is no evidence that τ^0 will yield interesting molecular information, but this is based upon analyses that are far from satisfactory. There is some thought that the intercept may, in fact, merely be an artifact of the way in which data are handled [18]. If, however, one does take the intercept seriously, it is important to note that it is sometimes positive, sometimes negative, with the negatives in the majority [19] (See Fig. 1.)

Recently, Glenn Evans and I have proposed [19] an explanation which we believe holds promise. By means of a simple Mori-type transformation, τ^0 can be written as the sum of time integrals over four correlation functions: one which is an autocorrelation function of torque-like terms, one which is an autocorrelation function of kinetic terms (bilinear in angular velocity), and two cross correlation functions between these. All the data for liquids have been taken in the "strong-torque regime", and it is widely assumed that it is the torque-torque correlation function which gives rise to the contributions which are proportional to viscosity η, i.e. to the SED terms. The purely kinetic terms, treated in the low-torque or inertial limit have been proposed as a possible source of the τ^0 intercept. Though this might indeed describe the <u>actual</u> low viscosity behavior, it cannot describe negative intercepts nor any aspect of the behavior <u>extrapolated</u> from the high to the low viscosity regime; this behavior is not inertial but is determined by the intermolecular torques and, consequently, it yields contributions to τ^0 that are not independent of viscosity but inversely proportional to it. It appears that this kinetic contribution to τ^0 is negligible in the viscosity regimes encountered in most experiments. The cross-correlations have usually been neglected but it is these that we propose are responsible for the intercept τ^0; they appear to be independent of viscosity. Cross-correlations are difficult to evaluate and we need the integral of these correlation functions over all time, which is why we cannot, as yet be sure that it is indeed these cross-correlation terms that give rise to τ^0; we can, however, evaluate the first few terms in a time expansion of the cross-correlation function, and we find that it vanishes at zero time and is negative for short times. We look forward to molecular dynamics simulations to help evaluate these cross-correlation functions.

Determination of Collective and Single Particle Correlation Times

The single molecule correlation function is $<Y_{\ell m}(j,t)Y_{\ell m}(j,0)^*>$; where $Y_{\ell m}(j,t)$ is the normalized spherical harmonic describing the orientation of the j-th molecule at time t. How can the single molecule rotational correlation time be determined? In nmr measurements one usually measures the single particle correlation time τ_2 as defined in Eq. (1), but it has to be extracted from spin relaxation data. Thus one determines the quantity sought, but only as a second order effect.

When applicable, the single particle τ_2 can also be obtained from Raman spectra. Here one actually can determine the rotational correlation function; therefore, more detailed information can be obtained than in most nmr studies. However, the extraction of rotational information from Raman spectroscopy is not always straightforward, because it is often intertwined with vibrational relaxations.

Fluorescence experiments in the time-domain yield direct information concerning the rotational correlation function, but as suggested at this conference by Dorfmüller and others, the correlation times obtained by this technique are often shorter than those obtained by other techniques. The suggestion has been made that in such experiments sufficient energy is transferred from the electronically excited tagged molecule to its nearest neighbors, or even to its own translational and rotational modes, that the effective local temperature may be considerably higher than the ambient temperature.

Dielectric relaxation, depolarized light scattering, Kerr effect measurements all yield collective rotational correlation information, i.e. $\Sigma <Y_{\ell m}(1,t)Y_{\ell m}(j,0)*\exp\{i\mathbf{k}\cdot[\mathbf{r}^1(t)-\mathbf{r}^j(0)]\}>$, where $\mathbf{r}^j$ is the position of the j-th particle, and the sum is taken over all (j) particles. One must then extract the single molecule relaxation times from the collective ones. If the single particle correlation function relaxes diffusionally (exponentially), with decay time τ_ℓ, it can then be shown that the collective correlation function is also exponential, and that the collective correlation time $\tau_\ell^{(c)}$ is given by the relation [20]

$$\tau_\ell^{(c)} = (g_\ell/\dot{g}_\ell)\tau_\ell \tag{4}$$

where g_ℓ is the static correlation factor,

$$g_\ell = \sum_j <Y_{\ell m}(1,0)Y_{\ell m}(j,0)*\exp\{i\mathbf{k}\cdot[\mathbf{r}^1(0) - \mathbf{r}^j(0)]\}> , \tag{5}$$

and $\dot{g}_\ell$ is a dynamic correlation factor. Most dielectric studies are carried out at very low k. Eq. (4) is useful because it appears that $\dot{g}_\ell$ is usually very close to one, and g_ℓ can sometimes be obtained by independent techniques such as measurements of dielectric relaxation, light scattering intensities [22] and various electro-optical experiments [23]. Generalizations of Eq. (4) for non-diffusional motion have also been discussed [21,24]. Careful studies of the relationship of τ_ℓ and $\tau_\ell^{(c)}$ for light scattering have been reported at this conference by Dorfmüller and Versmold. For solutions that are dilute in the tagged particle, the collective and molecular motions are often quite similar.

For dielectric phenomena, rather large discrepancies between collective and molecular rotational times are expected. Though considerable theory has been developed to handle this problem [21], there are continuing controversies over how this should be done. The uncertainty arises because of the subtle role of dipole-dipole interactions, interactions that vary as r^{-3} and are thus so long-ranged that the relevant correlation functions are dependent upon the shape of the sample. The analysis given by Madden and me [21] leads to the following conclusions. If the individual dipoles $\mathbf{\mu}^j$ reorient diffusionally, then

$$<\mathbf{\mu}^j(t)\cdot\mathbf{\mu}^j(0)> = <\mathbf{\mu}^j(0)\cdot\mathbf{\mu}^j(0)>\exp(-t/\tau_1), \tag{6}$$

and in the low k-limit, the collective dipole, $\mathbf{M}(\mathbf{k},t)$, where

$$\mathbf{M}(\mathbf{k},t) = \Sigma\mathbf{\mu}^j(t)\exp[i\mathbf{k}\cdot\mathbf{r}^j(t)] \tag{7}$$

also reorients diffusionally. It follows that [21]

$$\langle M(k,t)M(-k,0)\rangle = \hat{k}\hat{k}\cdot\langle M(k,0)M(-k.0)\rangle\cdot\hat{k}\hat{k}\, \exp(-t\varepsilon/\tau_T) \tag{8}$$

$$+ (1-\hat{k}\hat{k})\cdot\langle(M(k,0)M(-k,0)\rangle\cdot(1-\hat{k}\hat{k})\, \exp(-t/\tau_T)$$

where $\hat{k}$ is the unit vector along the propagation vector k, τ_T is the Debye or transverse dielectric relaxation time, and ε is the dielectric constant. These results hold for correlation functions taken as averages over infinite samples; this is important to note because of the long-range character of the correlation functions, as discussed above. The g_1 specified in Eqs. (4) and (5), therefore also depends upon the shape of the sample, but a shape-independent relation between the collective and single dipole correlation times can be obtained [21]:

$$\tau_T = \tau_1 3\varepsilon(g^S/\dot{g}_1)[2\varepsilon + 1]^{-1} \tag{9}$$

where g^S can be obtained from the Kirkwood relation

$$(2\varepsilon +1)(\varepsilon-1)/\varepsilon = (\mu^2\rho/k_B T\varepsilon_0)g^S, \tag{10}$$

and μ is the magnitude of the dipole, ρ is the density, k_B is the Boltzmann constant, and ε_0 is the permittivity in vacuo. The Kirkwood g-factor g^S differs from g_1 in that it is short-range and so independent of sample shape:

$$g^S = \sum_j \langle\mu^1(0)\cdot\mu^j(0)\rangle/\langle\mu^1(0)\cdot\mu^1(0)\rangle . \tag{11}$$

Eq. (9) is a generalization of the Glarum-Powles relation.

Non-Diffusional Motion; Intermolecular Contributions

As stated in the Introduction, I shall say little about the problem of non-exponential relaxation, or, equivalently, non-lorentzian spectra. Such phenomena occur both because the rotational relaxation itself may be non-diffusional and because the quantity measured, e.g. the dielectric permittivity, may have direct (primary) contributions associated with intermolecular interactions, e.g. dipole-induced-dipole effects. This is an important problem, both because it must be unraveled in order to understand the rotational data properly and because the intermolecular or collisional effects are interesting in and of themselsves. Here, I shall merely outline one approach to the study of this problem.

In depolarized light scattering one studies fluctuations of the off-diagonal dielectric tensor $\varepsilon(t)$, i.e. the correlation function $\langle\varepsilon(t)\varepsilon(0)\rangle$. We can write $\varepsilon(t)$ as a sum of microscopically well-specified dynamical variables, $Q_i(t)$:

$$\varepsilon(t) = \sum_i a_i Q_i(t). \qquad (12)$$

Quite generally, we can describe correlation functions of $Q_i(t)$ as sums of exponentials, or equivalently, as obeying the equations of motion,

$$\partial <Q_i(t)Q_j(0)>/\partial t = -\sum_k M_{ik}<Q_k(t)Q_j(0)>. \qquad (13)$$

We are interested only in the "slow" motions, i.e. those that corresponds to low frequency spectra; therefore we are interested only in "slow" hydrodynamic variables $Q_i(t)$, variables whose relaxation takes place on a relatively long time-scale [25]. But the basis set of "slow" variables $Q_i(t)$ must be "complete" in that they describe <u>all</u> relevant slow motions. It is not easy to decide <u>a priori</u> how many dynamical variables are needed, and what they must be. The number of necessary variables can be determined empirically by specifying the number of exponentials in time, or lorentzians in frequency, that are needed to describe the observed relaxation or spectrum. There will be one variable $Q_i(t)$ for each needed exponential or lorentzian. Note that the same variables must be included in both Eqs. (12) and (13), though in some cases a number of a_i's and M_{ij}'s may vanish. All $Q_i(t)$'s for which $a_i \neq 0$ are called "primary variables" [20].

The choice of variables $Q_i(t)$ is not unique because even if a complete set of slow variables is identified, linear combinations of these would serve equally well. We could choose $Q_1(t) = \varepsilon(t)$, in which case all the a_i's in Eq. (12), except $a_1 = 1$, would vanish. But we would still need the full complement of $Q_i(t)$'s in Eq. (13). And because $\varepsilon(t)$ is not a basic molecular quantity, it is not a well-chosen primary variable.

Instead, we choose an orientation variable, either $Y_{\ell m}(j,t)$ or $\sum Y_{\ell m}(j,t)\exp[i\mathbf{k}\cdot\mathbf{r}^j(t)]$, as one of the primary variables, i.e. as $Q_1(t)$. This is reasonable both because it is a well-specified quantity and one in which we are interested. We know that we must also include some intermolecular or collisional quantities as a primary variable; in particular, we know that there must be a primary variable $Q_2(t)$ which represents dipole-induced-dipole (DID) effects [26]. There may be other primary variables representing both inter- and intra-molecular motions, but for the moment we will limit ourselves to these two. Both these variables must also be included in Eq. (13). And there may be other variables required in Eq. (13) to yield a proper dynamical description of the slow motions, i.e. other "secondary" variables [20].

Let us examine a few examples of the above. Suppose the spectrum can be described in terms of two lorentzians. We choose $Q_1(t)$ to be an orientational and $Q_2(t)$ an intermolecular slow variable. We can then interpret the spectrum in three alternative ways:

1) If the dynamical couplings, $M_{12} = M_{21} = 0$, then $<Q_1(t)Q_1(0)>$ and $<Q_2(t)Q_2(0)>$

each relax exponentially, and at different rates, while cross-correlations $<Q_1(t)Q_2(0)> = 0$. Here we see that both the rotational and intermolecular contributions to the dielectric relax independently and exponentially, and the spectrum consists of the sum of these two effects. In this case, one could apply one of the theories discussed above to interpret the rotational problem and the theory of Madden [27] for the intermolecular problem.

2) If the intermolecular contribution to the dielectric relaxation is very small, i.e. if $a_2 = 0$, then the two lorentzians must both be associated with the rotational motion $Q_1(t)$. This non-exponential dynamical behavior arises because of the interaction of some as yet unspecified intermolecular variable Q_2, such as the intermolecular torque, with the primary rotational variable $Q_1(t)$. This means that though the static coupling a_2 vanishes, the dynamical coupling M_{12} does not vanish: and though the spectrum is dependent only upon the single correlation fucntion $<Q_1(t)Q_1(0)>$, this correlation function decays as the sum of two exponentials. Theoretical descriptions of this effect do exist [1,21], but they are not completely satisfactory.

3) It is possible that both a_2 and M_{12} are sizeable, in which case the two-lorentzian spectrum is a sum of the four correlation functions, $<Q_1(t)Q_1(0)>$, $<Q_2(t)Q_2(0)>$, $<Q_1(t)Q_2(0)>$, and $<Q_2(t)Q_1(0)>$. Each of these correlation functions is the sum of two exponentials.

In the first example, both the rotations and the intermolecular motions are exponential, and both can readily be determined. In the second case the direct contribution of the intermolecular effects to the dielectric properties are negligible, and the non-exponential character of the relaxation can be attributed entirely to the rotations. In this third case there is no way to distinguish exactly what is taking place. And unfortunately, there is as yet no reliable way to determine which situation applies. Fortunately, if the third and most troublesome situation applies, we know that the same two variables must enter Eqs. (12) and (13), or else we would need three "slow" variables, and we would then predict a three-, rather than a two-lorentzian spectrum.

Spectra that are described by more than two lorentzians are correspondingly more difficult to describe than those discussed above. However, in describing VH depolarized spectra one [28] always includes the conserved collective momentum variable in the slow set of $Q_i(t)$'s. The signature of this variable is a very distinct dependence upon scattering angle. The well-specified and well-defined presence of the momentum variable actually provides insight into the nature of the various dynamical coupling parameters M_{ij} [29]. But ultimately, it appears that identification of the members of the minimum complete set of slow $Q_i(t)$'s will require molecular dynamics simulations. In contrast to light scattering experiments, where it is the fluctuations of $\varepsilon(t)$ which are measured, in simulations one can focus directly on the correlation function of the orientational variables; few such simulations have been carried out [30].

References

[1]   D. Kivelson and T. Keyes, J. Chem. Phys. 57, 4599 (1972)

[2]   J. Dote, D. Kivelson, and R. Schwartz, J. Phys. Chem. 85, 2169 (1981)

[3]   G. Alms, D. Bauer, J. Brauman and R. Pecora, J. Chem. Phys. 59, 5310, 5321 (1973)

[4]   A. Masters and P. Madden, J. Chem. Phys. 74, 2450, 2460 (1981)

[5]   P. Debye, "Polar Molecules", (Dover, 1929)

[6]   F. Perrin, J. Phys. Radium 5, 497 (1934)

[7]   J. Dote and D. Kivelson, J. Phys. Chem. 87, 3889 (1983)

[8]   C. Hu and R. Zwanzig, J. Chem. Phys. 60, 4354 (1974)

[9]   G. Youngren and A. Acrivos, J. Chem. Phys. 63, 3846 (1975); J. Fluid Mech. 69, 377 (1975)

[10] T. Nee and R. Zwanzing, J. Chem. Phys. 52, 6353 (1970)

[11] P. Madden and D. Kivelson, J. Phys. Chem. 86, 4244 (1982)

[12] A. Gierer and K. Wirtz, Z. Naturforsch. A8, 532 (1953)

[13] S. Zager and J. Freed, J. Chem. Phys. 77, 3360 (1982)

[14] D. Kivelson, M.G. Kivelson and I. Oppenheim, J. Chem. Phys.

[15] P. Colonomos and P. Wolynes, J. Chem. Phys. 71, 2644 (1979)

[16] J. Hubbard and P. Wolynes, J. Chem. Phys. 69, 998 (1978)

[17] D. Kivelson and K. Spears, J. Phys. Chem. 89, 1999 (1985)

[18] J. Freed, These proceedings

[19] G. Evans and D. Kivelson, J. Chem. Phys. 84, 385 (1986)

[20] T. Keyes and D. Kivelson, J. Chem. Phys. 56, 1057 (1972)

[21] P. Madden and D. Kivelson, Advances in Chemical Physics, Vol. LVI, Edit. by I. Progogine and S. Rice (J. Wiley, N.Y.,1984)

[22] A. Burnham, G. Alms, and W. Flygare, J. Chem. Phys. 62, 3289 (1975); S. Bertucci, A. Burnham, G. Alms, and W. Flygare, J. Chem. Phys. 66, 605 (1977)

[23] M. Battaglia, T. Cox, and P. Madden, Molec. Phys. 37, 1413 (1979)

[24] D. Kivelson and P. Madden, J. Phys. Chem. 88, 6557 (1984)

[25] H. Mori, Prog. Theor. Phys. 37, 502 (1967)

[26] T. Keyes, D. Kivelson and J. McTague, J. Chem. Phys. 55, 4096 (1971)

[27] P.A. Madden, Molec. Phys. 36, 365 (1978); Chem. Phys. Lett. 47, 174 (1977); in: "Molecular Liquids", ed. by A. Barnes, W. Orville-Thomas, and J. Yarwood (Reidel, Lancaster, 1983), pp. 431-474

[28] B. Berne and R. Pecora, Dynamic Light Scattering (Wiley, N.Y., 1976)

[29] P. Chappell, M. Allen, R. Hallem, and D. Kivelson, J. Chem. Phys. 74, 5929 (1981)

[30] M. Allen and D. Kivelson, Molec. Phys. 44, 945 (1981)

STATISTICAL STUDY OF FREELY ROTATING ASYMMETRIC TOP MOLECULES.
CASE OF PLANAR TOPS.

J.-Cl. Leicknam, Y. Guissani and M. Aguado-Gómez[*]
Laboratoire de Physique Théorique des Liquides[**]
Université Pierre et Marie Curie, Tour 16
4, Place Jussieu, 75252 Paris Cedex 05, France

Abstract

Rotational correlation functions of freely rotating asymmetric-top molecules have been obtained by the help of functional series. The convergence of these series is examined in this paper, for the planar rotors. The convergence is found to be very fast. The first two terms approach the exact result to less than 1.10^{-3} for all types of planar molecules.

1. Introduction

The purpose of this paper is to illustrate a recent theory of rotational correlation functions of freely rotating asymmetric-top molecules. Four types of methods [1-4] exist to calculate vectorial correlation functions of rigid asymmetric rotors. In three of them [1-3], the computational time is largely excessive and involves heavy numerical integrations. This is not the case for the most recent theory [4] in which only one numerical integration has to be performed. The theory employs functional series built on a basis of free spherical-top rotor correlation functions.

In what follows, the convergence of these series will be examined for the case of planar rotors. This problem merits a special attention since a great number of molecules belong, or are similar, to this class of tops; one can cite, for example, H_2O, SO_2, monochlorethylene, dichlorethylene, pyridine, ... etc. It will be shown that the functional series conver e very rapidly. More than ninety-six per cent of the exact value is given by the first term alone. Including the second term gives the exact result within an error smaller than 1.10^{-3}; this is largely more than required in majority of cases. For general reviews on rotational spectroscopy of liquids, see e.g. Refs. [5] and [6].

---

[*]Present Address: Departamento de Fisica, Facultad de Ciencias, Universidad de Baleares, 07071 Palma de Mallorca, Spain

[**]Unité Associée au C.N.R.S.

II. Theory

A. Vectorial Correlation Functions

If collective effects are negligible, vectorial correlation functions for isotropic fluids of molecular rotors can be expressed as follows

$$G(t) = <\vec{u}(0) \cdot \vec{u}(t)> . \tag{1}$$

The vector $\vec{u}(t)$ is a unitary vector fixed in the molecule at the time t and rotating with it. In the case of a free rotation, the average $< >$ is over a canonical ensemble and implies only a probability density for the initial angular momentum $\vec{J}(0)$. This result is obtained by considering the following two points. (i) Scalar products like $\vec{u}(0) \cdot \vec{u}(t)$ are independent of orientation of axes on which they are expressed; in the subsequent discussion, these axes will always be molecular axes at the zero time. (ii) For isotropic fluids, all orientations of these axes are equiprobable. Thus, averaging over initial orientations is useless. Next, a rotation matrix $\mathbb{R}(t)$ is introduced such that the vector $\vec{u}(t)$ is obtained from the vector $\vec{u}(0)$ by the relation

$$G(t) = \sum_{i,j} u_i u_j <\mathbb{R}_{ij}(t)>_{\vec{J}} \tag{2}$$

where u_i are the three components of $\vec{u}(0)$.

B. Symmetry Considerations

In the general case of asymmetric-top molecules the choice of the axes is not necessarily unique. For example, the principal axes of the rotational diffusion tensor $\underline{D}$ may not be those of the inertia tensor $\underline{I}$. However, the principal axes of $\underline{I}$ should be adopted, if the free rotation is to be studied. Without loss of generality, the axes x, y and z can be chosen such as to verify the conditions:

$$I_x \leq I_y \leq I_z$$

where I_i's designate the principal values of the inertia tensor. Then, symmetry arguments prove that, in this frame, the matrix $<\mathbb{R}(t)>_{\vec{J}}$ is diagonal [1].

In the free rotation regime, two types of motion can be observed depending on the initial value of the angular momentum $\vec{J}(0)$. In the molecular frame, $\vec{J}(t)$ can rotate either around the x-axis of least moment of inertia, or around the z-axis of major moment of inertia [7]. In order to handle these two cases conveniently, a transformation is introduced [1], namely

$$(J_x(0), J_y(0), J_z(0)) \longrightarrow (J, \vartheta, \varphi) .$$

The limiting value of ϑ taken at the point when the regime of motion changes, is denoted by ϑ_m; the averaging over $\vec{J}(0)$ is then divided in two parts. (i) For $0 \leq \vartheta \leq \vartheta_m$, the integration over J and φ can be obtained analytically in the form of an infinite series [4]. However, the integration over ϑ remains numerical and will be denoted by the symbol $< >_I$ (average over the first region). (ii) For $\vartheta_m \leq \vartheta \leq \pi/2$, the integration over J and φ is given similarly by an analytical series. Likewise, the integration over ϑ remains numerical and will be denoted by the symbol $< >_{III}$ (average over the third region). For the justification of these notations, see Reference [1]. Symmetry arguments [1] then prove that

$$G(t) = u_x^2 G_x(t) + u_y^2 G_y(t) + u_z^2 G_z(t) \tag{3}$$

$$G_x(t) = <R_{xx}(t)>_I + <R_{zz}(t)>_{III} \tag{4}$$

$$G_y(t) = <R_{yy}(t)>_{I+III} \tag{5}$$

$$G_z(t) = <R_{zz}(t)>_I + <R_{xx}(t)>_{III} \tag{6}$$

where $R_{ii}(t)$ is the average over φ and J of the expression for $2\text{Re}[\mathbb{R}(t)_{ii}]$ in the first region.

C. Functional Series

According to the results of the preceding Section, the diagonal elements of $R(t)$ appear in the form of a functional series. In what follows, the time t is always a dimensionless quantity, the reducing factor is $(k_B T/I_y)^{1/2}$. The following formulas can be obtained [4]:

$$R_{xx}(t) = A_1^+(t) + A_3^+(t) + A_5^+(t) + \ldots \tag{7}$$

$$R_{yy}(t) = A_1^-(t) + A_3^-(t) + A_5^-(t) + \ldots \tag{8}$$

$$R_{zz}(t) = A_0^+(t) + A_2^+(t) + A_4^+(t) + \ldots \tag{9}$$

$$A_n^\pm(t) = \sum_{\sigma=-1}^{+1} \Lambda_{n\sigma}^\pm \, g(n,\sigma,t) \tag{10}$$

$$\Lambda_{n\sigma}^\pm = \frac{2 \, \Phi_n^\pm(Q_x,Q_y,Q_z)}{(1+\delta_{on})(1+\sigma^2)} \cdot \frac{q^n \, \exp(\sigma\pi a/K)}{[1\pm(-1)^\sigma q^n \exp(\sigma\pi a/K)]^2} \tag{11}$$

$$g(n,\sigma,t) = (1-\mu_{n\sigma}t^2) \, \exp\{-\mu_{n\sigma}t^2/2\} . \tag{12}$$

The functions $A_n^{\pm}(t)$ appearing in the series involve two types of terms: a time-independent term $\Lambda_{n\sigma}^{\pm}$ responsible for the convergence of the series and a function of time $g(n,\sigma,t)$ such that $g(n,\sigma,0) = 1$. These two quantities deserve the following comments. (i) The factors $\Lambda_{n\sigma}^{\pm}$ depend essentially on the so-called "nome q", a quantity well-known in the theory of Elliptic Functions [8]. In most cases, the value of q is very small as compared to unity. The above series are then found to be rapidly converging. The function $\Phi_n^{\pm}(Q_x,Q_y,Q_z)$ represents Q_z for an even value of n; when n is odd and the sign is positive, Φ indicates Q_x; finally, when n is odd and the sign is negative, Φ takes the value Q_y. The remaining quantities Q_i, a and K appearing in Eq. (11), are also ϑ-dependent; their expressions are given in Refs. [1] or [4]. (ii) The functions $g(n,\sigma,t)$ represent the functional basis for the development of the $R_{ii}(t)$'s. These functions have the form of the correlation functions for free spherical-top rotors. After an initial decrease and a passage through a negative minimum, they increase monotonically and vanish in the long time limit. The coefficients $\mu_{n\sigma}$ depend only on ϑ and are found to be of the type

$$\mu_{n\sigma} = (nm^* + \sigma\lambda^*)^2 \tag{13}$$

$$m^* = [(I_y/I_z) \cos^2\vartheta + (I_y/I_x) \sin^2\vartheta]^{-1/2} \, \pi m / 2 K J \tag{14}$$

$$\lambda^* = [(I_y/I_z) \cos^2\vartheta + (I_y/I_x) \sin^2\vartheta]^{-1/2} \, \lambda / J \; . \tag{15}$$

The quantities mJ^{-1}, λJ^{-1} and K are independent of J; their expressions are given in Ref. [1]. The correlation function $G(t)$ is thus obtainable for all values of t, once the coefficients $\Lambda_{n\sigma}^{\pm}$ and $\mu_{n\sigma}$ have been determined. There results a gain of the order of thousand in computer time by comparison with previous theories [1-3].

III. Results and Discussion

A. Generalities

In the preceding Section, the correlation function, $G(t)$ has been given in the form of an infinite series on a functional basis of spherical top correlation functions. When a molecular rotor is far from a spherical rotor, the convergence of the series may be questionable. Our purpose here is to study the convergence for planar asymmetric rotors $(I_z = I_x + I_y)$, a class of rotors particularly distant from the spherical ones. This situation can be illustrated by using our triangular classification of asymmetric tops [1]. In a two-dimensional coordinate system, the abscissa and ordinate are chosen to be equal to $1-I_x/I_y$ and $I_z/J_y - 1$, respectively (Fig. 1). Then, all existent molecules can be represented by a point into, or on the border, of the triangle $s\ell r$. Spherical, linear and symmetric planar tops are located on the apexes s, ℓ and r. All planar asymmetric-tops are on the side ℓr, opposite to the point s representing

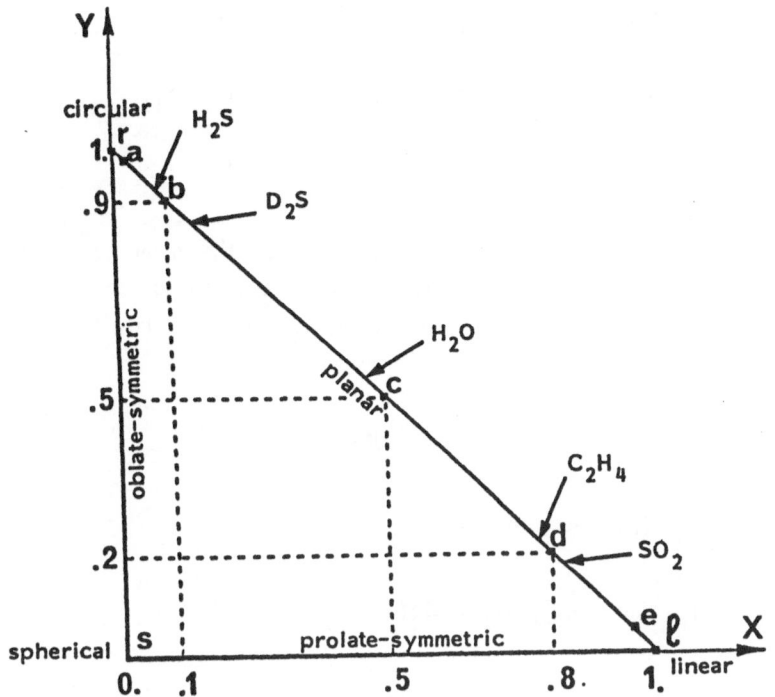

Figure 1: Representation of molecular rotors on the plane $X = 1-I_x/I_y$ and $Y = I_z/I_y-1$. The principal values I_i of the inertia tensor obey the conditions $I_x \leq I_y \leq I_z$. All existent molecules can be represented by a point in, or on the border, of the triangle $s\ell r$.

the spherical tops. This group of molecules may thus serve to examine the convergence of the series.

Results can be given either by studying various functions $G(t)$ or by analysing their Fourier transforms, $I(\omega)$. In view of possible spectroscopic applications, the subsequent discussion adopts this latter alternative. In what follows, the frequency is always a dimensionless quantity, the reducing factor is $(I_y/k_BT)^{1/2}$.

B. Convergence

In molecular spectroscopy of asymmetric-tops, the bands generated by the components u_x, u_y and u_z of the transition dipole moment are usually denoted by A, B and C bands [9]. Accordingly, the Fourier transforms of $G_x(t)$, $G_y(t)$ and $G_z(t)$ are here denoted by $I_A(\omega)$, $I_B(\omega)$ and $I_C(\omega)$, respectively. Applying Eqs. (4-9), the following formulas are readily obtained:

$$I_A(\omega) = <I_1^+(\omega) + I_3^+(\omega) + ...>_I + <I_0^+(\omega) + I_2^+(\omega) + ...>_{III} \tag{16}$$

$$I_B(\omega) = <I_1^-(\omega) + I_3^-(\omega) + \ldots>_{I+III} \qquad (17)$$

$$I_C(\omega) = <I_0^+(\omega) + I_2^+(\omega) + \ldots>_I + <I_1^+(\omega) + I_3^+(\omega) + \ldots>_{III} \qquad (18)$$

$$I_n^\pm(\omega) = (2\pi)^{-1} \sum_{\sigma=-1}^{+1} A_{n\sigma}^\pm \mu_{n\sigma}^{-3/2} \omega^2 \exp(-\omega^2/2\mu_{n\sigma}). \qquad (19)$$

The successive terms of these series differ by a factor of the order q^2; see Eq. (11). The contributions of elements of the same order of magnitude in q can then be collected to generate the first term, $I_i^{(1)}(\omega)$, second term, $I_i^{(2)}(\omega),\ldots$, respectively, of spectral densities. Proceeding this way, one finds

$$I_A^{(1)}(\omega) = <I_1^+(\omega)>_I + <I_0^+(\omega)>_{III} \qquad (20)$$

$$I_B^{(1)}(\omega) = <I_1^-(\omega)>_{I+III} \qquad (21)$$

$$I_C^{(1)}(\omega) = <I_0^+(\omega)>_I + <I_1^+(\omega)>_{III} \qquad (22)$$

$$I_A^{(2)}(\omega) = <I_3^+(\omega)>_I + <I_2^+(\omega)>_{III} \qquad (23)$$

$$I_B^{(2)}(\omega) = <I_3^-(\omega)>_{I+III} \qquad (24)$$

$$I_C^{(2)}(\omega) = <I_2^+(\omega)>_I + <I_3^+(\omega)>_{III} \qquad (25)$$

etc. ...

For the symmetric top case, only the first elements $I_i^{(1)}(\omega)$ survive and the results of St. Pierre and Steele [10] are found again. For asymmetric molecules, an estimation of the magnitude of successive terms $I_i^{(n)}(\omega)$ can be obtained by the help of their integrals, $M_i^{(n)}$. This quantity is always positive and is smaller than unity:

$$M_i^{(n)} = \int_{-\infty}^{+\infty} d\omega \, I_i^{(n)}(\omega) \qquad (26)$$

$$\sum_{n=1}^{\infty} M_i^{(n)} = 1 \,. \qquad (27)$$

Planar rotors verify the equation $I_z/I_y = 1 + I_x/I_y$. Thus, the value of $M_i^{(n)}$ depends only on the ratio I_x/I_y. This quantity is comprised in the interval [0,1]:

its lower limit corresponds to the linear rotor limit (point ℓ) and its upper limit represents the circular rotor limit (point r); see Fig. 1. The evolution of $M_A^{(1)}$, $M_B^{(1)}$ and $M_C^{(1)}$ has been studied for all values of I_x/I_y. These quantities $M_i^{(1)}$ are equal to unity at the linear rotor point, pass through a minimum for $0 < I_x/I_y < 1$ and increase again approaching the unity at the circular rotor point. The minima of $M_A^{(1)}$, $M_B^{(1)}$ and $M_C^{(1)}$ are attained at the points b, c and d, respectively. The values taken by I_x, I_y and I_z are given in Table I, and the corresponding values of $M_i^{(1)}$ are detailed in Table II. The results deserve the following comments.

Table I: Values of the standardized elements of the inertia tensor for the symmetric planar (r), asymmetric planar (a,b,c,d,e) and linear (ℓ) rotors presented on Fig. 1.

Point	I_x	I_y	I_z
ℓ	0.00	1.0	1.00
e	0.05	1.0	1.05
d	0.20	1.0	1.20
c	0.50	1.0	1.50
b	0.90	1.0	1.90
a	0.99	1.0	1.99
r	1.00	1.0	2.00

Table II: Values of the integral $M_i^{(1)} = \int_{-\infty}^{+\infty} d\omega \, I_i^{(1)}(\omega)$.

In all cases, the sum of the first two terms approaches unity, the exact value of the integral of the infinite series, to within a difference smaller than 1.10^{-3}.

Point	$M_A^{(1)}$	$M_B^{(1)}$	$M_C^{(1)}$
e	0.999	0.997	0.974
d	0.999	0.994	0.959
c	0.992	0.993	0.967
b	0.980	0.996	0.995
a	0.990	0.998	0.999

(i) From Table II, one observes that the first term, $I_A^{(1)}(\omega)$ gives at least 98% of the exact result, $I_A(\omega)$. Similarly, $I_B^{(1)}(\omega)$ and $I_C^{(1)}(\omega)$ give at least 99% and 96% of the exact results for $I_B(\omega)$ and $I_C(\omega)$. On Fig. 2, the exact profile $I_i(\omega)$ and the corresponding first term, $I_i^{(1)}(\omega)$ have been superimposed for the three bands A, B and C and for the various points a-e; the profiles for the linear and circular limits are also given. It appears that in the majority of the cases the first order term is practically indistinguishable from the exact profile. Only slight differences appear at the center of the band C for the points c, d and e. One concludes that in the majority of cases the first term of the series is sufficient for all practical purposes. This remarkable convergence of the series justifies partially an old model used to calculate band profiles [9]: a profile for an asymmetric

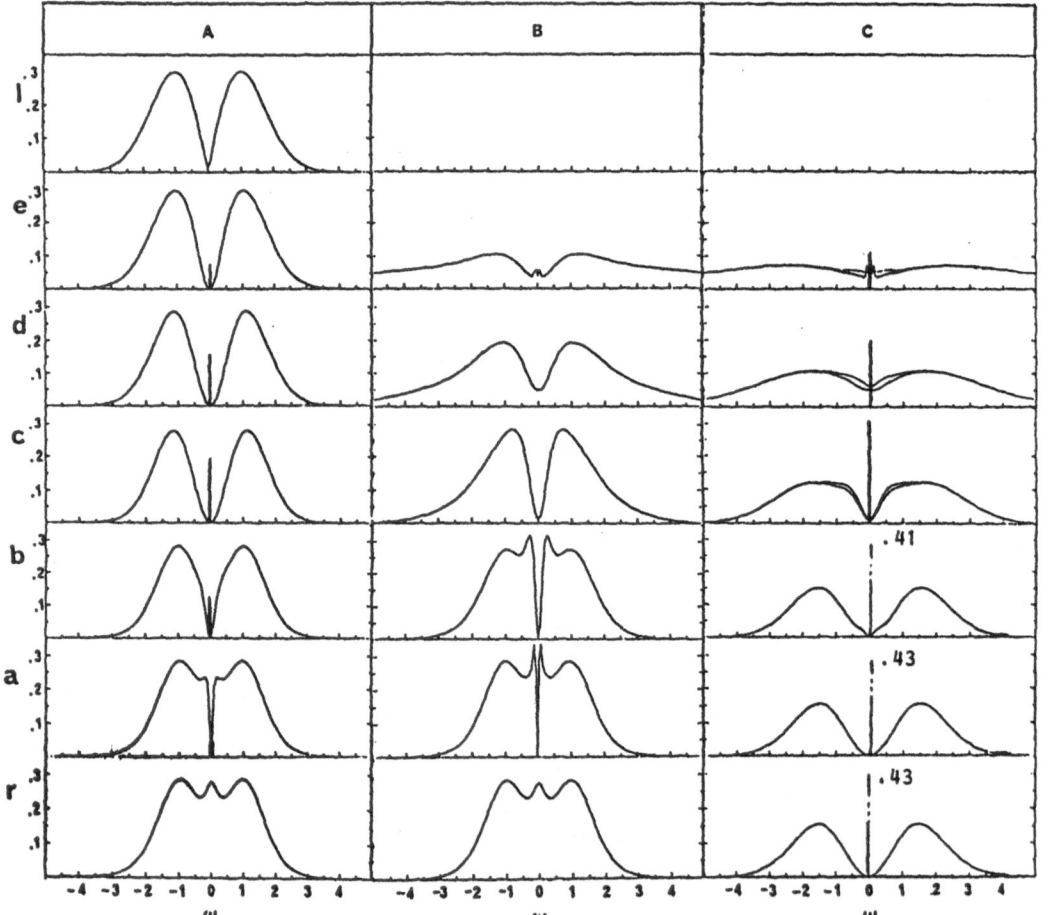

Figure 2: Comparison of exact spectral densities of A, B and C bands with those obtained from the first terms $I_i^{(1)}(\omega)$ of the series given in the text; see Eqs. (20-22). The points r, a, b, c, d, e and ℓ represent planar rotors situated on the segment ℓr of Fig. 1. The lower curves represent the first terms of the series and the upper curves are the exact results. In majority of cases, these two curves are indistinguishable from each other.

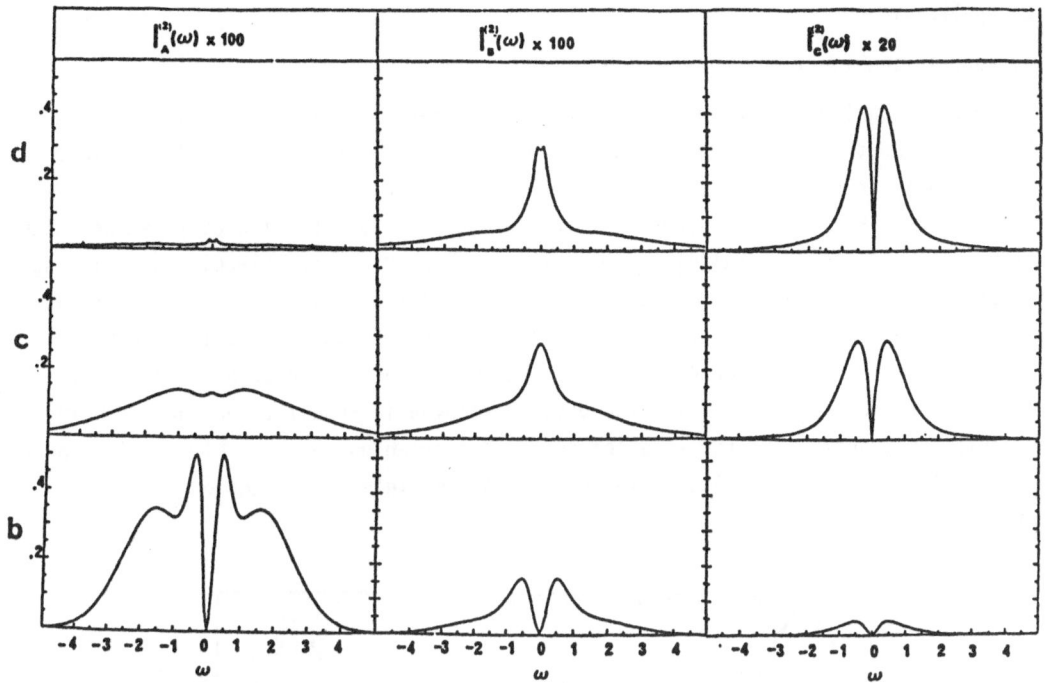

<u>Figure 3</u>: Contribution of the second terms $I_i^{(2)}(\omega)$ of the series given in the text; see Eqs. (23-25). The points b, c and d represent asymmetric-planar rotors situated on the segment lr of Fig. 1. Note the different scales for $I_A^{(2)}(\omega)$, $I_B^{(2)}(\omega)$ and $I_C^{(2)}(\omega)$.

top is approximated by the profile for its nearest symmetric top. However, this crude approximation is greatly improved if in the functional forms, $\omega^2\exp(-\omega^2/2\mu_{n\sigma})$ with $n = 0$ or 1, the correct coefficients $\mu_{n\sigma}$ are introduced.

(ii) The sum of the two first integrals, $M_i^{(1)} + M_i^{(2)}$ approaches the unity to within a difference less than 10^{-3} and the third term $I_i^{(3)}(\omega)$ is always of order 10^{-4}. The profiles $I_i^{(2)}(\omega)$ are represented on Figure 3. The term $I_A^{(2)}(\omega)$ is relatively important at the point b, that is for a rotor close to a planar-symmetric top such as H_2S or D_2S. On the contrary, this term is comparatively small for a rotor near a linear top (poind d). At the point b, the first region is large as compared with the third one (ϑ_m approaches $\pi/2$). Moreover, a detailed analysis of the results proves that, in this first region, the term corresponding to $\sigma = -1$ gives the major part of $I_A^{(2)}(\omega)$. This relative importance comes from the small values taken by $\mu_{3,-1} = (3m^*-\lambda^*)^2$; compare with Eqs. (13) and (19). Similar comments can be done on the evolution of $I_C^{(2)}(\omega)$ by exchanging the points b and d and the first and third regions. $I_B^{(2)}(\omega)$ is everywhere of the same order. One concludes that the major contribution to the second term comes from regions where the vectorial component, u_i, is perpendicular to the axis of the rotation.

C. Applications

a. Correlation functions for liquids at short times

Generally speaking, autocorrelation functions of rigid rotors in the liquid phase
behave at short times like correlation functions of free rotors [11]. This effect is
particularly pronounced if the intermolecular potential is spherical. Planar molecules do not escape to this general rule. On Figure 4 is shown the correlation function of a unit vector parallel to the axis of symmetry of the molecule SO_2 in the
liquid phase. The function is obtained by a Molecular Dynamics simulation over 256
particles in an elementary cell with periodic boundary conditions. The particles
interact via a three center L.J. potential with dipolar and quadrupolar potential
terms determined such as to give the best thermodynamic results. The corresponding
correlation function for the free motion is also drawn on the same Figure. It appears
clearly that the two functions coincide over a time interval of the order of 0.1
picosecond.

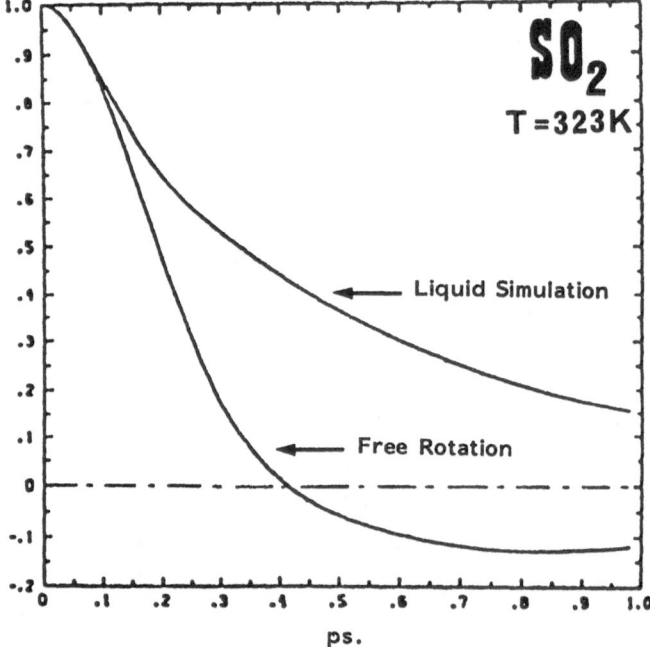

Figure 4: Rotational correlation function $\langle \vec{u}(0) \cdot \vec{u}(t) \rangle$. The vector $\vec{u}$ is along
the symmetry axis of the molecule SO_2. Molecular Dynamics simulation of the liquid
is performed along the coexistence curve at $323^\circ K$. The free rotation correlation
function has been calculated for the same temperature.

b. Models

Among all models describing molecular rotations in liquids, the extended diffusion model proposed by Gordon has been most frequently employed to study rotations of symmetric-tops; for a review, see for example Refs. [5] or [6]. In this model, the molecules are assumed to rotate like free rotors between collisions. The latter are instantaneous and randomize the angular momentum $\vec{J}(t)$. These models are connected with hard sphere theories of fluids. Planar molecules are far from being spherical. Nevertheless, these models can still give a good description of the rotation of planar rotors in liquids; see, for example, the rotations in the liquid state of the molecule C_2H_4 [12]. For all types of molecules, the calculations can conveniently be performed by employing Laplace transformation techniques. For asymmetric tops, the basic functions of the present theory, $g(n,\sigma,t)$, have Laplace transforms expressible in terms of elementary functions [4]. The functional series employed in this paper for planar rotors are thus well-adapted for practical computations. In the same context, more sophisticated models merit attention and will be analyzed in a near future.

Acknowledgments

The authors are grateful to Professor S. Bratos for fruitful discussions. The calculations in this work have been carried out with the support of the "Centre Interrégional de Calcul Electronique" (CIRCE, Orsay, France).

References

[1] Y. Guissani, J.-Cl. Leicknam and S. Bratos, Phys. Rev. A16, 2072 (1977)

[2] D.R. Fredkin, A. Komornicki, S.R. White and K.R. Wilson, J. Chem. Phys. 78, 7077 (1983)

[3] T.E. Bull and W. Egan, J. Chem. Phys. 81, 3181 (1984)

[4] M. Aguado-Gómez and J.-Cl. Leicknam, Phys. Rev. A34, 4195 (1986)

[5] W.A. Steele, Adv. Chem. Phys. 34, 1 (1976)

[6] W.G. Rothschild, Dynamics of Molecular Liquids, (Wiley Interscience, New York, 1984)

[7] L.D. Landau and E.M. Lifshitz, Mécanique, p. 160 (MIR, Moscow, 1966)

[8] M. Abramowitz and I.A. Stegun, Mathematical Functions (Dover, New York, 1965)

[9] R.M. Badger and L.R. Zumwalt, J. Chem. Phys. 6, 711 (1938)

[10] A.G. St. Pierre and W.A. Steele, Phys. Rev. 184, 172 (1969)

[11] R.G. Gordon, J. Chem. Phys. 41, 1819 (1964)

[12] J. Soussen-Jacob, T. Nguyen Tan and J.-Cl. Leicknam, Mol. Phys. 46, 769 (1982)

RELAXATION OF RIGID AND NON-RIGID MOLECULES IN LIQUIDS

James Mc Connell
Dublin Institute for Advanced Studies
Dublin 4, Ireland

1. Introduction

In the present article a summary will be given of a theoretical study of the relaxation of molecules in liquids. Most of the discussion will centre round the behaviour of a single rigid molecule, and the mathematical treatment will be analytical.
The relaxation is due to steady state thermal motion and this gives rise to the use
of time correlation functions. In most cases these are correlation functions of
spherical harmonics of order 0, 1 or 2, or of functions closely related to them.

We shall in the next Section introduce the stochastic rotation operator and
relate it to correlation functions. In the following Section the values of the ensemble average of the operator for different molecular models will be collected together. In Section 4 attention is directed to special features of the calculations
for specific relaxation processes. Relaxation of non-rigid molecules will be discussed briefly in the final Section.

2. The Stochastic Rotation Operator

We define a laboratory coordinate system S as the set of rectangular cartesian axes
with origin at the centre of mass of a molecule under investigation and with axes in
directions fixed with respect to the laboratory. We next define a molecular coordinate system S' as the set with the same origin as before and with coordinate axes
coinciding with the principal axes of inertia through the centre of mass. The orientation of S' at time t relative to its orientation at time zero may be specified
by the Euler angles $\alpha(t)$, $\beta(t)$, $\gamma(t)$, which of course vanish for t equal to zero.
The operator $R(t)$ related to the rotation of the molecule from its orientation at
time zero to its orientation at time t is given by [1]

$$R(t) = \exp(-\alpha(t)J_x) \exp(-\beta(t)J_y) \exp(-i\gamma(t)J_z), \qquad (2.1)$$

where J_x, J_y, J_z are the infinitesimal generators of rotation. We see that $R(0)$
is the identity operator, which we shall denote by E. When we are concerned, as we
shall be, with rotations caused by the thermal motion of the environment of the molecule, the Euler angles are random variables and $R(t)$ is called a stochastic rota-

tion operator.

The operator $R(t)$ of (2.1) occurs in the study of various relaxation processes. Thus for dielectric relaxation the complex polarizability $\alpha(\omega)$ for orientational polarization is expressible by [2]

$$\alpha(\omega) = \frac{1}{3kT} M^+ \left[1 - i\omega \int_0^\infty <R(t)>_{j=1} e^{-i\omega t} \right] M,$$ (2.2)

where M is the column matrix with components M_1, M_2, M_3 along the axes of S' and M^+ is the adjoint of M. In (2.2) $<R(t)>$ is the ensemble average of $R(t)$ subject to the condition that $R(0) = E$ and $j = 1$ signifies that we take the appropriate three-dimensional matrix representation of $<R(t)>$.

The stochastic rotation operator plays a central role in the study of nuclear magnetic relaxation by intramolecular interactions. There the interaction Hamiltonian $\hbar G(t)$ is a rotational invariant expressible by [3]

$$\hbar\, G(t) = \hbar \sum_{q=-\ell}^{\ell} (-)^q H_{\ell,-q}(t)\, A_{\ell q} ,$$ (2.3)

where $A_{\ell q}$ is the q-th component of a spherical tensor operator or rank ℓ. It follows from the rotational invariance of $\hbar G(t)$ that $H_{\ell q}(t)$ is the q-th component of a spherical tensor of rank ℓ. If the $H_{\ell q}(t)$ are commuting functions of orientational variables only, we denote by $\underset{\sim}{a}$ a unit vector specifying a direction fixed with respect to the molecule and by $\underset{\sim}{a}(t)$ the unit vector specifying at time t the same direction with respect to the laboratory system S and we express $H_{\ell q}(t)$ as $H_{\ell q}(\underset{\sim}{a}(t))$. Then by the extension of a theorem established for spherical harmonics [2] it is found that the correlation function

$$<H_{\ell q}^*(\underset{\sim}{a}(0))H_{\ell q'}(\underset{\sim}{a}(t))> = \frac{\delta_{qq'}}{2\ell+1} \sum_{n,n'=-\ell}^{\ell} <R^+(t)>_{nn'}^{\ell}\, H_{\ell n'}^*(\underset{\sim}{a})H_{\ell n}(\underset{\sim}{a}).$$ (2.4)

In this equation the subscripts nn' indicate the matrix representative with respect to the basis $Y_{\ell,-\ell}$, $Y_{\ell,-\ell+1}$, $\ldots$ $Y_{\ell\ell}$ associated with the molecular coordinate system, which is usually taken to be S'. Equation (2.4) shows that the cross-correlation functions of the $H_{\ell q}(\underset{\sim}{a}(t))$ vanish and that the autocorrelation functions are all equal to $<H_{\ell 0}^*(\underset{\sim}{a}(0))H_{\ell 0}(\underset{\sim}{a}(t))>$.

In order to see the relevance of (2.4) we introduce the operator $\tau(\omega)$ by

$$\tau(\omega) = \int_0^\infty <R^+(t)>\, e^{-i\omega t}dt .$$ (2.5)

Since spherical tensor components satisfy $H_{\ell,-q} = (-)^q H_{\ell q}^*$, it follows that $H_{\ell 0}$ is real and therefore that the autocorrelation function of $H_{\ell q}(\underset{\sim}{a}(t))$ is real. Moreover, since the motion is in a steady state,

$$<H_{\ell 0}(\underset{\sim}{a}(0))H_{\ell 0}a(-t)> \;=\; <H_{\ell 0}(\underset{\sim}{a}(t))H_{\ell 0}(a(0))> \;=\; <H_{\ell 0}(\underset{\sim}{a}(0))H_{\ell 0}(a(t))>, \qquad (2.6)$$

so that the autocorrelation function is an even function of the time. We define the spectral density $j(\omega)$ of $H_{\ell q}(\underset{\sim}{a}(t))$ by

$$j(\omega) = \frac{1}{2} \int_{-\infty}^{\infty} <H_{\ell 0}(\underset{\sim}{a}(0))H_{\ell 0}(\underset{\sim}{a}(t))> e^{-i\omega t}\, dt \;. \qquad (2.7)$$

We then deduce from (2.6) and (2.7) that

$$j(\omega) = \frac{1}{2} \int_{0}^{\infty} <H_{\ell 0}(\underset{\sim}{a}(0))H_{\ell 0}(\underset{\sim}{a}(t))> e^{-i\omega t} + \text{c.c.}, \qquad (2.8)$$

where c.c. denotes complex conjugate. We now define the operator $\sigma(\omega)$ by

$$\sigma(\omega) = \tau(\omega) + \tau(\omega)^{+} \;. \qquad (2.9)$$

Substituting from (2.4) into (2.8) and employing (2.5) and (2.9) we obtain

$$j(\omega) = \frac{1}{2(2\ell+1)} \sum_{n,n'=-\ell}^{\ell} H_{\ell n}(\underset{\sim}{a})\sigma(\omega)_{nn'}^{\ell} \, H_{\ell n'}(\underset{\sim}{a})^{*}. \qquad (2.10)$$

The nuclear magnetic relaxation times are expressed in terms of $j(\omega)$, so if we know $<R(t)>$ we can find the relaxation times.

We next consider the calculation of correlation times. By definition the correlation time is the integral from zero to infinity of the normalized autocorrelation function. Thus for the spherical harmonic $Y_{\ell q}(\theta(t),\phi(t))$ the correlation time

$$\tau_{\ell} = \frac{\displaystyle\int_{0}^{\infty} <Y_{\ell q}^{*}(\theta(0),\phi(0))\, Y_{\ell q}(\theta(t),\phi(t))>dt}{<Y_{\ell q}^{*}(\theta(0),\phi(0))\, Y_{\ell q}(\theta(0),\phi(0))>} \;. \qquad (2.11)$$

Since $R(0)$ is the identity operator, we deduce from (2.4) that

$$<Y_{\ell q}^{*}(\theta(0),\phi(0))Y_{\ell q}(\theta(0),\phi(0))> = \frac{1}{2\ell+1} \sum_{n=-\ell}^{\ell} \left| Y_{\ell n}(\theta',\phi') \right|^{2} = \frac{1}{4\pi} \;, \qquad (2.12)$$

where θ',ϕ' are the angles specifying the fixed direction of $\theta(t),\phi(t)$ with reference to the molecule and we use a well-known result for the sum [4]. Hence from (2.4), (2.5), (2.11) and (2.12)

$$\tau_{\ell} = \frac{4\pi}{2\ell+1} \sum_{n,n'=-\ell}^{\ell} Y_{\ell n}(\theta',\phi')\, \tau(0)_{nn'}^{\ell} \, Y_{\ell n'}^{*}(\theta',\phi'). \qquad (2.13)$$

It is important to distinguish τ_{ℓ} from another orientational correlation time τ_{ℓ}', which is the correlation time of $H_{\ell q}(\underset{\sim}{a}(t))$. Proceeding as for

$Y_{\ell q}(\theta(t),\phi(t))$ we deduce that

$$\tau_\ell' = \frac{\displaystyle\sum_{n,n'=-\ell}^{\ell} H_{\ell n}(\underset{\sim}{a})\tau(0)_{nn'}^{\ell} H_{\ell n'}(\underset{\sim}{a})^*}{\displaystyle\sum_{n=-\ell}^{\ell} \left|H_{\ell n}(\underset{\sim}{a})\right|^2} \quad . \tag{2.14}$$

If the direction of the vector $\underset{\sim}{a}$ is specified by the angles θ',ϕ', we see from (2.13) and (2.14) that a sufficient condition for τ_ℓ and $\tau_{\ell'}$ to be equal is

$$H_{\ell q}(\underset{\sim}{a}) = c\, Y_{\ell q}(\theta',\phi'), \tag{2.15}$$

where c is a constant independent of q .

When relaxation times are being expressed in terms of the spectral density, it will usually be found adequate to put the argument of j equal to zero. This is called the extreme narrowing approximation. We see from (2.10) that

$$j(0) = \frac{1}{2(2\ell+1)} \sum_{n,n'=-\ell}^{\ell} H_{\ell n}(\underset{\sim}{a})\sigma(0)_{nn'}^{\ell} H_{\ell n'}(\underset{\sim}{a})^* \quad . \tag{2.16}$$

Now it will be found in the next section that $\tau(0)_{nn'}^{\ell}$ is a real and symmetric matrix for all the molecular models that we encounter. Thus from (2.9) and (2.16)

$$j(0) = \frac{1}{2\ell+1} \sum_{n,n'=-\ell}^{\ell} H_{\ell n}(\underset{\sim}{a})\, \tau(0)_{nn'}^{\ell}\, H_{\ell n'}(\underset{\sim}{a})^* \quad . \tag{2.17}$$

Comparing (2.17) with (2.14) we deduce that

$$j(0) = \frac{\tau'_\ell}{2\ell+1} \sum_{n=-\ell}^{\ell} \left|H_{\ell n}(\underset{\sim}{a})\right|^2 \quad . \tag{2.18}$$

The ratio of $j(0)$ to τ_ℓ' depends only on the $H_{\ell q}(\underset{\sim}{a})$: it depends on the interaction causing the relaxation and it is independent of the molecular model. Hence to find the ratio for any molecule it will suffice to calculate it for a spherical molecule. In the extreme narrowing approximation the nuclear magnetic relaxation rates are constant multiples of $j(0)$, and so the ratio of any relaxation rate to τ_ℓ' does not depend on the shape of the molecule.

3. Molecular Models

In the previous section we have seen that the ensemble average $<R(t)>$ of the stochastic rotation operator plays a central role in the calculation of correlation times and of nuclear magnetic relaxation times. We shall now report the values of $<R(t)>$ for different models of a rigid molecule referring the reader elsewhere [5] for calculational details and for approximations to a higher order than those provided here.

The calculations were performed by a theory of rotational Brownian motion based on Langevin-type equations. The theory is more general than rotational diffusion theory in that it includes the effects of the inertia of the molecule. The inertial theory is very important for the study of dielectric relaxation but not for nuclear magnetic relaxation. The rotational diffusion theory is treated as a limiting case of the inertial theory [6], and the values of $<R(t)>$ for the former are deduced from those of the latter theory.

The calculation of $<R(t)>$ was performed in the molecular coordinate system S' described at the beginning of the previous section. The principal moments of inertia of the molecule are denoted by I_1, I_2, I_3. The components of angular velocity at time t are $\omega_1(t)$, $\omega_2(t)$, $\omega_3(t)$, and it is assumed that the components of the frictional torque resisting the motion may be expressed as $I_1 B_1 \omega_1(t)$, $I_2 B_2 \omega_2(t)$, $I_3 B_3 \omega_3(t)$. The infinitesimal generators of rotation are denoted in S' by J_1, J_2, J_3.

We now consider several molecular models.

3a. The Spherical Molecule

For the spherical model

$$I_2 = I_3 = I_1 , \quad B_2 = B_3 = B_1 = B , \quad \text{say,}$$

and in the inertial theory

$$<R(t)> = \left[E + \kappa J^2 (1-e^{-Bt}) + \kappa^2 \left\{ J^2 \left[\frac{5}{4} - (Bt+1)e^{-Bt} - \frac{1}{4} e^{-2Bt} \right] \right. \right.$$

$$\left. \left. + J^4 [\frac{1}{2} - e^{-Bt} + \frac{1}{2} e^{-2Bt}] \right\} + \ldots \right]$$

$$\times \exp[-\kappa B(1 + \frac{1}{2}\kappa + \frac{7}{12} \kappa^2 + \ldots)J^2 t], \tag{3.1}$$

where $J^2 = J_1^2 + J_2^2 + J_3^2$ and $\kappa = kT/(I_1 B^2)$, a small dimensionless quantity. For a $2\ell+1$ - dimensional representation $J^2 = \ell(\ell+1)E$, and it follows from (2.5) and (2.9) that

$$\tau(\omega)_{mn}^{\ell} = \delta_{mn} \left\{ \frac{1}{B(G'+i\omega')} + \frac{\ell(\ell+1)\kappa}{B} \frac{1}{(G'+i\omega')(1+G'+i\omega')} + \ldots \right\}$$

$$\sigma(\omega)_{mn}^{\ell} = \delta_{mn} \left\{ \frac{2G'}{B(G'^2+\omega'^2)} + \frac{2\ell(\ell+1)\kappa}{B} \frac{G'(1+G') - \omega'^2}{(G'^2+\omega'^2)[(1+G')^2+\omega'^2]} + \ldots \right\},$$

$$\tag{3.2}$$

where

$$\omega' = \omega/B , \quad G' = \ell(\ell+1)\kappa(1 + \frac{1}{2}\kappa + \frac{7}{12} \kappa^2 + \ldots). \tag{3.3}$$

Since $<R(t)>_{mn}^{\ell}$ is proportional to the Kronecker delta, we deduce from (2.12) and (2.13) that

$$\tau_{\ell} = \tau(0)_{oo}^{\ell} \; . \tag{3.4}$$

The rotational diffusion results corresponding to (3.1) - (3.4) are obtained by performing the limiting process

$$I_1 \to 0, \quad B \to 0, \quad I_1B \text{ finite,} \tag{3.5}$$

from which it follows that $\kappa \to 0$ but that

$$\kappa B \to \frac{kT}{I_1 B} \; ,$$

the diffusion coefficient for the sphere. We also deduce that

$$<R(t)> = \exp\left[-\frac{kT}{I_1 B} J^2 t\right] \tag{3.6}$$

$$\tau_{\ell} = \frac{I_1 B}{\ell(\ell+1)kT} \tag{3.7}$$

$$\sigma(\omega)_{mn}^{\ell} = \frac{2\delta_{mn} I_1 B}{\ell(\ell+1)} \; \frac{1}{1 + \left(\frac{I_1 B \omega}{\ell(\ell+1)kT}\right)^2} \; . \tag{3.8}$$

3b. The Linear Molecule

We take the third coordinate axis along the molecule, so that the other two axes are perpendicular to the molecule and to one another. Then $I_2 = I_1$, $B_2 = B_1 = B$ and in the inertial theory [7]

$$<R(t)> = \left\{E + \kappa(J^2 - J_3^2)(1 - e^{-Bt}) + \kappa^2\left[(2J^2 - 5J_3^2)(\frac{5}{4} - Bte^{-Bt} - e^{-Bt} - \frac{1}{4} e^{-2Bt})\right.\right.$$
$$\left.\left. + (J^2 - J_3^2)^2(\frac{1}{2} - e^{-Bt} + \frac{1}{2} e^{-2Bt})\right] + \dots\right\} \exp(-BGt), \tag{3.9}$$

where

$$G = \kappa\left\{(J^2 - J_3^2) + \kappa(J^2 - \frac{5}{2} J_3^2) + \kappa^2\left[\frac{8}{3} J^2 - \frac{29}{3} J_3^2 + 4(J^2 - J_3^2)J_3^2\right] + \dots\right\} \; . \tag{3.10}$$

In the $2\ell+1$-dimensional representation $J^2 = \ell(\ell+1)E$ and the matrix representation of J_3^2 is the diagonal matrix with consecutive elements $\ell^2, (\ell-1)^2, \dots 1, 0, 1, \dots (\ell-1)^2, \ell^2$.

The matrix representative of $\tau(\omega)^\ell_{mn}$ will be diagonal but will not be expressible in a simple form as in (3.2).

Let us therefore restrict our considerations to rotational diffusion results for $\ell = 2$. Then from (3.5), (3.9) and (3.10)

$$<R(t)> = \exp\left[-\frac{kT}{I_1 B}(J^2 - J_3^2)t\right] . \tag{3.11}$$

It may then be deduced that

$$\tau(0)^2_{00} = \frac{I_1 B}{6kT}$$

$$\tau(0)^2_{11} = \tau(0)^2_{-1,-1} = \frac{I_1 B}{5kT}$$

$$\tau(0)^2_{22} = \tau(0)^2_{-2,-2} = \frac{I_1 B}{2kT} \tag{3.12}$$

$$\sigma(\omega)^2_{00} = \frac{I_1 B}{3kT\left[1 + \left(\frac{I_1 B\omega}{6kT}\right)^2\right]}$$

$$\sigma(\omega)^2_{11} = \sigma(\omega)^2_{-1,-1} = \frac{2I_1 B}{5kT\left[1 + \left(\frac{I_1 B\omega}{5kT}\right)^2\right]}$$

$$\sigma(\omega)^2_{22} = \sigma(\omega)^2_{-2,-2} = \frac{I_1 B}{kT\left[1 + \left(\frac{I_1 B\omega}{2kT}\right)^2\right]} , \tag{3.13}$$

the other elements of $\tau(0)^2_{mn}$ and $\sigma(\omega)^2_{mn}$ vanishing. We see that the values of $\tau(0)^2_{00}$ and $\sigma(\omega)^2_{00}$ in (3.12) and (3.13) agree with those in (3.7) and (3.8) for the sphere. We note that (3.4) does not hold for the linear molecule: it holds for the sphere because $<R(t)>$ is then a multiple of the identity operator.

3c. The Asymmetric Molecule

The rotational Brownian motion of an asymmetric top was investigated by Ford, Lewis and Mc Connell [2]. Many of their results are too lengthy to reproduce and we shall just recall a few of the basic ones. In the inertial theory

$$\langle R(t) \rangle = \left\{ E + \sum_{i=1}^{3} \frac{kT}{I_i B_i^2} (1 - e^{-B_i t}) J_i^2 + \dots \right\}$$

$$\times \ \exp \left\{ \left[-\sum_{i=1}^{3} D_i \ J_i^2 + \frac{i}{3} P \frac{(kT)^2}{I_1 I_2 I_3} \sum_{i=1}^{3} \frac{B_j - B_k}{B_j^2 B_k^2} + \dots \right] t \right\} ,$$

$$(3.14)$$

where i, j, k is a cyclic permutation of 1,2,3, the diffusion coefficient $D_i = kT/(I_i B_i)$ plus a small correction, and

$$P = J_1 J_2 J_3 + J_1 J_3 J_2 + J_2 J_3 J_1 + J_2 J_1 J_3 + J_3 J_1 J_2 + J_3 J_2 J_1 .$$

Explicit expressions for $\tau(\omega)_{mn}^{\ell}$ and $\sigma(\omega)_{mn}^{\ell}$ may be deduced from (3.14).

The rotational diffusion limit of (3.14), obtained by extending (3.5) to three dimensions, is

$$\langle R(t) \rangle = \exp \left\{ - \left[D_1 J_1^2 + D_2 J_2^2 + D_3 J_3^2 \right] t \right\} ,$$

$$(3.15)$$

where the diffusion coefficients are now given by

$$D_1 = \frac{kT}{I_1 B_1} , \quad D_2 = \frac{kT}{I_2 B_2} , \quad D_3 = \frac{kT}{I_3 B_3} .$$

$$(3.16)$$

3d. The Symmetric Molecule

When the third axis of the molecule is an axis of rotational symmetry C_n with $n \geq 3$, we have

$$I_2 = I_1 , \quad B_2 = B_1 , \quad D_2 = D_1 .$$

$$(3.17)$$

Then the matrix representatives of $\tau(\omega)$ and $\sigma(\omega)$ are diagonal. On substituting from (3.17) into (3.15) we have in the rotational diffusion limit

$$\langle R(t) \rangle = \exp \left\{ - \left[D_1 J^2 + (D_3 - D_1) J_3^2 \right] t \right\} .$$

$$(3.18)$$

In applications we shall be interested chiefly in five-dimensional representation of $\tau(\omega)$ and $\sigma(\omega)$. For these we obtain

$$\tau(\omega)_{00}^{2} = (6D_1 + i\omega)^{-1}$$

$$\tau(\omega)_{11}^{2} = \tau(\omega)_{-1,-1}^{2} = (5D_1 + D_3 + i\omega)^{-1}$$

$$\tau(\omega)_{22}^{2} = \tau(\omega)_{-2,-2}^{2} = (2D_1 + 4D_3 + i\omega)^{-1} .$$

It then follows from (2.9) that

$$\sigma(\omega)^2_{mm} = \tau(\omega)^2_{mm} + \tau(\omega)^{2^*}_{mm} .$$

3e. Comparison of Results for Different Models

We shall find in the next section that for some relaxation processes the only element $<R(t)>^{\ell}_{mn}$ required in the course of calculation is that with $m = n = 0$. If now we compare the values of $<R(t)>$ in (3.6), (3.11) and (3.18) and remember that $(J^2_3)^{\ell}_{00}$ vanishes, we deduce that in rotational diffusion theory $<R(t)>^{\ell}_{00}$ has the same values for the spherical, linear and symmetric rotator models. Consequently we shall obtain for the above processes the same expressions for $\tau(\omega)^{\ell}_{00}$, $\sigma(\omega)^{\ell}_{00}$, $j(\omega)$ and therefore for the correlation times τ_{ℓ}, τ'_{ℓ} and the nuclear magnetic relaxation times, if they appear in the process.

It should, however, be pointed out that this result cannot be extended to a circular plate molecule [8].

4. Application to Relaxation Processes

4a. Dielectric Relaxation [5]

We had in (2.2) a relation between complex polarizability and $<R(t)>$. It is usually more convenient for the discussion of experiments to employ complex permittivity $\varepsilon(\omega)$ rather than polarizability. When we have a very dilute solution of a liquid dielectric in a nonpolar solvent,

$$\frac{\varepsilon(\omega) - \varepsilon_{\infty}}{\varepsilon(\omega) - \varepsilon_{\infty}} = 1 - i\omega \int_0^{\infty} <R(t)>^1_{00} e^{-i\omega t} dt. \tag{4.1}$$

If we write

$$\varepsilon(\omega) = \varepsilon'(\omega) - i\varepsilon''(\omega),$$

the refractive index $n(\omega)$ and the absorption coefficient $a(\omega)$ are given by

$$n(\omega) = \left[\frac{|\varepsilon(\omega)| + \varepsilon'(\omega)}{2} \right]^{1/2} \tag{4.2}$$

$$a(\omega) = \frac{2^{1/2}\omega}{c} \left[|\varepsilon(\omega)| - \varepsilon'(\omega) \right]^{1/2} . \tag{4.3}$$

We see from Section 3e. and (4.1) that in rotational diffusion theory (4.2) and (4.3) will provide the same expressions for $n(\omega)$ and $a(\omega)$ when the molecule is spherical, linear or a symmetric rotator. It is found that the graph of $a(\omega)$ as a function of the angular frequency ω has a plateau in the far infrared region. This Debye plateau, as it is called, of the absorption curve disappears, if in place of (3.6) we employ (3.1) for the value of $<R(t)>$ associated with the spherical molecule. We cannot immediately take over this result to linear and symmetric top molecules in the inertial theory, but in fact it may be proved by direct calculation that the shape of the absorption curve is the same for the three molecular models [9].

4b. Nuclear Magnetic Relaxation [3]

We shall indicate how results from Section 2 may be applied to nuclear magnetic relaxation by intramolecular interactions. The simplest interaction from the mathematical viewpoint is dipolar [10]. For two spins I and S distant r apart the interaction Hamiltonian is given by

$$\hbar\, G(t) = -\left(\frac{6\pi}{5}\right)^{1/2} \gamma_I \gamma_S\, \hbar^2 \sum_{q=-2}^{2} (-)^q\, r^{-3}\, Y_{2,-q}(\theta(t),\phi(t)) A^{(q)}, \quad (4.4)$$

where γ_I, γ_S are the gyromagnetic ratios of the nuclei,

$$A^{(0)} = \frac{2}{6^{1/2}} \left[3 I_0 S_0 - (\underset{\sim}{I} \cdot \underset{\sim}{S}) \right]$$

$$A^{(\pm 1)} = \mp (I_0 S_\pm + I_\pm S_0); \quad A^{(\pm 2)} = I_\pm S_\pm \ .$$

Hence, from (2.3) and (4.4)

$$H_{2q}(t) = -\left(\frac{6\pi}{5}\right)^{1/2} \gamma_I \gamma_S\, \hbar^2 r^{-3}\, Y_{2q}(\theta(t),\phi(t)). \qquad (4.5)$$

Since $H_{2q}(t)$ is proportional to $Y_{2q}(\theta(t),\phi(t))$, Eq. (2.15) will be obeyed in the molecular system S' and therefore the correlation time for $H_{2q}(t)$ is equal to that for $Y_{2q}(\theta(t))$, which is denoted by τ_2.

We calculate $j(\omega)$ from (2.10), (4.5) and the values of $\sigma(\omega)^2_{nn}$, given in Section 3 for different molecular models. The spin-lattice relaxation time T_1 and the spin-spin relaxation time T_2 are expressed in terms of $j(\omega)$ by

$$\frac{1}{T_2} = I(I+1)\left\{ \frac{4}{3} j(\omega_0) + \frac{16}{3} j(2\omega_0)\right\}$$

$$\frac{1}{T_2} = I(I+1)\left\{ 2j(0) + \frac{10}{3} j(\omega_0) + \frac{4}{3} j(2\omega_0)\right\}$$

for like spins and by a more complicated set of equations for unlike spins. In the extreme narrowing approximation the calculation for like spins may be shortened by putting

$$\frac{1}{T_1} = \frac{1}{T_2} = \frac{20}{3} I(I+1) j(0)$$

and employing (2.18) to find j(0).

An example of a relaxation mechanism, whose $H_{2q}(t)$ is not proportional to $Y_{2q}(\theta(t),\phi(t))$, is anisotropic chemical shift [11]. We begin the study by taking a molecular coordinate system S'' with origin at the nucleus and axes in directions that will diagonalize the tensor associated with the relaxation process. In S'' we have

$$H''_{20} = \frac{1}{2} \gamma\delta_{z''} \; , \; H''_{2,\pm1} = 0, \; H''_{2,\pm2} = \frac{\zeta}{2\,(6)^{1/2}} \; \gamma\delta_{z''} \; , \qquad (4.6)$$

where $\delta_{x''}$, $\delta_{y''}$, $\delta_{z''}$ are the diagonal elements of the tensor, ζ is the asymmetry parameter defined by

$$\delta_{x''} = -\frac{1}{2} (1-\zeta)\delta_{z''} \; , \; \delta_{y''} = -\frac{1}{2} (1+\zeta)\delta_{z''}$$

and γ is the gyromagnetic ratio of the nucleus.

In order to employ the results of the previous section we must transform from S'' to S'. While these two systems have different origins, we may regard the transformation as a rotation about a common origin, if we agree to neglect rototranslational effects. Indeed the rotation about any axis is equivalent to an equal rotation about a parallel axis together with a translation. If we go from S' to S'' by rotating through the constant Euler angles α, β, γ, then from (4.6)

$$H'_{2m} = \sum_{S=-2}^{2} D_{mS}^{2*}(\alpha,\beta,\gamma)H''_{2S}$$

$$= (\frac{\pi}{5})^{1/2} \gamma\delta_{z''} Y^*_{2,-m}(\beta,\alpha) + \frac{\zeta\gamma\delta_{z''}}{2(6)^{1/2}} \left\{ D_{m2}^{2*}(\alpha,\beta,\gamma) + D_{m,-2}^{2*}(\alpha,\beta,\gamma) \right\}. \quad (4.7)$$

When the asymmetry parameter is negligible, Eq. (4.7) reduces to

$$H'_{2m} = (-)^m(\frac{\pi}{5})^{1/2} \gamma\delta_{z''} Y_{2m}(\beta,\alpha). \qquad (4.8)$$

Apart from the multiplying factor $(-)^m$, which from the explicit value of $\tau(\omega)_{mn}^2$ for molecular models may be shown to be irrelevant, Eq. (4.8) agrees with (2.15). Hence τ'_2 is equal to τ_2. The calculation of relaxation times may be completed in the manner summarized for dipolar interaction.

When the relaxation arises from electric quadrupole interaction, we choose the molecular coordinate system S'' to be that with origin at the centre of mass of the relaxing nucleus and with axes in the directions that make the field gradient tensor diagonal [12]. Then

$$H''_{20} = \frac{1}{2} eq \,, \quad H''_{2,\pm 1} = 0 \,, \quad H''_{2,\pm 2} = \frac{\eta eq}{2(6)^{1/2}} \qquad (4.9)$$

where eq is the field gradient of greatest magnitude and η is the asymmetry parameter for quadrupole interaction. On comparing (4.9) with (4.6) we see that we can
deduce many of the results for quadrupolar relaxation from those derived in the study
of anisotropic chemical shift.

We now draw attention to a theorem related to the above three nuclear magnetic
relaxation mechanisms, when the calculations are made according to rotational diffusion theory [13]. The formulae for $j(\omega)$, and consequently for relaxation times, obtained in the case of spherical molecules are applicable to linear and symmetric rotator molecules, if and only if the asymmetry parameters ζ and η vanish so that
the relevant tensors have axes of cylindrical symmetry. For dipolar interaction the
line of dipoles is itself the axis of cylindrical symmetry. In addition these axes
must coincide with the linear molecule or they must be parallel to the axis of rotational symmetry of the symmetric molecule. When these conditions are satisfied, it
will be seen that the only value of $\langle R(t) \rangle^2_{mn}$ required for the calculation of correlation and relaxation times is that where $m = n = 0$. The theorem then follows from
Section 3e.

The investigation of nuclear magnetic relaxation that results from spin-rotational interactions presents certain features that make calculations more difficult
[14]. In Eq. (2.3) we now have for the molecular coordinate system S'

$$H_{1q}(t) = \hbar^{-1} \sum_{\nu=1}^{3} I_\nu \sum_{m=-1}^{1} b^*_{m\nu} D^{1*}_{qm}(\alpha(t),\beta(t),\gamma(t))\omega_\nu(t), \quad A_q = I_q, \qquad (4.10)$$

where the notation is explained in ref. 3. The $H_{1q}(t)$ is a function of both the
angular variables $\alpha(t),\beta(t),\gamma(t)$ and the components $\omega_\nu(t)$ of angular velocity of
the molecule. Hence we can no longer apply (2.4). Employing the values of $\langle R(t) \rangle$
given in Section 3 for the various molecular models we must first calculate $R(t)$
and then $\langle R(t)\omega_\mu(t)\omega_\nu(0) \rangle$. It is found that

$$j(0) = \frac{kT\tau'_1}{3\hbar^2} \sum_{\mu=1}^{3} \sum_{m=-1}^{1} (-)^m b_{m\mu} b_{-m,\mu} I_\mu \,. \qquad (4.11)$$

On account of the presence of the moment of inertia I_μ in (4.11), the ratio of $j(0)$
to τ'_1 is no longer independent of the molecular model, as it was in (2.18). Analytical expressions for correlation and relaxation times have been derived for spherical,
linear and symmetric rotator molecular models in both inertial and rotational diffusion theories [7,14,15].

5. Relaxation of Non-Rigid Molecules

While we have so far regarded as a rigid body the molecule in whose relaxation we are interested, it must be realized that such a picture is unphysical for the large major- ity of molecules. It is true that methane, ethylene, benzene and their simple deriva- tives may properly be considered as rigid bodies. Most organic molecules contain these rigid bodies as "bricks". The bricks are rotating relative to each other in the organic molecule, the axes of rotation usually being C-C, C-N, C-O or C-S bonds [16]. These internal rotations should influence the expressions for the nuclear magnetic relaxation rates T_1^{-1}, T_2^{-1} .

A basic contribution to the study of internal rotations was made by Woessner [17], who considered relaxation by dipole-dipole interaction. The line of dipoles ro- tates about an axis fixed in the molecule, with which it makes a constant angle. The molecule itself rotates with respect to the laboratory system. It is assumed that the reorientation of the line of dipoles about the axis of rotation is not influenced by the random rotation of the molecule, and vice versa. This assumption of the mutual independence of the two motions is less acceptable nowadays [16].

When discussing nuclear magnetic relaxation by anisotropic chemical shift and by quadrupolar interaction we employed the laboratory coordinate system S and two molecular coordinate systems S' and S". In order to discuss internal rotations let us simplify the picture for dipolar relaxation by supposing that the axis of rotation and the dipole-dipole axis both pass through the centre of mass of the molecule. We now take S and S' as before and for S" we take a molecular coordinate system with third axis in the direction of the axis of internal rotation. We denote by α', β', γ' the constant Euler angles which specify the orientation of S" with respect to S'. Let θ be the constant angle which the line of dipoles makes with the axis of internal rotation and let $\phi(t)$ be the azimuthal angle in S". Then a spherical harmonic $Y_{2S}''(\theta,\phi(t))$ in S" transforms to H_{2m}' in S' given by

$$H_{2m}' = \sum_{S=-2}^{2} D_{mS}^{2*} (\alpha',\beta',\gamma') \, Y_{2S}''(\theta,\phi(t)).$$

We see from this equation that H_{2m}' is time dependent, and so it cannot be used for substitution into the right hand side of (2.4). Hence the treatment based on the stoch- astic rotation operator and employed in the preceding sections for rigid molecules is not applicable here.

Progress in the investigation of internal rotations has been made by employing probability density functions [18]. Working in a rotational diffusion theory Hertz [19] assumed that the conditional probability density function for the system is the product of that for the rigid molecule and that for the internal rotation. Taking the molecule to be a symmetric top, the first coordinate axis being the axis of symmetry, he deduced that the autocorrelation function $G_2(t)$ of $Y_{2m}(\theta(t),\phi(t))$ is given by

$$G_2(t) = \frac{1}{16\pi} \left\{ \frac{1}{4}(3\cos^2\vartheta-1)^2(3\cos^2\beta-1)^2 e^{-6D_1 t} + \frac{9}{8}\sin^2 2\vartheta \sin^2 2\beta\, e^{-(6D_1+D^*)t} \right.$$

$$+ \frac{9}{8}\sin^4\vartheta \sin^4\beta\, e^{-(6D_1+4D^*)t} + \frac{3}{4}(3\cos^2\vartheta-1)^2\sin^2 2\beta\, e^{-(5D_1+D_2)t}$$

$$+ \frac{3}{2}\sin^2 2\vartheta(4\cos^4\beta-3\cos^2\beta+1)e^{-(5D_1+D_2+D^*)t}$$

$$+ \frac{3}{4}\sin^4\vartheta \sin^2\beta(1-\cos^2\beta)e^{-(5D_1+D_2+4D^*)t}$$

$$+ \frac{3}{4}(3\cos^2\vartheta-1)^2\sin^4\beta\, e^{-(2D_1+4D_2)t}$$

$$+ \frac{3}{2}\sin^2 2\vartheta \sin^2\beta(1+\cos^2\beta)e^{-(2D_1+4D_2+D^*)t}$$

$$\left. + \frac{3}{8}\sin^4\vartheta(1+6\cos^2\beta+\cos^4\beta)e^{-(2D_1+4D_2+4D^*)t} \right\}.$$

In this equation β is the angle between the axis of internal rotation and the third axis of S', ϑ is the fixed angle which the line of dipoles makes with the axis of symmetry, D_1 and D_2 are the diffusion coefficients for the rotation of the molecule as a whole and D^* is the diffusion coefficient for the internal rotation.

Further discussion of theory and experiment of internal rotations may be obtained in the papers of Hertz and his collaborators [16,19,20,21]. It appears that the formulation of a theory of the relaxation of non-rigid molecules in liquids is still at a preliminary stage.

References

[1] M.E. Rose, Elementary Theory of Angular Momentum (Wiley, New York, 1957)

[2] G.W. Ford, G.T. Lewis and J. Mc Connell, Rotational Brownian motion of an asymmetric top. Physical Review A19, 907-919 (1979)

[3] J. Mc Connell, The Theory of Nuclear Magnetic Relaxation in Liquids (Cambridge Univ. Press, in course of publication)

[4] Ref. 1, eq. (4.28)

[5] J. Mc Connell, Rotational Brownian Motion and Dielectric Theory (Academic Press, London, 1980)

[6] J. Mc Connell, Debye limit of the stochastic rotation operator, Physica 128A, 611-630 (1984)

[7] J. Mc Connell, Nuclear magnetic spin-rotational relaxation times for linear molecules, Physica 112A, 488-504 (1982)

[8] Ref. 3, section 6.4

[9] Ref. 5, Chapter 14

[10] J. Mc Connell, Theory of nuclear magnetic relaxation by dipolar interaction,

Physica 135A, 38-62 (1986)

[11] J. Mc Connell, Theory of nuclear magnetic relaxation by anisotropic chemical shift, Physica 127A, 152-172 (1984)

[12] J. Mc Connell, Nuclear magnetic relaxation by quadrupole interactions in non-spherical molecules, Physica 117A, 251-264 (1983)

[13] J. Mc Connell, Correlation and nuclear magnetic relaxation times, Physica 138A, 367-381 (1986)

[14] J. Mc Connell, Stochastic differential equation study of nuclear magnetic relaxation by spin-rotational interactions, Physica 111A, 85-113 (1982)

[15] J. Mc Connell, Nuclear magnetic spin-rotational relaxation times for symmetric molecules, Physica 112A, 479-487 (1982)

[16] T. Frech and H.G. Hertz, Rotational, internal rotational and translational motion of acetic acid and dimethyl sulfoxide in their liquid mixture (with some results for DMSO dissolved in Me OH), Journal of Molecular Liquids 30, 237-282 (1985)

[17] D.E. Woessner, Spin relaxation processes in a two-proton system undergoing anisotropic reorientation, The Journal of Chemical Physics 36, 1-4 (1962)

[18] H. Versmold, Time correlation functions for internal and anisotropic rotational motion of molecules, Zeitschrift für Naturforschung 25a, 367-372 (1970)

[19] H.G. Hertz, The problem of intramolecular rotation in liquids and nuclear magnetic relaxation, Progress in Nuclear Magnetic Resonance Spectroscopy 16, 115-162 (1983)

[20] M.C. Ansari and H.G. Hertz, An NMR investigation of internal molecular motion in liquids ethanol, Zeitschrift für Physikalische Chemie Neue Folge 137, 187-220 (1983)

[21] T. Frech and H.G. Hertz, Rotational, internal rotational and translational motion of liquid i-propanol, Berichte der Bunsen-Gesellschaft für Physikalische Chemie 89, 948-958 (1985)

THE PROBLEM OF INTERNAL MOTION OF MOLECULES· IN THE LIQUID
AS SEEN FROM NMR RELAXATION STUDIES

H.G. Hertz

Institut für Physikalische Chemie und Elektrochemie der
Universität Karlsruhe, Kaiserstr. 12, D-7500 Karlsruhe

Abstract

As a brief introduction the most important formulas are given and the crucial physical aspects will be worked out. It is essentially the Woessner theory which gives the framework of the facts to be treated. A number of experimental results are presented for the following molecules: Methanol, acetic acid, DMSO, ethanol, i-propanol, toluene and propylene carbonate. It will be shown that the proton-proton distances in the molecule which result from the proton relaxation data obtained in the dispersion range together with the "classical" theory do not agree with the generally accepted molecular geometry. Moreover, the general nature of the resulting motion in its qualitative and basic features does not correspond to the pattern generally accepted. A very general description of the molecule is given which is free of the difficulties which have appeared.

1. Introduction

Generally, a molecule representing a system of atoms which are chemically bound is drawn as a set of letters to symbolize the elements from which the compound is built up and these letters are partially connected by single or double bars. They represent the "chemical bonds" between the atoms. These two-dimensional schemes sometimes also indicate certain features of the shape of the molecule in three-dimensional space, but whether these stereo-chemical aspects are introduced or not, in any case the objects "molecule" are thought to exist as mechanical systems in three-dimensional space. Some very important small molecules are thought to be <u>rigid bodies</u>. A macroscopic material object is a rigid body if, with the translational and rotational motion it undergoes, all point distances within the object remain unaltered. Thus, for instance, a stone or a piece of metal are rigid bodies. If we mark any points on the surface of these bodies, the respective distances remain the same if we throw the body in an arbitrary way through the air. Of particular importance for the results to be discussed below is the following property of a rigid body. If it performs a rotation about an axis with angular velocity $\vec{w}$ then, as all the distances are constant, the velocity of a point i belonging to the rigid body is

$$\vec{v}_i = \vec{r}_i \times \vec{w} \qquad\qquad (1)$$

where $\vec{r}_i$ is the vector connecting a point on the axis with the i-th point of the system.

If we now consider a molecule to be a microscopic rigid body, then the definition given above should remain valid. But one fundamental difficulty appears. This is due to the fact that the atoms in the molecule perform vibrations. Certainly any of these normal vibrations distorts the molecule, the point distances vary with time, and thus, at least partly, this feature invalidates our definition of a rigid body, namely the requirement that the distances between all points in the system remain constant. How large can the amplitudes of the vibrations be so that we would still be able or willing to recognize the molecule as a rigid body? The answer to this question can only be some kind of convention, no "objective" property of the system can be established. One argument in favour of the picture of a molecule as a rigid body will be that in the gas phase the properties of the rigid body are fairly well-established by microwave and electron diffraction studies. In the solid state the question we are asking does not arise because molecules as a total entity do not rotate or do not perform translational motion in the lattice. In the isotropic liquid the existence of a molecule as a rigid body has not yet been proven explicitly. Whether the optical activity connected with chirality is a proof for the existence of partial rigidity of molecules in the strict sense probably has not yet been investigated. In an anisotropic liquid rigid body properties of molecules have definitely been verified [1-3].

If we turn now to molecules which contain 10-20 atoms, then in some cases one or more vibrational modes appear which certainly cause drastic distortions of the system from the ideal rigid body picture. These are the torsional vibrations of certain groups. These movements may even be so strong that they attain zero frequency character, leading to practically free rotation which is hindered more by intermolecular than by intramolecular forces. However, if one admits that in the molecule free or almost free rotation of certain groups like CH_3 or $_{\backslash C/}H_2$ occurs, then it is implicitly assumed that these rotating groups themselves are microscopic rigid bodies. In contrast to those experiments in which rigid body properties of whole molecules were investigated and which to the author's knowledge are extremely scarce [4,5], there is now a fair amount of experimental material available which regards the movement of these bricks of molecules, which, although rotating relative to one another, should themselves be identifiable to form such microscopic rigid bodies as the generally accepted models tell us. On the following pages a number of these experimental results will be described and from the difficulty of their interpretation it is concluded that so far we are not able to show that we are justified to picture small flexible molecules as chains of microscopic rigid bodies which are linked by rotational axes with atom-atom distances along the axes (the C-C axes or O-C axes) which are constant. The question whether molecules are really rigid bodies or systems of rigid bodies of course is of great importance for a large number of treatments of statistical mechanics and of computer simulations in which the attempt is made to deduce macroscopic properties of liquids from the nature of the microscopic entities which build up the system.

2. Conventional Theoretical Basis of the Pertinent Investigations

The experimental quantity which yields the starting point of the investigations to be described is the longitudinal nuclear magnetic relaxation rate $1/T_1$. It is a macroscopic rate constant and according to Eq. (2) describes the approach of the z-component of the nuclear magnetization M_z (in the direction of the static magnetic field) towards its equilibrium value M_z^0, if by an appropriate operation $M_z - M_z^0 \neq 0$ has been achieved

$$\frac{dM_z}{dt} = \frac{1}{T_1} (M_z^0 - M_z). \tag{2}$$

In the microscopic picture each of the magnetic nuclei interacts with its neighbours via the magnetic field they produce at the position of the respective reference nucleus selected at random. Let this interaction energy be H(t). As all the surrounding nuclei are moving, it is understandable that we have written H(t). H(t) fluctuates with time; it is a random function of the time. We may take the product of the interaction energy at two times, $t = 0$ and $t = t$. Now t is a microscopic time. Then the theory shows [6,7,8] that the macroscopic rate constant $1/T_1$ and the microscopic quantities H(0), H(t), are connected by the relation

$$\frac{1}{T_1} \sim \int_0^\infty <H(0)H(t)> \cos\omega t \, dt \tag{3}$$

where < > represents the ensemble average and ω is the NMR resonance frequency. We said that H(t) is caused by many neighbours, i.e.

$$H(t) = \sum_{i=1}^{N} H_i(t)$$

where N is the number of nuclei in the system.

For $H_i(t)$ we write

$$H_i(t) = \vec{\mu} \cdot \vec{H}_{i \text{ local}}(t)$$

where $\vec{\mu}$ is the magnetic moment of the reference nucleus and $\vec{H}_{i \text{ local}}$ is the local magnetic field at the reference nucleus caused by the i-th neighbour. The last lines refer to the magnetic dipole-dipole interaction. This interaction will be the main subject of our treatment. In certain cases the quadrupole interaction plays a role as the relaxation mechanism. Here we have, using an abbreviated form of writing,

$$H_i(t) = Q \cdot \nabla \varepsilon_{i \text{ local}}$$

where Q is the electric quadrupole moment of the reference nucleus and $\nabla \varepsilon_{i \text{ local}}$ is the electric field gradient at the reference nucleus caused by the i-th atomic or molecular species.

We have

$$H_{i \text{ local}} = \frac{\mu}{r_i^3} \qquad (4)$$

r_i is the distance between the reference nucleus and the i-th neighbour nucleus. One sees that here a quantity r_i appears which in an essential way entered in our definition of a rigid body. Therefore the interpretation of relaxation rates caused by magnetic dipole-dipole interaction will be of particular importance for our investigation.

$$\nabla \varepsilon_{i \text{ local}} = 2 \frac{e}{r_i^3} + \text{ electron contribution}$$

Because the electric field gradient contains electron contributions, statements of geometry cannot be given in a direct way and therefore the respective experimental results are only of secondary importance for us.

The time correlation function of the magnetic dipole-dipole interaction occurring in the integrand of Eq. (3) is particularly simple to treat if among the N nuclei in the system there is one with a distance so short that all other neighbours give minor or even negligible contributions. In this case the simplest expression is

$$\langle H(0)H(t) \rangle \sim \frac{1}{r^6} g(t) \qquad (5)$$

with

$$g(t) = e^{-t/\tau_c} .$$

Now $r \equiv r_1$, and the time constant τ_c is called the rotational correlation time of the vector connecting the two nuclei. It must be stressed that in Eq. (5) r is considered to be a constant quantity, this fact defines the rotational motion and it is clear that in the overwhelming majority of cases these two nuclei will be members of the same molecule. Fourier transformation of Eq. (5) according to Eq. (3) yields the result

$$\frac{1}{T_1} \sim \frac{1}{r^6} \frac{\tau_c}{1 + \omega^2 \tau_c^2} . \qquad (6)$$

It should be noted that Eq. (6) is again an abridged form of the correct equation, however it contains the physical essence of the treatment to be presented. For τ_c we write down the simplest form of temperature dependence

$$\tau_c = \tau_c^0 \, e^{E_A/RT} \tag{7}$$

where E_A is the activation energy of the respective rotational process.

Eq. (5) refers to the simplest case, a vector performing isotropic rotational diffusion. The material object which is adequate to cause such a motion of a nucleus-nucleus vector is a molecule of spherical shape, indeed a rigid body of spherical symmetry. Clearly, the property of rigidity does not show up explicitly.

However, we wish information about the shape of the molecule in the sense that the spatial extension of the molecule deviates from spherical symmetry. The shape may be cigar-like or pancake-like. The theory shows that in such a case the time correlation function $g(t)$ is of a more complex form [9-13]

$$g(t) = a_1 e^{-t/\tau_1} + a_2 e^{-t/\tau_2} + a_3 e^{-t/\tau_3} \tag{8}$$

with the coefficients a_i being functions of ϑ which is the angle between the "main axis" of the molecule and the vector connecting the two nuclei or atoms between which the interaction is effective (see Fig. 1a). $D_{||}$ is the rotational diffusion coefficient about the main axis, $D_\perp$ is the rotational diffusion coefficient of this axis, $D_{||} \geq D_\perp$. Then, as shown by the theory [11-13], the time constants appearing in Eq. (8) are given by the relations

$$\frac{1}{\tau_1} = 6D_\perp \,, \quad \frac{1}{\tau_2} = 5D_\perp + D_{||} \,, \quad \frac{1}{\tau_3} = 2(2D_{||} + D_\perp) \,. \tag{9}$$

Eq. (8) is derived from the solution of the rotational diffusion equation of a rigid body having <u>one</u> axis of symmetry, as shown in Fig. 1a. As all three coefficients a_1, a_2 and a_3 are functions of the same angle ϑ there must hold two equations between these coefficients. In the corresponding way, there must be one equation between the time constants τ_1, τ_2 and τ_3. The respective relations are

$$a_2 = 3(\tfrac{4}{3} a_3)^{1/2} (1 + 2a_1^{1/2}) \tag{10}$$

$$a_1 + a_2 + a_3 = 1 \tag{11}$$

$$\frac{1}{\tau_3} = \frac{4}{\tau_2} - \frac{1}{\tau_1} \,. \tag{12}$$

Now we return to the problem of rigidity of a microscopic body. In Eq. (5), as before, the constant atom-atom distance r occurs together with the new time corre-

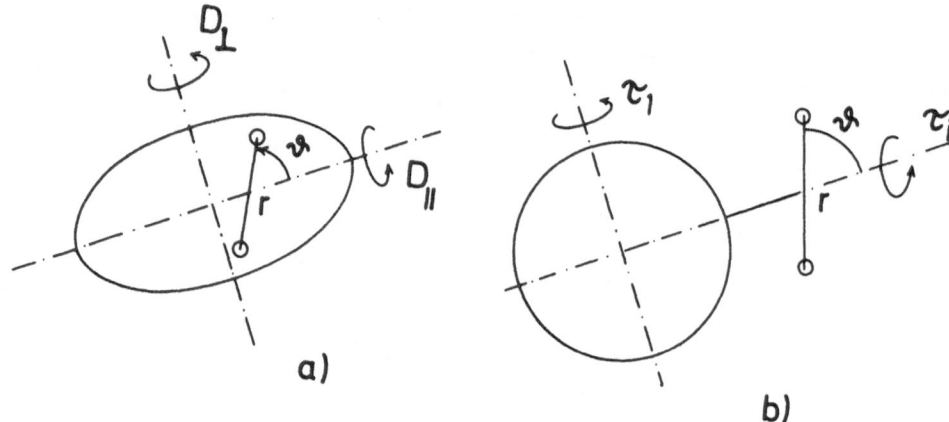

Fig. 1: a) A rigid body of non-spherical shape which performs anisotropic rotational diffusive motion. b) A rigid body of spherical shape to which an axis with rotating group is attached.

lation function g(t) according to Eq. (8). In the coefficients a_i the angle ϑ occurs. Constant distance and constant angle are properties of a rigid body. But we stated that generally rigidity is defined by the constancy of all point distances in the object. For a microscopic body like a molecule this cannot be experimentally veri-fied. Since Eqs. (8) - (12) are derived from the diffusive property of a rigid body, we consider Eqs. (10) - (12) as the rigidity conditions. We say, if the relations (10) - (12) are fulfilled, then we are dealing with a microscopic rigid body, if not, we must be doubtful whether the molecule can really be treated as a rigid body.

Now it may be shown that a molecule in which one group rotates relative to the larger remaining part of the molecule is dynamically equivalent to the symmetric top rigid body for which Eqs. (8) - (12) have been derived (see Fig. 1b) [9-13]. Conse-quently, all the same relations hold with the one difference that two of the Eqs. (9) have to be replaced by

$$\frac{1}{\tau_2} = \frac{1}{\tau_1} + \frac{1}{\tau_i} \; ; \quad \frac{1}{\tau_3} = \frac{1}{\tau_1} + \frac{4}{\tau_i} \tag{13}$$

where τ_i is the time constant of the rotation of the group around the axis, now ϑ is the angle which the nucleus-nucleus interaction vector forms with the axis of in-ternal rotation.

Thus, the molecular model is a rigid body as the carrier of a rotational axis, about the axis a nucleus-nucleus vector of constant length r rotates under a fixed angle ϑ, this is the part of the system we see directly, then according to classical molecular geometry other vectors as well belonging to the group should rotate leaving the angles relative to the reference vector unchanged. In this sense we may call Eqs. (10) - (13) partial rigidity conditions, by presupposition total rigidity of the mole-cule is absent in the case of interest here. We shall see below that one fails to identify experimentally such system of rigid bodies linked by fixed rotational axes.

With Eq. (8), as an abbreviated expression Eq. (3) takes the form

$$\frac{1}{T_1} \sim \frac{1}{r^6}\left(\frac{a_1\,\tau_1}{1+\omega^2\tau_1^2} + \frac{a_2\,\tau_2}{1+\omega^2\tau_1^2} + \frac{a_3\,\tau_3}{1+\omega^2\tau_3^2}\right) \quad . \tag{14}$$

Each of these terms has its maximum at a certain τ_1 value which is given by the relation $\tau_i \approx \omega^{-1}$. If in Eq. (14) we insert a temperature dependence of the τ_i according to Eq. (7), then as we lower the temperature, $1/T_1$ passes through two maxima, if τ_1, and τ_2, τ_3 are sufficiently different. τ_2 and τ_3 according to Eq. (13) are so close to one another that the two maxima connected with τ_2 and τ_3 cannot be separated. Eq. (14) also yields the desired information about the proton-proton distance r. Fig. 2 gives a schematic representation of the behaviour

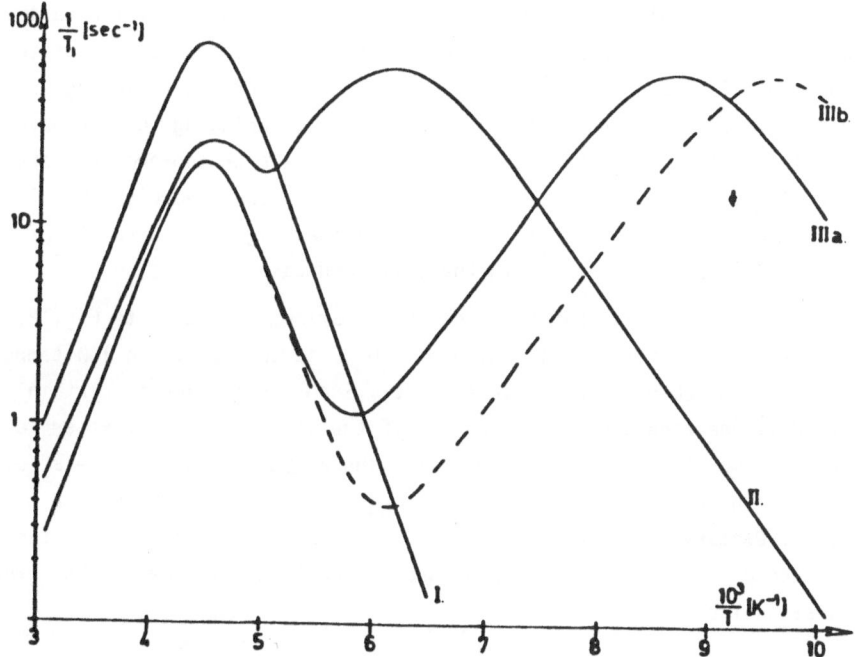

Fig. 2: Schematic representation of $1/T_1$ as a function of the reciprocal temperature when the nuclear interaction vector rotates about an axis and "the molecule" undergoes isotropic reorientation. I: $\tau_i \to \infty$, i.e. no internal rotation, in cases II, IIIa, IIIb, τ_i becomes gradually shorter (after Refs. [13], [14]).

to be expected which may be summarized in the following form. If a molecule contains a rotating group, then this system of two rigid bodies connected by an axis should be "pictured" in the plot $\log 1/T_1$ as a curve with two maxima. If the curve has only one maximum, then the whole molecule should be a rigid body. Our main question on the following pages will be, do we find such a double maximum and do we find the rigidity conditions of Eqs. (10) - (13) fulfilled? When we fail to find the expected behaviour, what is the reason for the discrepancy between prediction and experiment and what is the correct way to describe a flexible molecule in the liquid in the sight of the nuclear magnetic relaxation method?

3. Some Typical Experimental Results

3.1 Methanol

Our idealized model is a rigid isotropically rotating carrier, "the molecule", to which a rotating group is attached. In methanol we have the rotating methyl group but the rigid carrier certainly is not realized satisfactorily [15]. We must consider the H-bonded OH-group as acting as the carrier. Pure methanol cannot be cooled down sufficiently in order to reach the dispersion range in which $\omega\tau_i \approx 1$, i = 1,2,3. Mixtures with DMSO, water, and glycerol are appropriate. In all cases in the temperature range so far accessible only one maximum of $\log 1/T_1$ was found, not two maxima as expected for a two-rigid-bodies-system, in which one part rotates fast relative to the other one [16,17]. There are two possibilities to interpret our finding. Either, contrary to our expectation, the whole methanol molecule including CH_3 behaves like a rigid body. But in this case the data evaluation yields a proton-proton distance in the CH_3 group of $r_{HH} = 2.7$ Å. This is much too large when compared with the classical H-H-distance in the methyl group $r_{HH} = 1.78$ Å, the latter figure is a result of X-ray and electron diffraction measurements in the solid and gaseous state.

The other possibility is that the right-hand maximum according to Fig. 2, which should represent the fast rotational motion of CH_3, is shifted to so low temperatures that we were not able to detect this effect. Such a situation is certainly unsatisfactory. Will one find the second maximum if the temperature is further lowered? The proof for the two-rigid-bodies-system is lacking as yet. Even with the assumption of extremely fast CH_3-rotation, the resulting r_{HH} in the methyl group is 2.1 Å, thus still too large. Deviation of the angle ϑ from 90° would make the r_{HH} distance smaller, but then one would have to give up the trigonal symmetry of the CH_3 group about the rotational axis.

3.2 Acetic Acid

This molecule should be a better representation of a two-rigid-bodies-system than methanol because the carboxyl group is bulkier than the OH group together with one or two H-bonded neighbour hydroxyl groups. Also here from microwave studies in the gas phase it is known that CH_3 performs fast rotation due to the low torsional potential barrier [18]. The nuclear magnetic relaxation studies yielded only one maximum of $\log 1/T_1$ in the accessible temperature range [19]. Again we have the choice: either CH_3COOH as a whole is a rigid body, or CH_3 rotation is so fast that it becomes detectable at temperatures lower than ≈ 140 K. Thus, to yield an answer further experiments at lower temperatures are needed, however, from the very scarce information available at this time it seems that a distinct second maximum does not occur. The proton-proton distance in CH_3 again comes out too large (2.6 or 2.1 Å without and with internal rotation, respectively) which is unsatisfactory in any case. The corresponding behaviour one observes for the deuteron relaxation rate as well, only one maximum exists in the temperature range so far studied and a quadrupole coupling constant results which is too small, see Fig. 3 [19].

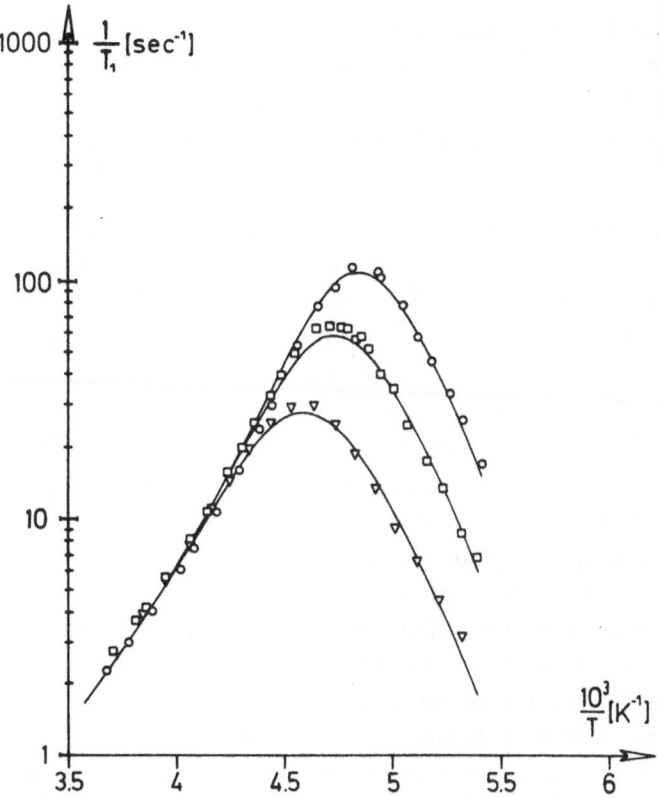

Fig. 3: Deuteron magnetic relaxation rates of the system 66.7 mole % CD_3COOH + 33.3 mole % DMSO at the frequencies (top to bottom) 12, 22, and 46.06 MHz (after Ref. [19]).

Non-linearities of the time dependence of the logarithm of the nuclear magnetization which have been observed for the methyl proton relaxation add further problems [19].

3.3 Dimethylsulfoxide (DMSO)

In order to investigate DMSO in the dispersion range one has to use binary mixtures, a most favourably with a polar substance. We studied three such mixtures, DMSO/methanol, DMSO/acetic acid, and DMSO/water [19,20]. The mixtures with MeOH have already been mentioned when the motion of methanol was described. Let us continue to consider this mixture. The logarithm of the proton relaxation rate of DMSO mixed with CD_3OD when plotted against 1/T shows only one maximum. According to our scheme of argumentation this tells us that the whole DMSO molecule behaves as a rigid body [19]. In contrast to the situation with the molecules described before, this is really an acceptable statement because the potential barrier hindering methyl rotation is 11.9 kJ/mol [21]. In the former cases it was much lower, 2-4 kJ/mol. From the observed pattern of the relaxation curves one might conclude that the rotational diffusion of the molecule is approximately isotropic, in fact, the molecule has approximately spherical shape. The oxygen points out of the plane formed by $H_3C\diagup S \diagdown CH_3$, the lone electron pair occupies the other side of this plane. The instantaneous rotational axis, if we use the picture of a rigid body, fluctuates somewhere in the molecule. It is reasonable to assume that the S-atom mostly is in the neighbourhood of this axis. However, the H-H distance in the methyl group which results from the experiments is definitely too large (2.1 Å compared with 1.78 Å, the classical distance) [19]. Thus, again the result is not that which one would expect for the molecule as a rotationally diffusing rigid body.

Now there is a possibility to improve the resulting value for the H-H distance in the methyl group. This is to introduce a time correlation function for the dipole-dipole or quadrupole interaction which is different from the simple form given by Eq. (5). One usually applies time correlation functions which involve a continuous distribution of correlation times. Note that Eq. (8) as well contains a distribution of correction times, we have three correlation times which, however, are connected by the one Eq. (12). The distribution of correlation times mostly used is the Cole-Davidson distribution. The main feature of the resulting correlation function, which we may call $g_{CD}(t)$, is that at short times $g_{CD}(t)$ decays faster than the exponential, at longer times it decays fairly slowly. Transcribing this fact in the frequency language it means that in the rotational motion high frequency Fourier components are more abundant than they should be according to the rigid body rotational diffusion model. Indeed, with the use of the Cole-Davidson distribution (of correlation times) one can achieve almost the correct methyl proton-proton distance. One has to choose the proper parameters of the distribution. But it is very important to keep in mind that there is no physical argument from which the use of the Cole-Davidson distribution could be justified. So using the Cole-Davidson distribution means that one gives up the rigid body model of a molecule in the isotropic liquid.

Next we consider the mixtures of DMSO with acetic acid and with water. In both cases one now observes the indication of a double maximum (see Fig. 4) [19,20]. The degree of separation of these maxima depends on the NMR frequency belonging to the measurement, for the higher frequencies only a shoulder or an inflection of the curve log $1/T_1$ vs. $1/T$ can be observed, for lower frequencies the separation is better. As may be seen from a strong increase of the viscosity in these two mixtures relative to the pure DMSO liquid, there is an appreciable interaction between DMSO and acetic acid or water. As a consequence of this interaction, the rotation about the fluctuating axis passing through the molecule somewhere close to the S-atom becomes much slower. Were the total molecule a rigid body in the strict sense, then the methyl proton-proton vectors as well would have to perform a slow rotational motion because

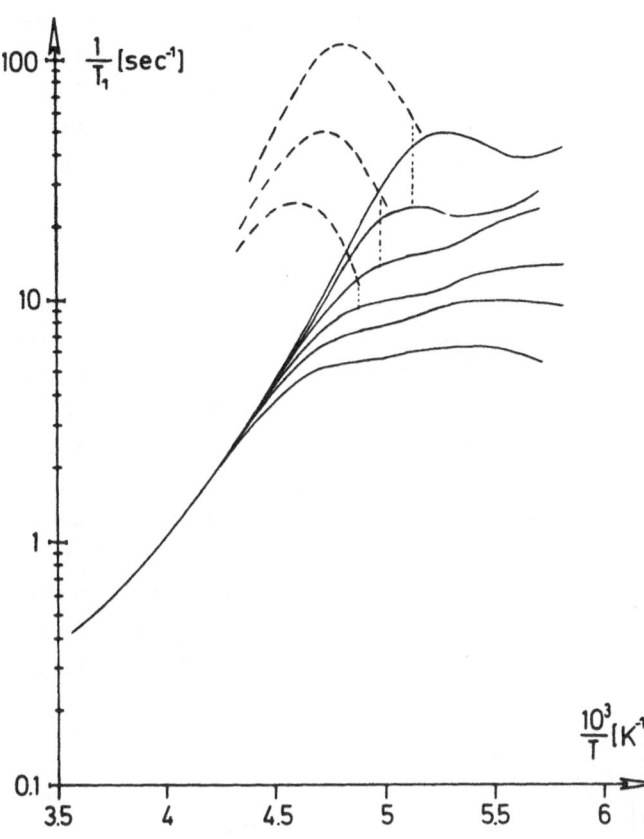

Fig. 4: (Intramolecular) proton magnetic relaxation rates of DMSO in the mixture 33.3 mole % DMSO + 66.7 mole % CD_3COOD at the resonance frequencies (top to bottom) 6, 12, 30, 60, 90 and 144 MHz. The dashed curves are the OD relaxation rates of the solvent AcD at 12, 30 and 60 MHz divided by ten (after Ref. [19]).

the instantaneous value in the rigid body is the same everywhere. But this is not the case, two new rotational axes appear, namely the S-C axes, the methyl protons move about these two intramolecular axes. However, again the H-H distance, even if the fast internal motion is taken into account, comes out too large, about 2.0 Å [19,20]. One can try to improve the resulting proton-proton distance in two possible ways (apart from the attempt of introducing a continuous distribution of correlation times), (1) to change the angle ϑ such that r_{HH} comes out correctly. But in the present case ϑ has to be equal to 90°, because otherwise one would give up the rigid body property: trigonal symmetry with respect to the internal rotational axis S-C. (2) One could make the assumption that the "molecule" itself performs an anisotropic rotational motion. This means that a third rotational axis appears in the molecule, this axis being a property of the rigid body carrier part of the total molecular system (see Fig. 1a). Around this axis fast diffusive motion occurs as described in connection with Eqs. (8) and (9). But firstly, considering the geometry of the molecule, it is difficult to see where this axis of fast motion should be placed inside the molecule and, secondly, in order to obtain the correct proton-proton distance one has to assume a ratio $D_{||}/D_{\perp} \approx 1000$ which is physically unacceptable. Thus, we are not able to understand our data in terms of a system of rigid bodies (methyl groups and skeleton C $\diagup^S\diagdown$ C) which are connected by rotational axes. However, one result is of great interest. If we consider the deuteron relaxation rates of DMSO-d$_6$ in the same systems, then the shoulder on the log $1/T_1$ vs. $1/T$ curve, forming the residuum of the slow "molecule" motion is developed to a lesser degree. This is in agreement - at least qualitatively - with the Woessner theory which predicts that a_1 in Eq. (8) should be proportional to $(3\cos^2\vartheta - 1)^2$ [9-13].

3.4 Ethanol

In analogy to other substituted hydrocarbons where spectroscopic information concerning the gas phase is available, the rotational barrier of CH$_3$ relative to CH$_2$ is comparatively high, i.e. $\approx$14.5 kJ/mol. On the other hand, as in methanol, the barrier of OH rotation relative to the CH$_2$ group is only about 4 kJ/mol. Thus, one might expect that the ethyl group represents the "molecule" as the rigid body and that the hydroxyl proton simply rotates around the oxygen atom. On the other hand, from the studies of methanol there is already the information that OH moves slowly due to H-bonding. Early investigations on EtOH in the extreme narrowing range also showed that the OH group has the slowest motion when compared with the hydrocarbon part as may be seen from Fig. 5 [22,23]. So, how does the ethanol molecule decide to behave?

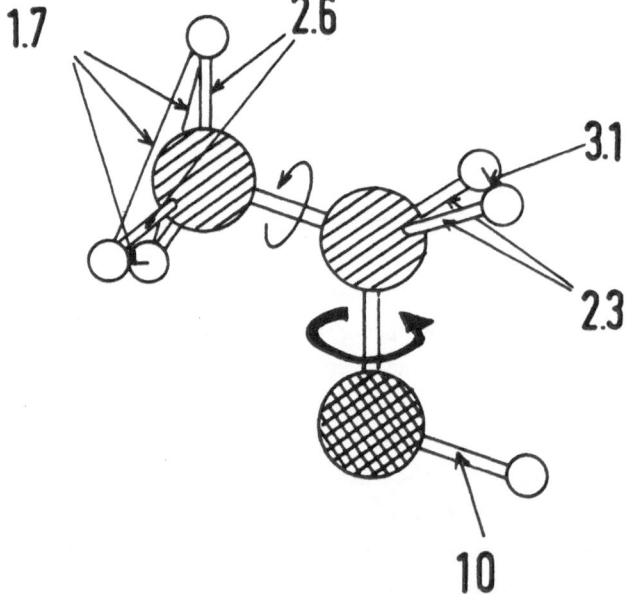

Fig. 5: The classical picture of the ethanol molecule as a system of rigid bodies linked by rotational axes of constant length. Numbers give effective correlation times of vectors indicated in picoseconds. The figure refers to the liquid at 25°C, this corresponds to the extreme narrowing range of the relaxation rate (after Ref. [13]).

Let us begin with the methyl proton relaxation rates. As may be seen from Fig. 6, the curve $\log(1/T_1)$ vs. $(1/T)$ is of the simplest type one could imagine [24]. One con-

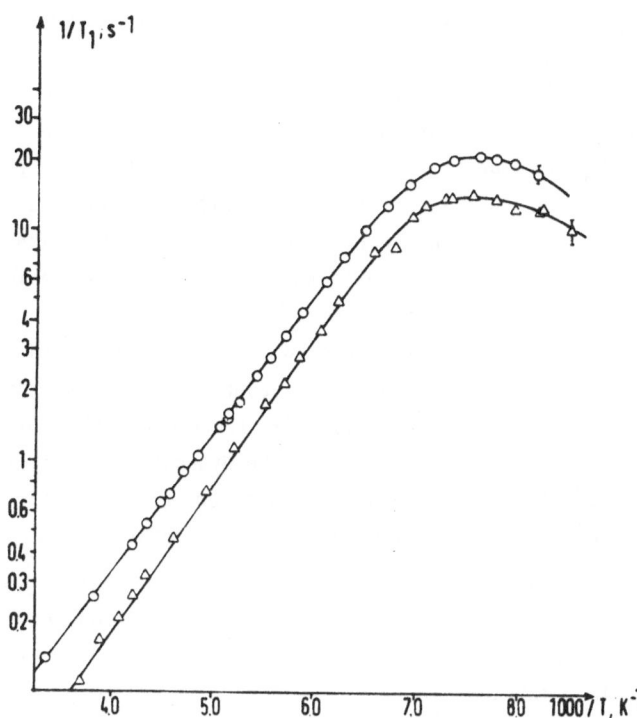

Fig. 6: Proton magnetic relaxation rates of the liquid
CH_3CD_2OD (o) and the mixture 10 mole % CH_3CD_2OD +
90 mole % CD_3CD_2OD (Δ)
as a function of the reciprocal absolute temperature,
ν = 30 MHz (after Ref. [24]).

cludes that OCH_2CH_3 behaves like a rigid body. The rotational axis for the overall
molecular reorientation fluctuates isotropically, according to the experimental information described so far it will pass through the methylene carbon. If the EtOH
system is a rigid body reorienting isotropically, then all correlation times for
the other intramolecular atom-atom vectors should be equal. The experiment tells us
that <u>this is not the case</u>. The methylene proton-proton vector and the methylene car-
bon-deuteron vector undergo a slower reorientation than the corresponding vectors in
the methyl group. Considering this finding the next step of the argumentation should
be that the rigid body "ethanol" has an anisotropic rotational motion. Then the axis
moving slowly would have the direction of the methylene H-H vector, i.e. it would be
perpendicular to the C-C-O plane. This is difficult to imagine, it also contradicts
the observation that $\log 1/T_1$ is a strict linear function of $1/T$ on the left hand
side of the relaxation maximum (see Fig. 6). The methylene proton-proton distance according to the conventional theory comes out much too large, r_{HH} = 2.5 Å.

Furthermore, the OH vector moves still slower than the methylene group, which
firstly confronts us with the difficulty where to place the axis of slowest motion,
and secondly is in contradiction to our previous statement that the hydroxyl proton
rotates around the CO axis. Thus, one is forced to reject the picture that the ethanol
molecule or the CH_2CH_2O skeleton might form a rigid body. The next attempt is to consider the H-bonded OH network to be the effective "molecule", molecule in the sense
that it represents the part of slowest motion. Thus now OH...OH is the carrier of

the rotating axis and the whole ethyl group rotates around the O-C axis. In order to get the correct proton-proton distance in the methylene group, CH_2 has to move very rapidly around the OC axis and, moreover, the OH system should undergo anisotropic motion. But from what physical property of a network of H-bonds should follow such an anisotropy? Another severe difficulty shows up if we return to Eq. (1). Let $\vec{w}$ be the instantaneous angular velocity around the O-C axis. It fluctuates with time and, as was just indicated, assumes very high values. Then, in accordance with Eq. (1), each point in the methyl group, if it is rigidly connected with the methylene group, would have to undergo fluctuations of the velocity which are coherent with those of the vector $\vec{w}$ and thus also must reach very high values. It seems to be very difficult to accept such a picture. As a consequence, it seems adequate to reject the proposal that the ethyl group as a rigid body rotates about the OC axis. Still there are certain observations which are in favour of this approach to understand the dynamics of ethanol and some other molecules. Before these effects are described, another variant of the possible interpretations of the experimental relaxation curves must be mentioned.

So far we had considered models which involve one axis of internal rotation and one instantaneous external axis of the whole molecule rotation, perhaps this one axis may split into two axes in the case of anisotropic motion. We have now briefly to discuss the model of a flexible molecule which contains two internal axes and one external axis. The Woessner theory has been generalized to this case [12, 25-27], for ethanol the "rigid" body methylene rotates relative to OH...OH, and the rigid body methyl rotates relative to CH_2. And in fact the experimental relaxation rate of CH_3 can be described properly by such a model, however the respective dynamical description of the CH_3 group implies also a prediction of the dynamical behaviour of the CH_2 group which in our case of ethanol, and similarly for i-propanol [27], are not in agreement with the experimental findings. Therefore, the two internal axes model has to be rejected as well. We return now to the rigid body model of ethanol and show that C_2H_5OH exhibits similar features as the "rigid body" molecule DMSO.

Components have been added to ethanol which execute a strong intermolecular interaction with the molecule [28-31]. Admixed components of this kind are glycerol and mixtures of LiCL and D_2O. The result is that now the methyl proton relaxation curve splits to yield a double maximum according to Fig. 7 if conditions are chosen properly. Thus, under these circumstances an internal rotational axis appears around which the methyl protons move. One sees that the ethyl group behaves as a rigid body in the sense that it is possible to break it in two parts connected by a rotational axis if at the OH end the intermolecular interaction is strong enough. But the methyl proton-proton distance again comes out too large. Thus the result conflicts with the two-rigid-body model. One could introduce the correct distances, but then the a_i-values according to Eqs. (8), (10), and (11) do not assume the values given by the geometry (angle $\vartheta = 90°$) thus, again conflicting with the rigid body requirement.

The methylene proton relaxation curve under the same conditions first broadens strongly, then the relaxation maximum of these protons coincides precisely with that

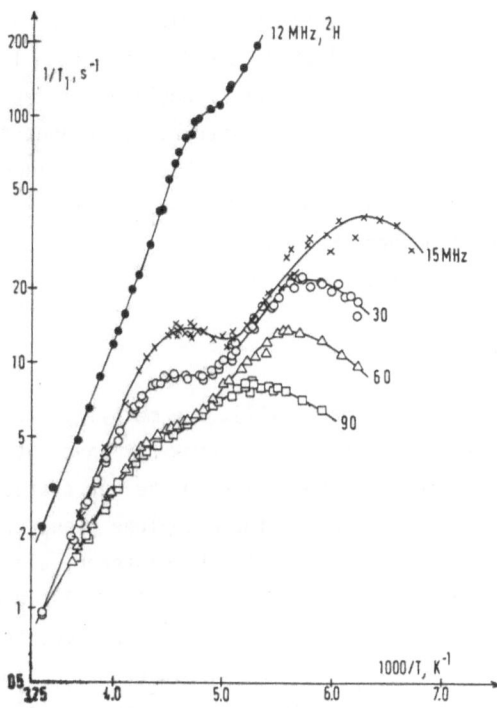

Fig. 7: Proton magnetic relaxation rates of the mixture 19.8 mole % CH_3CD_2OD + 80.2 mole % D_2O containing 14.5 $\bar{m}$ LiCl at various resonance frequencies ($\bar{m}$ = aquamolarity). The experimental points (•) refer to the deuteron magnetic relaxation rate of the mixture 19.6 mole CD_3CH_2OH + 80.4 mole % H_2O containing 13.67 $\bar{m}$ LiCL at a frequency ν = 12 MHz (after Ref. [30]).

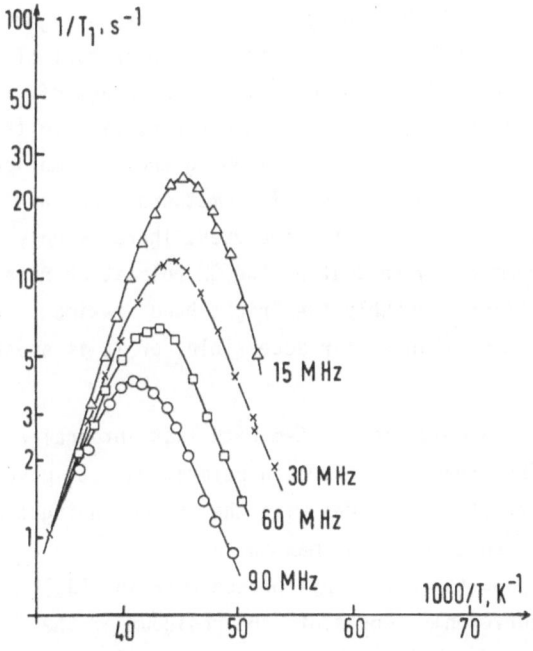

Fig. 8: Proton magnetic relaxation rates of the mixture 19.9 mole % CD_3CH_2OD + 80.1 mole % D_2O containing 14.42 $\bar{m}$ LiCl at a number of frequencies (after Ref. [30]).

of the OH vector as shown in Fig. 8 [30]. But this is only the residuum of the slow
OH motion, it represents the movement of the rotational axis, the OC axis. The flow
of the methylene protons around this axis is very fast so that it could not be detected in the temperature range so far accessible to our experiments. It is doubtful
whether there will be a distinct maximum at all. This would mean that even at lowest
temperatures there are Fourier components dominant in the correlation function $g(t)$,
which are of higher frequency than the NMR resonance frequency. Note, that all these
statements refer to the supercooled or glassy state.

3.5 i-Propanol, Toluene, and Propylene Carbonate

i-propanol shows essentially the same behaviour as ethanol [27]. The methyl proton
relaxation curve $\log 1/T_1$ vs. $1/T$ is entirely smooth, i.e. it does not have any
particular features which are the residuum of the slow OH motion in the correlation
function of CH_3. Again the understanding of the motion of the methylene group or in
the methylene group offers the greatest difficulties. Here only the deuteron relaxation was studied. The quadrupole coupling constant came out much too low [27]. When
the two internal axes approach involving the OC and the CC axes was applied, then
the methyl proton relaxation could be interpreted satisfactorily, however this fitting of the experimental data necessarily implies statements of the motion of the CD
vector in the methylene group which were contrary to observation. As in ethanol addition of glycerol as a strongly interacting component caused the appearance of a
shoulder of the $\log 1/T_1$ vs. $1/T$ curve which is the residuum of the slow OH group
motion [29]. So, here again the i-propanol group behaves as something like a rigid
body in the absence of a component which strongly interacts with the polar part of
the molecule. In the presence of such an admixture, the methyl motion breaks off in
two parts, one being connected with a fast decay of correlation corresponding to the
H_3 motion around the CC axis, and a slower decay of $g(t)$ following the slow motion
of OH. It should be mentioned that the addition of glycerol to methanol does not
cause the splitting of the $\log 1/T_1$ vs. $1/T$ curve in two features. There is only
one relaxation maximum coinciding approximately with that of the OH relaxation rate,
thus being the residuum of the slow OH motion, probably the "right-hand" maximum
(see Fig. 2) is shifted to lower temperatures than so far accessible, or it is smeared
out to a very broad feature [29].

Almost a classical representative of a molecular system with fast internal rotation is toluene [32], see also [33]. The study of the proton relaxation rate gave
a $\log 1/T_1$ vs. $1/T$ curve which almost exactly coincides with the corresponding curve
for the ring protons, there is no indication of a second maximum at lower temperatures (down to 125 K) which would correspond to fast methyl proton rotation [32].
Thus the maximum of the CH_3 relaxation curve only represents the residuum of the
slower motion of the ring. For the ring and methyl protons proton-proton distances
are resulting from the relaxation measurements which again are larger than the clas-

sical generally accepted values. Application of a Cole-Davidson distribution of correlation times yields better geometrical results for C_6H_5 and CH_3, however, as
was already mentioned, when using such a continuous distribution we are leaving the
domain of a microscopic rigid body approach.

Finally, the proton relaxation of propylene carbonate

$$
\begin{array}{ccccc}
& D & & D & \\
& | & & | & \\
H_3C & - C & - & C & - D \\
& | & & | & \\
& O & & O & \\
& \backslash & & / & \\
& & C & & \\
& & \| & & \\
& & O & &
\end{array}
$$

was measured as a function of the temperature [32]. In the pure liquid a double maximum according to Fig. 2 was observed. Thus, here we have an example where the carrier
of the methyl group is so large and so strongly interacting with its surroundings that
already in the pure liquid the correlation function of the CH_3 protons breaks off in
two parts. The fast part, corresponding to CH_3 motion, yields the maximum at lower
temperatures, the slower part corresponding to the motion of the rotational axis fixed
in the bulky carrier $C_3O_3D_3$ leads to the maximum at higher temperatures. Addition of
$LiClO_4$ increases the distinctness of maximum separation [32]. It should be noted that
in this example as well the methyl-proton-proton distance comes out too large, within
the temperature range studied the behaviour of the relaxation curve is not yet fully
understood. Propylene carbonate is the first case for which the proton relaxation
curve maximum separation has been observed for the pure liquid. In a very brief and
preliminary communication Versmold reports on an indication of beginning maximum splitting of the CD_3 deuteron relaxation rate in pure n-propanol [34].

4. Formal Description of Proton-Proton Vector Motion in a Flexible Molecule Like Ethanol

From the details described in Section 3 it has become obvious that we have to abandon
the rigid body approach (or system of rigid bodies approach) as a starting point which
involves constant distances, constant angles, instantaneous rotational axes, and axes
of internal rotation. The only dynamical facts with which we are really dealing are
flows of hydrogen nuclei relative to one another. In our samples prepared by proper
isotopic substitution effectively there is a flow of one methylene H-atom relative to
a given methylene proton and likewise, there is a flow of two methyl protons relative
to a given methyl proton. In the case of relaxation by quadrupole interaction a flow
of electrons being connected with the movement of the nuclei also becomes effective.

If in the following we introduce the flow picture we give up at the outset the
concept of rotation. Rotation is defined as a motion for which the distance between

the points of interest remains constant during the movement. Thus we consider the proton-proton vectors in the methylene and methyl group as quantities the <u>magnitude</u> and <u>direction</u> of which fluctuate with time. As a consequence we have to replace Eq. (5) by the relation

$$<H(0)H(t)> \sim \iint p(\vec{r}_0) \frac{Y_2(\Omega_0)}{r_0^3} \cdot \frac{Y_2(\Omega(t))}{(r(t))^3} P(\vec{r}_0, \vec{r}, t) \, d\vec{r}_0 \, d\vec{r} . \tag{15}$$

$p(\vec{r}_0)d\vec{r}_0$ is the probability that the proton which acts as the interaction partner is at $\vec{r}_0$ relative to the reference proton, $\vec{r}_0 = \{r_0, \vartheta_0, \varphi_0\}$, where ϑ_0 and φ_0 are the polar and azimuthal angle, respectively, of the proton-proton vector in the laboratory coordinate system. The $Y_2(\Omega)$ are the spherical harmonics of second degree, the order m is of no importance, $\Omega_0 = \{\vartheta_0, \varphi_0\}, \Omega(t) = \{\vartheta(t), \varphi(t)\}$ is the orientation of the proton-proton vector at time t, then Ω_0 has the meaning of the orientation at any time zero. $r = r(t)$ is the proton-proton distance at $t > 0$. $P(\vec{r}_0, \vec{r}, t)$ is the probability density of finding the proton at $\vec{r} = \vec{r}(t)$ at time t if it was at $\vec{r}_0$ at $t = 0$. The local probability density $P(\vec{r}_0, \vec{r}, t)$ is connected with the relative proton flow $\vec{j}_{rel}$ by the relation

$$\frac{\partial P(\vec{r}_0, \vec{r}, t)}{\partial t} = -\text{div}(\vec{v}_{rel} \, P(\vec{r}_0, \vec{r}, t))$$
$$= -\text{div} \, \vec{j}_{rel}(\vec{r}_0, \vec{r}, t) \tag{16}$$

where $\vec{v}_{rel}$ is the mean drift velocity of the interacting proton relative to the reference proton. In the methyl group, in first approximation both proton flows are assumed to be independent. It will be seen from Eq. (15) that the distance between the protons, r, is time-dependent, it is no longer constant as required for a rotational process. At $t = 0$, $P(\vec{r}_0, \vec{r}, t)$ is a δ-function localized at $\vec{r}_0$. Then, for $t > 0$ at this position $\vec{r}_0$, j_{rel} has the value zero. For long times $P(\vec{r}_0, \vec{r}, t) = p(r_0)/4\pi$ thus $\partial P(\vec{r}_0, \vec{r}, t)/\partial t = 0$ and also $j_{rel} = 0$ everywhere. The quantity which is transported by the flows is the magnetic dipole-dipole interaction $Y_2(\Omega_0)/r_0^3 \cdot Y_2(\Omega)/r^3$. The time $t = 0$ is given by the NMR pulse experiment. One $\vec{r}_0$-value (methylene) or two $\vec{r}_0$-values (methyl) correspond to one molecule. To take into account the motion in the total isotropic liquid, according to the integration over $\vec{r}_0$ in Eq. (15) all the fluxes $\vec{j}_{rel}(\vec{r}_0, \vec{r}, t)$ have to be added with the proper weight factor $p(\vec{r}_0)$. This gives the correct average relaxation effect. Formally Eq. (15) has the same appearance as the corresponding expression for the correlation function to be used for the calculation of the intermolecular relaxation rate. However, when we integrate in Eq. (15) with respect to $\vec{r}(t) = \vec{r}$, there must be a surface on which the radial components of all the $\vec{j}_{rel}(\vec{r}_0, \vec{r}, t)$ vanish, i.e. $P(\vec{r}_0, \vec{r}, t) = 0$ for all $\vec{r}$ outside this surface. This means that there is no proton exchange between the methylene and methyl group. We may also say that the zero frequency Fourier component of the relative velocity correlation function between the interacting protons vanishes [35]

$$\int_0^\infty <\vec{v}_{rel}(0) \cdot \vec{v}_{rel}(t)>dt = 0 \ . \tag{17}$$

Eq. (17) essentially has the physical significance that the mean square displacement after a sufficiently long time is exactly the same for all nuclei in the respective groups (here CH_2 and CH_3).

In many cases $\vec{J}_{rel}(\vec{r}_0,\vec{r},t)$ is given by Fick's first law

$$\vec{J}_{rel}(\vec{r}_0,\vec{r},t) = -\tilde{D} \nabla P(\vec{r}_0,\vec{r},t)$$

or $\tag{18}$

$$\vec{v}_{rel}(\vec{r}_0,\vec{r},t) = -\tilde{D} \nabla \ln P(\vec{r}_0,\vec{r},t)$$

where $\tilde{D}$ is the diffusion coefficient. But this is clearly not the correct relation in the present situation. Firstly, as a consequence of the validity of Eq. (17), and, secondly, the wide extension to high frequencies of the Fourier transform of $<H(0)H(t)>$ is due to marked deviations from Eq. (18), the almost coherent vibrational modes in the liquid contribute to the fast initial decay of the correlation function of the interaction energy.

The quantity $\vec{J}_{rel}(\vec{r}_0,\vec{r},t)$ in Eq. (15) refers to the relative proton-proton flow in the methyl and methylene group. There are also relative carbon hydrogen flows, $\vec{J}_{rel}(CH)$ and $\vec{J}_{rel}(CD)$ which are the microdynamic representations of the ^{13}C and deuteron relaxation rates, respectively. As already mentioned, the deuteron relaxation rate is also connected with a partial flow of electrons relative to the reference nucleus. Information regarding the carbon-carbon relative flows is not available.

Thus, as a consequence of our analysis, the picture of the partial rigid body molecule ethanol as shown in Fig. 5 is transformed to a set of relative nucleus-nucleus flows with fairly high frequency Fourier components and with the very important property as given by Eq. (17), i.e. vanishing proton exchange. Note that for the relative velocity between 0 and H Eq. (17) is not valid, there is in fact a slow proton exchange such that a feeble radial component of $\vec{J}_{rel}(OH)$ exists over the entire space.

Let us now consider the relative atomic flow system in ethanol a little more in detail. The flow vector distribution in the surroundings of an $\vec{r}_0$-value selected at random is shown schematically in Fig. 9a. The arrows represent the flow vectors $\vec{v}_{rel}P(\vec{r},\vec{r}_0,t)$ at a number of $\vec{r}$-values in the neighbourhood of $\vec{r}_0$. It will be seen that the flow has a radial component, in general protons move away and approach towards the reference proton. But the distribution of the flow vectors has rotational symmetry with respect to the angle ϕ. In this situation we say that the motion is isotropic. The methyl protons in pure ethanol behave in this way. The motion of the methylene proton is similar, but the correlation function according to Eq. (15) is not identical.

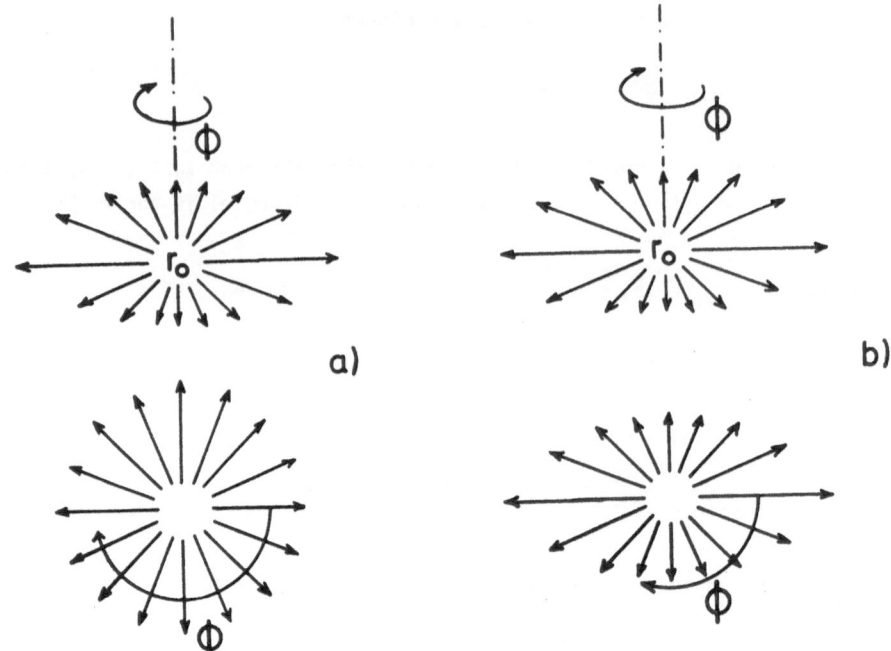

Fig. 9: Schematic representation of relative proton-proton flow distribution in the neighbourhood of a starting position r_o a) isotropic, b) anisotropic or internal motion. For further details see text.

The long time diffusive part of the motion is slower than that of the methyl protons, the flows exist a little longer. Moreover, as far as we know today, the fast "vibrational" short time part contributes to a greater degree. Thus, with respect to long times methylene is the part of slow motion and with respect to short times it is the fast motion part of the molecule.

Now it may happen that the flow distribution is not of rotational symmetry with respect to the angle ϕ. In the direction $\phi = 0^o$, say, the flow $\vec{j}_{rel}(\vec{r}_o, \vec{r}, t)$ is much greater than for $\phi = 90^o$. In this event there must be something which impedes the motion in the direction $\phi = 90^o$. This may be a part of a bulky molecule or a strongly interacting component in a mixture, as we have shown to exist for ethanol. The situation is shown in Fig. 9b, we are now dealing with anisotropic motion or internal motion depending on the nature of the molecule. For instance, in ethanol the fast part of the flow in the direction $\phi = 0$ represents the "internal rotation" of the methyl protons which arises when the differences between the flow velocities in the molecule, caused by some interacting species, exceed a certain critical value [31]. It should be noted that $\vec{j}_{rel}$ is the average flow, there is no instantaneous rotational axis. The slow motion in the direction $\phi = 90^o$ reflects the slow flows in the carrier part of the molecular system, it gives rise to the slow motion residuum in the correlation function and this in turn leads to a shoulder or second maximum at higher temperatures when $\log 1/T_1$ is plotted against $1/T$. for the methylene protons

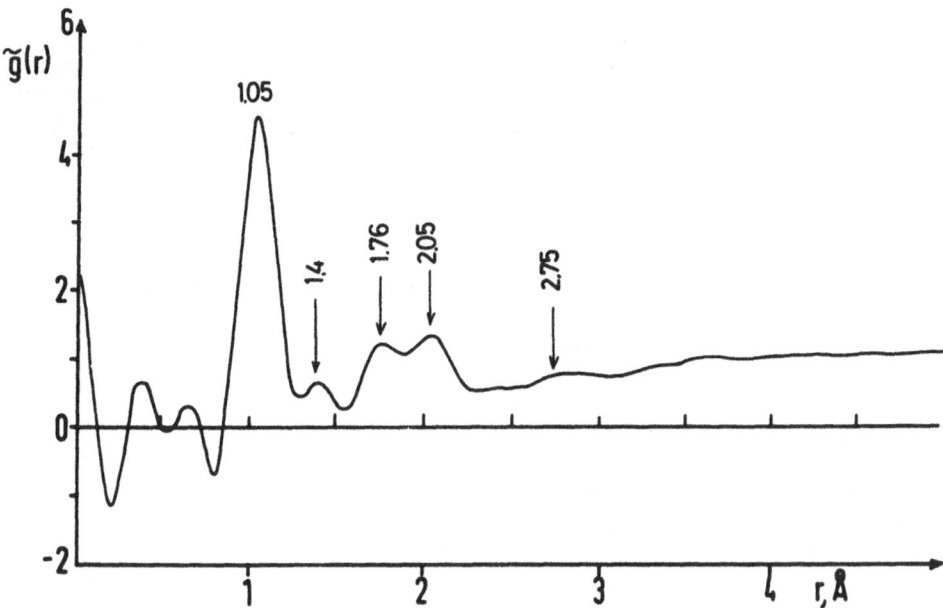

Fig. 10: Weighted sum of atom-atom pair correlation functions of liquid methanol obtained from neutron diffraction study. Curve for distances below 0.9 Å has no physical significance. T = 20°C [36].

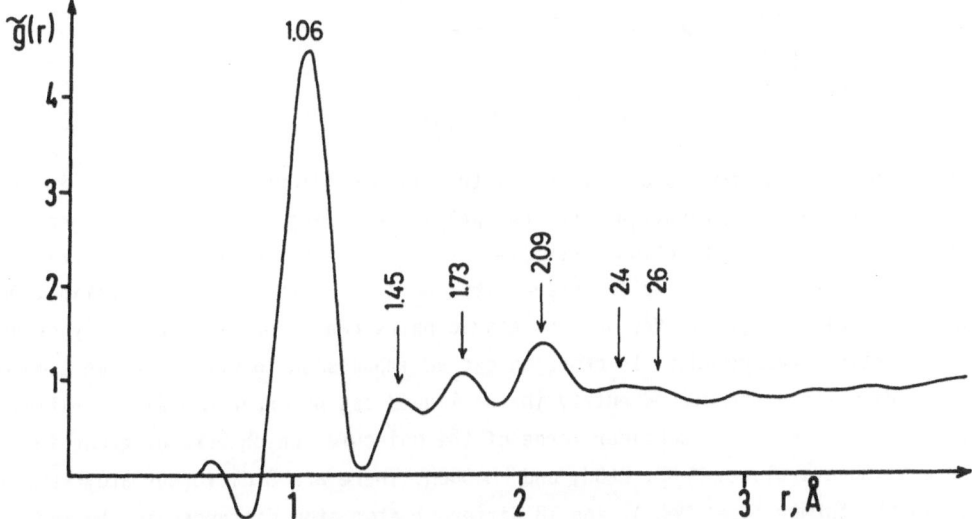

Fig. 11: Weighted sum of atom-atom pair correlation functions of liquid ethanol obtained from neutron diffraction study. T = 20°C [37].

of EtOH in certain mixtures, according to the experimental results so far available
(see Section 3.4), the fast part of the relative proton-proton flow seems to be so
fast that the resulting Fourier components of the correlation function even at very
low temperatures remain beyond the range of NMR frequencies, i.e. $> 10^8$ Hz.

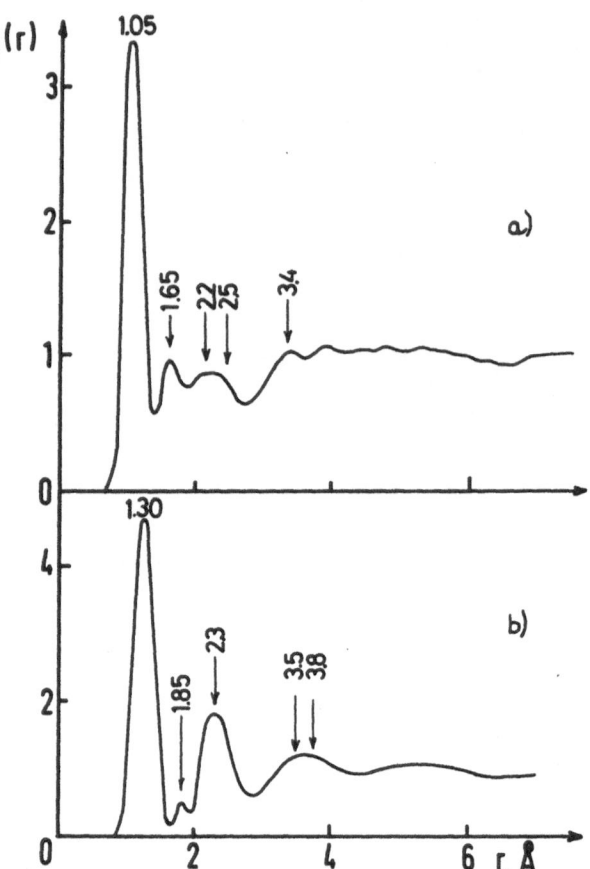

Fig. 12: Weighted sum of atom-atom pair correlation functions of liquid acetic acid obtained from neutron a) and X-ray b) diffraction studies. T = 20°C [38].

Of course, of crucial importance for the distribution of flows $\vec{j}_{rel}(\vec{r}_0,\vec{r},t)$ in the molecule is the pair distribution function $p(\vec{r}_0)$ occurring in Eq. (15). In general the radial part of these functions $p(r_0) = c' \cdot g(r_0)$ are not quantities directly accessible to experiment. c' is the mean number density of the atoms in question. Only weighted sums of the $g(r_0)$ for various atomic pairs can be obtained by X-ray or neutron diffraction measurements. In order to get an impression to what degree a molecule as a discretely structured entity in the liquid can be experimentally verified by these techniques let us consider three of the molecules which were of great importance in this article: MeOH, EtOH, and MeCOOH. There are 8 different atom-atom distances in the former case (MeOH), and 18 different atom-atom distances in the two latter cases. As may be seen from Figs. 10-12 the number of maxima which can be detected and which possibly includes those for intermolecular atom-atom distances are only 5, 6, and 6 for MeOH, EtOH and MeCOOH, respectively. One sees that smearing out of atom-atom distances, which usually is ascribed to intramolecular vibrations and which in our approach reflects the existence of lively relative atomic flows, is appreciable.

Acknowledgement:

I wish to thank Professor Dr. H. Bertagnolli, Würzburg, for supplying me with Figures 10 and 11.

References

[1] A. Saupe, Angew. Chemie 80, 99 (1968)

[2] S. Meiboom and L.C. Snyder, Science 162, 1337 (1968)

[3] P. Diehl and J. Jokisaari, in "Appl. of NMR Spectroscopy to Problems in Stereochemistry and Conformational Analysis", Eds. Y. Takeuchi and A.P. Marchand, Methods in Stereochemical Analysis, Vol. 6, 1986

[4] W. Müller-Warmuth and W. Otte, J. Chem. Phys. 72, 1749 (1980)

[5] M. Will, Karlsruhe, research work in progress

[6] A. Abragam, "The Principles of Nuclear Magnetism", Oxford 1961

[7] C.P. Slichter, "Principles of Magnetic Resonance", 2nd edition, Berlin 1980

[8] T.C. Farrar and E.D. Becker, "Pulse and Fourier Transform NMR", London 1971

[9] D.E. Woessner, J. Chem. Phys. 36, 1 (1962)

[10] D.E. Woessner, J. Chem. Phys. 37, 647 (1962)

[11] W.T. Huntress Jr., J. Phys. Chem. 73 103 (1969)

[12] H. Versmold, Z. Naturforsch. 25a, 367 (1970)

[13] H.G. Hertz, Progress in NMR Spectroscopy 16, 115 (1983)

[14] H. Bertagnolli, H.G. Hertz and H.A. Posch, Ber. Bunsenges. Phys. Chem. 85, 992 (1981)

[15] H.G. Hertz and M. Holz, Z. Phys. Chem., Neue Folge, 136, 81 (1983)

[16] B. Blicharska, H.G. Hertz and H. Versmold, J. Magn. Reson. 33, 531 (1979)

[17] H. Versmold, Ber. Bunsenges. Phys. Chem. 84, 168 (1980)

[18] L.C. Krisher and E. Saegebarth, J. Chem. Phys. 54, 4553 (1971)

[19] T. Frech and H.G. Hertz, J. Mol. Liquids 30, 237 (1985)

[20] B. Blicharska, T. Frech and G.H. Hertz, Z. Phys. Chem.. Neue Folge, 141, 139 (1984)

[21] W. Feder, H. Dreizler, H. Rudolph and V. Typke, Z. Naturforsch. A24, 266 (1969)

[22] E. v. Goldammer and H.G. Hertz, J. Phys. Chem. 74, 3734 (1970)

[23] H.J. Bender and H.G. Hertz, Ber. Bunsenges. Phys. Chem. 81, 468 (1977)

[24] M.S. Ansari and H.G. Hertz, Z. Phys. Chem., Neue Folge, 137, 187 (1983)

[25] D. Wallach, J. Chem. Phys. 47, 528 (1967)

[26] M.S. Ansari and H.G. Hertz, J. Soln. Chem. 13, 877 (1984)

[27] T. Frech and H.G. Hertz, Ber. Bunsenges. Phys. Chem. 89, 948 (1985)

[28] H. Versmold, Ber. Bunsenges. Phys. Chem. 78, 1318 (1974)

[29] H.G. Hertz, T. Wild, and H. Weingärtner, Z. Phys. Chem., Neue Folge, 140, 71 (1984)

[30] M.S. Ansari and H.G. Hertz, Z. Phys. Chem., Neue Folge, 146, 15 (1985)

[31] M.S. Ansari and H.G. Hertz, J. Soln. Chem. 15, 919 (1986)

[32] C. Meißner, Thesis, Karlsruhe 1986

[33] E. Rössler and H. Sillescu, Chem. Phys. Lett. 112, 94 (1984)

[34] H. Versmold, in "Molecular Liquids, Dynamics and Interactions", eds. A.J. Barnes,
      W.J. Orville-Thomas and J. Yrwood, p. 309, Reidel Publ. Co., Dordrecht 1984

[35] H.G. Hertz, Z. Phys. Chem. Supp. Issue $\underline{1}$, 117 (1982)

[36] D.G. Montague, J.P. Gibson, and J.C. Dore, Mol. Phys. $\underline{44}$, 1355 (1981)

[37] D.G. Montague, J.P. Gibson, and J.C. Dore, Mol. Phys. $\underline{47}$, 1405 (1982)

[38] H. Bertagnolli and H.G. Hertz, Phys. Stat. Sol. $(\underline{a})$ $\underline{49}$, 463 (1978)

A COMPARATIVE STUDY OF MOLECULAR ROTATION AS STUDIED BY
DYNAMIC LIGHT SCATTERING, FLUORESCENCE ANISOTROPY DECAY
AND RAMAN BANDWIDTH ANALYSIS

Th. Dorfmüller

Universität Bielefeld
Fakultät für Chemie
D-4800 Bielefeld, FRG

Abstract

Molecular rotation is studied by three different spectroscopic methods on diphenyl
polyenes, di- and terphenyl molecules, and some mono- and diphenyl substituted small
molecules. The differences obtained by the three techniques are shown to result from
the different components of the diffusion tensor observed by these methods. The simultaneous application of these techniques may allow us to better assess the rotational
correlation times about all three axes of the diffusion tensor.

1. Introduction

It is well known that rotation on the molecular level in liquids can be monitored by
several experimental techniques. However, due to the many-body character of liquid
interactions the theoretical modelling of molecular reorientation in liquids and its
connection to the experimental data is still an unsolved problem of statistical mechanics. Since no "exact" solution of this problem is in view it is very important to
assess the range of validity of the different approximations used in this context. It
appears that molecular reorientation correlation functions $C_r(t)$ and correlation times
τ_r besides being well measurable quantities are also sensitive probes to test the
validity of approximations used in statistical mechanical theory of liquids. These
quantities also seem to sensitively reflect what is going on in the immediate neighbourhood of rotating molecules. Thus, molecular reorientation correlation times are,
in principle, a rich source of information on effects of chemical interest like solvation, aggregation, hydrogen bonding, isomerization, photochemical processes and many
others. Unfortunately, this information is well encoded and we are still at the beginning of cracking the code.

In this contribution the motion of solute molecules "A" will be studied whose
reorientation can be monitored experimentally in a solvent consisting of the molecular

species "B", the case of neat liquids being a special case where solvent and solute molecules become identical.

The accumulated experimental evidence clearly indicates that the relevant factors affecting molecular reorientation in liquids are molecular interactions, the size of the molecules, their shape/symmetry and their flexibility.

2. Molecular Interactions

2.1. "AB" type interactions

"A" molecules interact with "B" molecules (and eventually with other "A" molecules) via events more or less randomly distributed in time which may be treated as instantaneous collisions or, alternatively, as fluctuating forces. Any motion of an "A" molecule in a disordered medium like a liquid will entail displacements of other molecules accompanied by kinetic energy randomization and hence dissipation phenomena. These effects are generally treated ignoring the details of molecular interaction and postulating a retarding force which is linearily connected to the molecular velocity and to a "friction constant" β. This is a phenomenological approach whereby the complex problem of random molecular interactions boils down to defining an appropriate friction constant which is supposed to properly average any irrelevant (or prohibitively complex) details of this interaction. Fast interactions, i.e. fast with respect to rotational correlation times, will also be averaged out and are generally neglected. The friction constant is connected to transport coefficients like the shear viscosity η and the translational and rotational diffusion coefficients D_t and D_r via the Stokes-Einstein relation which is formulated below for the rotational case, β_r being the rotational friction constant:

$$D_r = kT/\beta_r \tag{1}$$

The Stokes-Einstein-Debye (SED) equation:

$$\tau_r = \frac{\eta V}{kT} + \tau_0 \tag{2}$$

is in its application to molecular systems, especially for the case of small molecules, a highly controversial, imperfectly understood, but useful and widely used relation (see for example the article of Kivelson in this volume). The hydrodynamic volume V for spherical particles under stick boundary conditions can be identified with the molecular volume. The puzzling aspects are the applicability of macroscopic hydrodynamic concepts to the molecular scale and the appearance of a constant term τ_0 on the rh. side of Eq. (2). τ_0 is an additive term introduced in the original SED equation to account for the observed non-zero value of τ_r at the extrapolated zero-viscosity infinite-temperature limit. As this extrapolation seems at least problematic any physical interpretation of τ_0 should be considered with great caution. Eq. (2) can be thought of as an efficient and simple means to rationalize experimental data or as a first approximation to more involved theoretical relations still to be developed. Most of the improvements of these theories have proceeded by incorporating features like a frequency or a wavevector dependence in β_r [1, 2]. "AB" interactions or rather their relation to "BB" interactions are also important in determining whether the hydrodynamic approach to the problem is best approximated using slip, stick or intermediate boundary conditions. This state of affairs has also been described in terms of roughness of the rotating "A" molecules as a measure of the coupling of these molecules to the fluctuating torque exerted on the "A" molecule by its environment. From the molecular point of view we must ask for the origin of the frictional torque exerted on the surface of "A" molecules by intermolecular interactions with the surrounding molecules of the first layer.

2.2. "AA"-type interactions

Such interactions can severely modify the mobility of "A" molecules depending on their concentration, the temperature, and the nature of the interactions. These interactions determine to a large degree molecular and macroscopic properties in systems in which we have high concentrations of large molecules or particles. In the extreme case we may have molecular aggregation and, if the lifetime of the aggregates is longer or comparable to the time needed for a significant reorientation, the aggregates can be detected and characterized by their reorientational correlation times [3]. "AA"-type interactions will also directly influence light scattering and dielectric data since these methods being "coherent" methods monitor correlated intermolecular motion rather than single-molecule motion.

2.3. "BB"-type interactions

These are important in determining the value of the friction constant β for the motion of "A" molecules in the "B" environment. Thus in some liquids like e.g. glycerol β is extremely large due to the formation of a hydrogen-bonded network opposing both to liquid flow and to the reorientation of large dissolved probe molecules. On the other hand, we sometimes observe that small "A" molecules reorient easily in highly viscous liquids being apparently trapped in larger holes of an extended network.

3. Molecular Size

The important quantity in the present context is the relative size of "A" and "B" molecules i.e. whether the interaction of "A" molecules with "B" molecules can be considered as a sequence of small uncorrelated perturbations of the dynamics of the former or not. Consequently, we can distinguish between the folllowing cases:

3.1. "A" is large relative to "B"

In this case we feel intuitively justified in handling the situation with the same methods as the problem of a macroscopic body immersed in a structureless liquid. The latter is viewed as a structureless dissipative bath acting on the "A" molecules through random fluctuations and friction. This approach has proved quite accurate for the motion of Brownian particles. The simplifying feature which is exploited in this case is that most of the details of molecular interactions assumed to be uncorrelated can be ignored. What has to be considered is mainly the range of the interaction.

3.2. "A" is similar in size as "B"

The continuum approximation made in the previous case does not impose itself in this case too, however, due to its strong simplifying features, it may be profitable to assess how far we may use it. Typical for this case is the strong static and dynamic coupling of the motions of "A" molecules with the neighbouring molecules strictly speaking, the present case requires a molecular theory as opposed to the Brownian particle theory which applies only when the "A" particles are much larger than the "B" particles.

3.3. "A" is smaller than "B"

This case can be viewed as a motion of molecules in a more or less rigid, i.e. slowly moving matrix. We then often observe relatively fast motions that can be described as a rattling of an otherwise free molecule in a rigid cage. One can also view this behaviour as a result of the decoupling of the motion of the "A" and "B" molecules due to the large separation on timescale. Viscosity obviously is not the proper quantity to parametrize the dynamics in this case.

4. Molecular Symmetry

This is important since, depending on the experimental method used to monitor reorientation, its precise description requires the determination of the axis/axes of reorientation. In fact, the relevant quantities in this context are the diffusion tensor $\underset{\sim}{D}$ and its relative position with respect to either the polarizability $\underset{\sim}{A}$ or the Raman tensor $\underset{\sim}{R}$ for the detected vibrational line and, in the case of fluorescence polarization decay data, the electronic transition moments for absorption μ_a and fluorescence emission μ_e. In principle, all of these quantities may have different orientations within the molecule. It is however fortunate that the structural symmetry of the molecule generally also determines to a large extent the symmetry of the wave functions with the effect that the principal axes of the above mentioned $\underset{\sim}{D}$ and $\underset{\sim}{\alpha}$ tensors often coincide. In the case of flexible molecules, however, we may have serious deviations. Actually, the transition moment μ_e for fluorescence emission may differ significantly from the corresponding absorption transition moment μ_a as a consequence of configurational changes occuring in the excited state. All these geometrical relations must be carefully studied before interpreting the data in view of calculating rotational relaxation times.

4.1. Asymmetric rotors:

The calculations involving molecules with low symmetry are notoriously difficult not only because the mathematics are quite involved but also because the measured reorientational correlation functions are expected to display several reorientational correlation times which in most cases cannot be separated with the available experimental data. Under these conditions the observed stretched exponential correlation functions or the non-Lorentzian spectra have been properly analyzed in terms of a superposition of distinct processes only in few cases.

4.2. Symmetric tops:

The main body of interpretable rotation data was obtained from molecules which were described as symmetric tops with the additional feature that the location of the monitored quantity, like the optical transition moments μ_a and μ_e, the polarizability tensor or the dipole vector in the molecule is well-defined and lies parallel or normal to the symmetry axis. On the other hand, all the other requirements which must be met by a good molecular probe are so stringent that we very often must put up with "quasi symmetric tops" like for example diphenyl ether and diphenyl methane [4,5] consisting of two phenyl rings connected by an oxygen or a carbon atom respectively. Also dyes like ethidium bromide and acridine orange which are popular labels for biologic applications consisting of three fused aromatic rings with substituents are often considered as quasi symmetric tops [6,7]. The same considerations apply to diphenyl polyenes and poly phenyl molecules which will be discussed extensively below. We know that, strictly speaking, molecular symmetry can only be defined exactly and a "quasi symmetric" top is in fact an asymmetric top. This important point is stressed in the contribution of Leicknam in this volume. On the other hand, the error involved in approximating such molecules by symmetric tops is very difficult to assess quantitatively and apparently several experimental results involving different techniques seem to justify such an approximate procedure.

The extension of Eq. (2) to ellipsoidal molecules [8] results in an extended Stokes-Einstein-Debye equation (SED):

$$\tau_r = f(\rho) \cdot z(\rho) \cdot V \cdot \eta/kT + \tau_0 \tag{3}$$

In this equation $f(\rho)$ and $z(\rho)$ are a molecular shape factor and a factor describing the hydrodynamic boundary conditions respectively. In the case of ellipsoids of revolution with a symmetry axis and an axis normal to it of lengths a and b respectively $f(\rho)$ and $z(\rho)$ are functions of the axial ratio $\rho = b/a$. For such a rotor Perrin [8] has given values of $f(\rho)$ and Hu and Zwanzig [9] have calculated $z(q)$. Kivelson describes $z(\rho)$ in terms of the relative size of the rotating molecules and the solvent molecules (this volume). We can define a hydrodynamic volume for ellipsoids as $V_h = f(\rho) \cdot z(\rho) \cdot V$. V_h will depend on the actual molecular volume, on the shape and the prevailing boundary conditions. Unfortunately, it is not possible to disentangle these factors from experimental values of V_h alone although we usually know the molecular shape and size. Even admitting the validity of Eq. (3) a thorough and systematic study of many molecules under different conditions is still required. The value of the extended SED equation (3) lies in the possibility to calculate experimental values of the hydrodynamic volume V_h which can be compared to independent data on molecular shape and size although this raises a number of problems related to intermolecular interactions. SED plots of τ vs. η/T have been found to be in many cases linear indeed over a sufficiently large η/T scale. The information both on the size and the shape of the rotating scatterer has found interesting applications to biological macromolecules whose exact shape is known by electron microscopy, however, leaving the important question of their conformation in solution open [10].

Another symmetry which has been extensively used in model calculations is that of rigid rods [11,12]. Rodlike molecules can be considered as symmetric tops whose axial ratio ρ is very small. The hydrodynamic behaviour of rigid rods is particularly amenable to calculations and illustrates the effect of shape anisotropy very well. It has been found that the parallel study of rotational and translational diffusion was particularly useful as it allows us to determine the molecular parameters. However, when we compare the theoretical results to experimental data we find that it is very difficult to find real molecules whose shape and dynamical behaviour do comply to the rigid rod model. If molecules become sufficiently long they also become increasingly flexible. On the other hand, if they are to be rigid, for example as an effect of aligned conjugate double bonds, their solubility is extremely low as they tend to gain little entropy on dissolving.

4.3. Spherical tops

The interesting feature of rotating spherical tops in a liquid is that they do not require extra volume to rotate about an axis passing through the center and thus are less than other molecules hindered by friction with the environment. Actually, the viscous drag exerted on such molecules is only due to direct "AB" interactions or, in terms of hydrodynamic theory, to the stick boundary conditions prevailing at the surface of the "A" molecules. This effect may also be explained by invoking the non-spherical shape of molecular spherical tops, i.e. to the atomic structure of such molecules as is illustrated in the contribution of Kivelson in this volume. This approach was fruitfully used to interpret reorientational correlation times of liquid benzene [13]. The benzene molecule was shaped as a hexagonal solid with six protuberances for the hydrogen atoms. This proved much more adequate than using simply a disc-shaped molecule. The same considerations also apply to the rotation of symmetric top molecules around the main symmetry axis [14]. In such cases we find that the motion is little affected by the viscosity of the solvent if the latter does not display strong interactions with the rotor.

5. Intramolecular Motions

Strictly speaking, molecules display high internal mobility and the neglect of this feature may lead to serious errors. However, treating molecules as rigid objects with a well-defined shape is appealing for its simplicity and is common practice. In order to discuss the validity of the rigid-molecule assumption we must distinguish between molecular vibrations, internal rotations and conformational changes. The problem is to describe the coupling of molecular overall rotation with each of the mentioned internal degrees of freedom.

5.1. Intramolecular vibration

In most cases we can safely assume that the difference in timescales precludes significant coupling between overall rotation and internal vibration.

5.2. Intramolecular rotation

The coupling of overall to internal rotations must certainly be acknowledged since symmetry as well as the timescales will favour it. We distinguish the case of small rotors attached to large molecules or the case of equal or approximately equal rotors. In principle, the rate of internal rotations is determined both by the solvent viscosity and the potential barrier(s) along the angular degree of freedom. In most of the molecules discussed below the height of the potential barrier prevents full reorientational motions of the individual groups and the internal degree of freedom is a torsional libration about an equilibrium configuration in which for example phenyl groups form a small angle of 10 to 30 degrees. Under such conditions the relevance of this motion to the overall rotation is apparently not very large.

5.3. Conformational changes

In large molecules conformational changes with low energy barriers play an important role in determining the overall motion, especially rotation. The timescales of conformational changes are widely spread over many orders of magnitudes and their effect is, roughly speaking, to make the molecular shape time-dependent. If an external torque acts on such a molecule it not only suffers a change of its angular momentum but it also complies to the torque by changing its shape on a timescale comparable to the timescale of overall rotation. Mainly macromolecules either synthetic or biological like for example proteins display this effect. The fluctuations in overall shape and local configuration of proteins is important since for example enzymatic activity is controlled by the latter. In some cases it has been well established that the conformational reorientation of a particular functional group opens the way to a given substrate to interact with a particular enzymatic site.

6. Comparison of Experimental Techniques

In this contribution a few results will be presented obtained by application of depolarized light scattering (DLS), fluorescence anisotropy decay (FA) and Raman bandwidth analysis (RA) techniques.

6.1. The nature of the correlation functions

In all three techniques we monitor directly or indirectly the temporal evolution of a property $\Xi(t)$ of the whole system which is a function of the conditional probability $\xi_i[P(\Omega,t|\Omega_0,0)]$ that for the molecule i ξ_i has the orientation Ω at the time t when it had the orientation Ω_0 at the time origin t = 0. Ω represents the Euler angles describing the orientation of the pertinent quantity.

$$\Xi(t) = \sum_i A_i \cdot \xi_i [P(\Omega,t|\Omega_0,0)] \qquad (4)$$

The quantity $\xi_i[P(\Omega,t|\Omega_0,0)]$ is, depending on the technique discussed, the molecular polarizability, the Raman tensor or a transition moment between the excited and the ground state. The summation is, in principle, over all molecules involved, however, in the case of DLS light scattering proceeds coherently and the signal monitored is basically coherent containing additionally phase factors which modulate the factors A_i and which thus become functions of the position of each scatterer. In the two other methods the process of emission and hence the signal recorded from the molecules is incoherent and thus A_i does no more depend on i and the summation is over uncorrelated terms. In the process of calculating second order time correlation functions of the type:

$$C(t) = < \Xi(t) \cdot \Xi(0) > \qquad (5)$$

in DLS we must include the ij cross correlations:

$$C_d(t) = < \xi_i[P(\Omega,t|\Omega_0,0)] \cdot \xi_j[P(\Omega,t|\Omega_0,0)] > \qquad (6)$$

corresponding to two different scatterers while in the two other methods these will average out and we obtain only single-molecule ii terms in the correlation function:

$$C_s(t) = < \xi_i[P(\Omega,t|\Omega_0,0)] \cdot \xi_i[P(\Omega,t|\Omega_0,0)] > \qquad (7)$$

Despite this clear cut difference between DLS on the one hand and FA and RA on the other any physical coupling between the emission process of nearby molecules may, of course, lead to a non-zero value of the cross correlation function. On the other hand, if we are interested in studying single-molecule reorientation with DLS we must work at very low concentrations which for intensity reasons is only possible if we work with the resonance-enhanced light scattering technique.

The necessary expressions to interpret the obtained correlation functions are derived by assuming for example that the time evolution of $\xi_i[P(\Omega,t|\Omega_0,0)]$ obeys a diffusion equation. We then obtain:

$$\text{DLS:} \quad C(t) = K \cdot \sum_{i=1}^{5} B_i \cdot e^{-t/\tau_i} \qquad (8a)$$

RA: $\quad C(t) \;=\; L \cdot \displaystyle\sum_{i=1}^{5} C_i \cdot e^{-t/\tau_i}$ $\hspace{4cm}$ (8b)

FA: $\quad r(t) \;=\; 0.3 \cdot \displaystyle\sum_{i=1}^{5} D_i \cdot e^{-t/\tau_i}$ $\hspace{4cm}$ (8c)

The constants K and L involve the differential scattering cross section for Rayleigh and Raman scattering respectively. The correlation times τ_i correspond to the five eigenvalues of the rotational diffusion equation involving the diffusion tensor. The amplitude factors B_i, C_i and D_i are determined by the relative position of this tensor with respect to the polarizability tensor, the Raman tensor or to the transition vectors μ_a and μ_e. In the case of FA we usually obtain from the experimental decay curves of the vertical-vertical (VV) and the vertical-horizontal (VH) polarization geometries the anisotropy defined as:

$$r(t) \;=\; \frac{I_{VV}(t) - I_{VH}(t)}{I_{VV}(t) + 2 \cdot I_{VH}(t)} \hspace{4cm} (9)$$

The analytical form of r(t) [Eq. (8)] resulting from application of the the diffusion equation is similar to the form of the correlation functions C(t) in the DLS and the RA case as given in Eqs. (8a) and (8b). r(t) is connected to the rotational correlation function C(t) by Eq. 8d below involving, additionally to molecular rotation, the intramolecular dynamics of the transition vectors μ_a and μ_e [20]:

$$r(t) \;=\; < k_F(t) \cdot \rho(\omega,t) \cdot P_2[\mu_e(t) \cdot \mu_a(t)] > \hspace{3cm} (8d)$$

All three correlation functions in Eqs (8) describe the time evolution of a second-order Legendre polynomial $P_2(\cos \varphi)$.

It is important to notice that the summation over i from 1 to 5 in Eqs (8a-c) results in the occurence, in the general case, of five superimposed exponential decay curves with different time constants τ_i and amplitudes to which the experimental data must be fitted. It is well known that the non-linear fit procedures commonly used tend to become unstable in the case where the time constants are not very different as in most cases the resulting decay curves can hardly be distinguished from a single exponential decay with an average time-constant. New developments in experimental techniques may hopefully improve this situation by allowing for better signal to noise ratios of the spectra.

6.2. <u>DLS and FP correlation times of diphenyl polyenes and poly phenyl molecules</u> [15-17]

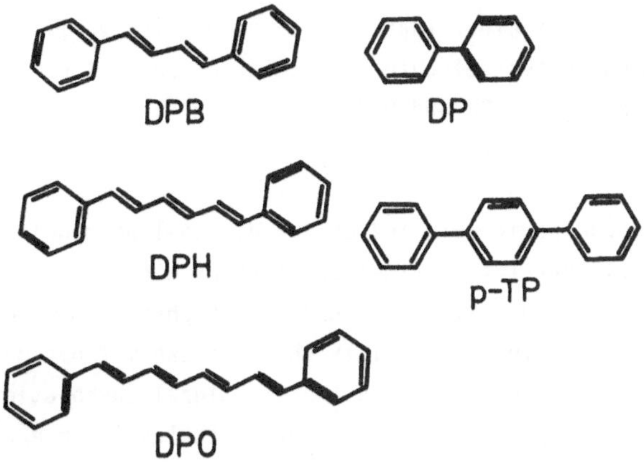

Fig. 1: Structure of the rodlike molecules studied in this article. DPB: diphenyl butadiene; DPH: diphenyl hexatriene DP: diphenyl; p-TP: para terphenyl

A number of elongated organic molecules display rotational correlation times which present some interesting features. Fig.1 illustrates the structure of the molecules discussed below. The DLS data presented below were performed in the preresonance region thus achieving at concentrations as low as 10^{-6}mol/l acceptable signal to noise ratios. The DLS spectra were fitted to single Lorentzians. Actually, the fits with two Lorentzians do not improve the correlation coefficients. The r(t) curves also do not give any indication of the occurence of more than one Exponential. The SED plots where the neat solvent viscosity was used were in most cases linear allowing us to determine unambigiously hydrodynamic volumina. The FA data were obtained under similar conditions and wherever possible in the same solvent. Thus, for PTB both sets of data were obtained in $CHCl_3$ solutions and the DPH spectra in 2-propanol. The experimental hydrodynamic volumina obtained with the two methods are displayed in Table 1

Table 1: Experimental and calculated hydrodynamic volumina

	V_h(DLS)	V_h(FA)	$V_h^{\parallel}$	$V_h^{\perp}/V_h^{\parallel}$	References
DP	72	-	37	1.7	[15]
p-TP	280	277	69	4.0	[22] [25]
DPB	292	-	65	4.1	[16]
DPH	390	130	53	7.7	[16] [26]
DPO	680	283	79	7.8	[16]

We see that while the experimental values of V_h(DLS) and V_h(FA) are similar for p-TP the same quantities display considerable differences for the two diphenyl polyenes DPH and DPO the hydrodynamic volumina of these molecules obtained by the two methods differing by roughly a factor of three. The ratio of the hydrodynamic volume parallel and normal to the symmetry axis displayed in column 5 is seen to increase in the series by a factor of 4. It also becomes clear that the DLS values for V_h fit quite well to the picture according to which the molecules rotate about an axis perpendicular to the symmetry axis. The FA data for DPH and DPO, on the contrary, yield significantly smaller values.

As this discrepancy in the hydrodynamic volumina of the diphenyl polyenes casts a doubt on the accuracy of rotational correlation times obtained by either of the two methods it is important to go into some detail of the underlying physics in order to understand its origin. We must consider the possibility that the two methods do not monitor the same species. In fact, while DLS measures the rotational correlation times of molecules in the electronic ground state, FA does so for molecules in an excited state which is a different species with respect to configuration and/or to charge distribution. This may go so far that we have photoisomerization processes conferring to the excited molecule a different shape. Furthermore, vibrational relaxation processes occuring in the excited state may heat the molecule, especially given the large Stokes shifts observed in diphenyl polyenes, thus increasing τ_r. In any case, it seems that the photophysics involved especially in the FA experiment must be fully taken into account.

First, we can discard the effect of increased dielectric friction in the excited state molecule and of local heating because the former would tend to decrease τ_r and the latter can be shown to be negligibly small. On the other hand, it is known that the photophysics of diphenyl polenes involve several processes which must be acknowledged in interpreting the data. Thus, the potential energy surface for the conformational degrees of freedom in the excited state of diphenyl polyenes are such that these molecules are able to undergo a rotatory conformational transition which is thermally unaccessible in their ground state. More precisely, absorption is a $S_0 \rightarrow S_2$ transition (A_g to B_u symmetry), the $S_0 \rightarrow S_1$ transition (A_g to A_g symmetry) being symmetry forbidden [18,19]. The absorbed energy is then transferred in a few picoseconds by virtue of vibronic coupling into the S_1 state. In this state the molecule undergoes a conformational change whose time constant is of the order of 50 ps. This state is then capable of emitting fluorescence with a time constant of the order of a few nanoseconds polarized in the direction of the transition moment μ_e which, as a consequence of the $S_1 \rightarrow S_2$ transition and the following conformational change entailing symmetry change, is significantly shifted with respect to μ_a. Thus, internal reorientation of μ_e is reflected in FA as well as molecular overall reorientation. In addition to this, the species whose rotatory motion is detected by FA has a different conformation than the ground state molecule. Note that these processes which determine the orientation of μ_e are also solvent dependent. Thus, in FA both external and internal dynamics are manifested requiring, stricly speaking, a full treatment of internal and external dynamics and their coupling. Under certain conditions, however, μ_e can be considered molecule-fixed and thus r(t) will depend on external rotation only.

Considering DLS, the situation is less complex since no excited states are involved. What must be known is the orientation of the polarizability tensor in the molecule and its symmetry. An additional complicating factor is the collective character of the measurement. This latter feature is of little relevance in the present case since we have worked at very low concentrations and rotational coupling of "A" particles will be negligible. Depending on the architecture of the molecule the deviation between the diffusion tensor and the polarizability tensor and their approximation by symmetric tops may be a source of error, but of minor importance given the difference by a factor of three between FA and DLS correlation times which must be explained. We thus can safely conclude that while DLS remains in many cases an excellent probe for molecular rotational relaxation we must exert much care in interpreting FA data. However, this result also indicates that extremely useful information on conformation and interactions of electronically excited molecules is contained in the FA data and that trying to extract it is worthwhile.

6.3. Comparison of DLS and RA correlation times of diphenyl [15]

Depolarized light scattering intensities are a function of the optical anisotropy $\beta = \alpha_\parallel - \alpha_\perp$ and the number of scatterers in the case of independent scatterers. When the scattered light comes from correlated scatterers we may learn something about these correlations by analyzing the data in terms of single-scatterer reorientation corrected for the effect of correlations between the scatterers. An approximate theory of the relation between single-molecule (τ_r), and collective (τ) rotational correlation times as monitored by DLS is given by Keyes and Kivelson [21]. These authors find a rather simple relation between the two correlation times:

$$\tau = \tau_r \cdot \frac{1 + N \cdot f}{1 + N \cdot g} \tag{10}$$

N is the number of scatterers in the scattering volume. The factors f, g are the static and dynamic correlation factors. These are defined as normalized correlation functions of the second Legendre polynomials $P_2(\cos \varphi)$ and their time derivatives respectively. It is usually assumed that g is sufficiently small so that $N \cdot g \ll 1$. With this approximation we can write the following expression for τ and for the total scattered intensity:

$$\tau = \tau_r \cdot (1 + N \cdot f) \tag{11a}$$

$$I_{VH} = A \cdot \beta^2 \cdot N \cdot (1 + N \cdot f) \tag{11b}$$

In Eq. 11b A is a constant and β the optical anisotropy of the single scatterer. This equation allows us to relate relative intensities to the static correlation factor f. The separation of single-scatterer rotational correlation times from the total rotational correlation times is based upon the difference between the scattering physics of DLS and vibrational Raman scattering. Whereas the former is a coherent process monitoring the time evolution of the coherent optical anisotropy $\beta(t)$ of all N scatterers in the scattering volume the latter registers the incoherent addition of light intensity emitted by the scatterers in the scattering volume. In Raman, depending on the symmetry species of the vibrational line chosen the axis of rotation can be unambiguously identified. Generally, totally symmetric vibrations are convenient and give correlation times which can be compared to those obtained from DLS.

$$\tau_r = \frac{1}{2 \cdot \pi \cdot \Gamma_a} \tag{12}$$

The half-width of the anisotropic Raman band $I_a(v)$ ($=I_{VH}(v)$) is a convolution of the orientational spectrum with the isotropic line shape

$$I_{iso} = I_{VV}(v) - \frac{4}{3} \cdot I_{VH}(v) \tag{13}$$

Thus the pure reorientational correlation time τ_r can be calculated from the corresponding bands obtained in the VV and VH scattering geometry, possibly contaminated by some collision-induced light.

For the rotation of symmetric tops, assuming diffusive dynamics, we expect Lorentzian bandshapes. On the other hand, although Lorentzian bandshapes are often registered it would be premature to consider this as a proof that rotating scatterer is indeed a symmetric top rotating with diffusional dynamics. In order to decide whether this is the case indeed, we must consider the time-window of our spectral analysis and the relative intensities expected from all pertinent components in relation to the noise level of the spectra. It appears that although very often one component seems to dominate the spectrum other components which may take the form of unshifted additional bands manifesting themselves as a band-broadening effect may appreciably contribute to the dynamics and still remain undetected.

In molecules displaying internal mobility the reorienting moieties can be considered as scatterers on their own whose motion is strongly correlated. In the case of an intramolecular configuration which is rigid in the pertinent timescale the static correlation factor will be 1. If the timescales of the internal and external correlations are of comparable size then Eqs. (11a) and (11b) can be replaced by the following:

$$\tau = \tau_r \cdot [1 + (n-1) \cdot f_i + N \cdot f_e] \tag{14a}$$

$$I_{VH} = A \cdot \beta^2 \cdot N \cdot [1 + (n-1) \cdot f_i + N \cdot f_e] \tag{14b}$$

n and N are the number of rotating moieties in the molecule and the number of scatterers in the scattering volume respectively. The factors f_i, f_e are the internal and external static correlation factors respectively. As Eq. (11) relies on the assumption of a weak coupling of different scatterers the validity of the extension illustrated in Eqs. (14a) and (14b) is limited to a more narrow range than that of Eqs. (11a) and (11b). Furthermore, dynamic correlations have also been ignored although in the case of internal correlations they may be significant. On the other hand linear relations like Eq. (14) represent experimental data quite well making it worthwhile to test their range of validity.

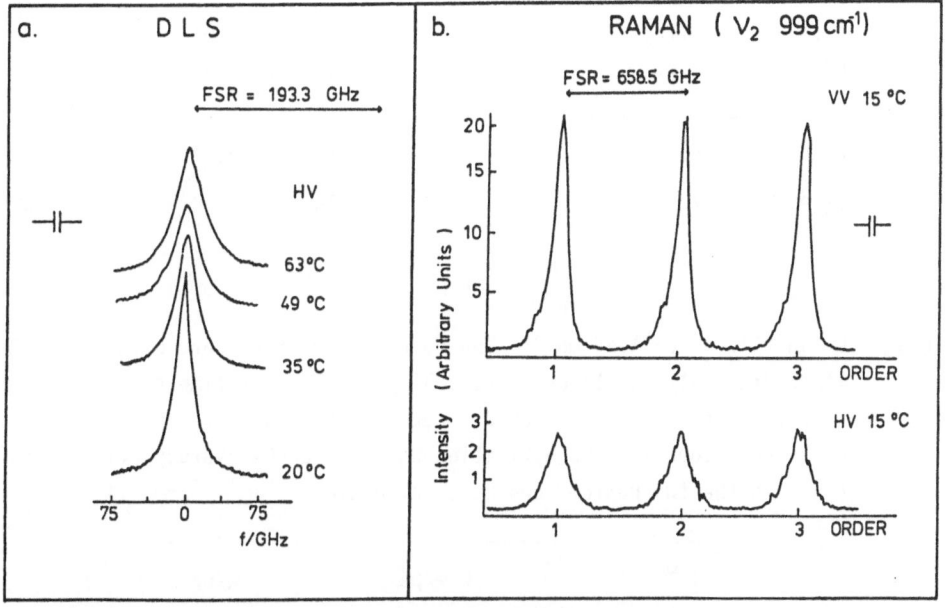

Fig. 2: The figure displays three orders of the Raman spectrum of the totally symmetric vibration of the phenyl ring in DP. The two scattering polarization geometries are displayed. The instrumental linewidt is indicated. The spectra were obtained by high resolution interferometric Raman spectroscopy.

Fig. 2 displays the DLS and the Raman spectra obtained with DP at different temperatures. The effect of temperature on the spectral shape both in DLS and in Raman are clearly visible. The Raman spectra at 992 cm^{-1} can be assigned to the totally symmetric breathing vibration of the phenyl ring belonging to the A_{2g} species. This line is free from overlap with other lines and the pure isotropic spectrum is very narrow. The slight asymmetry on the left side of the Raman lines is due to a combination line which could be removed by fitting with an additional shifted Lorentzian.

These conditions, which are necessary prerequisites for the calculation of reliable rotational correlation times, are met to a large extent in the used A_{1g} line of the phenyl group. Table 2 summarizes the results obtained with DP at different concentrations and temperatures in n-heptane.

Table 2: DP in n-heptane. The correlation times were obtained by DLS. From the data the values for the activation energies and the hydrodynamic volume of DP given in the two bottom rows were calculated.

	c=0.13mol/l			c=0.77mol/l	
$T/^{o}C$	τ/ps	η/cp		τ/ps	η/cp
20	11	0.63		15	0.66
34.8	9.4	0.54		11.7	0.58
49.1	7.7	0.48		9.3	0.51
63	6.7	0.43		7.5	0.45
$E^{\#}$/kJ	40	29.7		54	29.7
$V_h/Å^3$	72			110	

Table 3: DP in CCl_4 at c= 1.45 mol/l. The correlation times were obtained by Raman linewidth analysis. In order to achieve sufficient time resolution interferometric high resolution Raman spectroscopy was used [22]. From the data the values for the activation energy and the hydrodynamic volume of DP given in the two bottom rows were calculated.

$T/^{o}$	C—/ps	η/cp
15	7.5	.78
27	6.0	.68
40	5.8	.60
55	4.6	.50
$E^{\#}$/kJ/mol	35	29.7
$V_h/Å^3$	38	

The correlation times vary im the range between 5 to 15 ps depending on temperature, concentration and the technique used to monitor τ_r. Note the difference in the activation energies obtained from the DLS and the viscosity data. The activation energy for the rotation also depends on the concentration. This is remarkable since the concentration does not affect the viscosity as can be seen from the table. The concentration dependence of the hydrodynamic volumina is indicative of dynamic correlation effects.

The Raman data give an activation energy of 35 kJ/mol for the rotational relaxation which is lower than that obtained from DLS, although the concentration is higher. Likewise the hydrodynamic volume obtained by RA with only 38 A^3 is approximately one half of the value obtained with DLS at the lower concentration. Raman data at lower concentrations show that these values do not change with concentration of DP. This result is typical for the relation of RA and DLS hydrodynamic volumina as will be seen below. Fig. 3 visualizes these results for 15°C. We see that the DLS correlation times

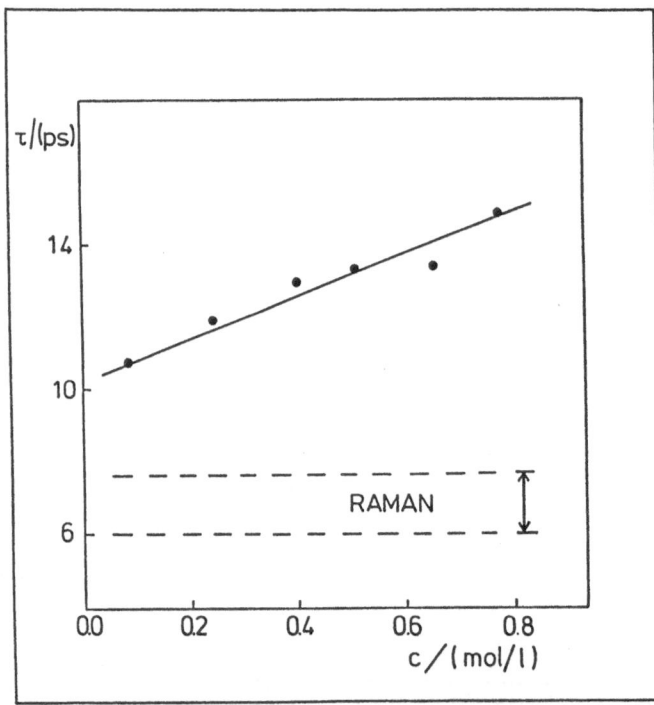

Fig.3: Concentration dependence of the DLS and the Raman rotational correlation times of diphenyl as a function of the concentration. The Raman data were obtained at high concentrations whereas the DLS data extend from 0.13 to 0.77 mol/l. The concentration dependence of the former is weak and is indicated by the error bar and the dashed lines.

increase linearily by 50% with concentration in this range, giving a correlation time of approximately 10 ps in the extrapolated zero-concentration limit. The Raman data could not be followed at so low concentrations due to the low intensity of the depolarized Raman line, but the value for a concentration of 0.8 mol/l is 7.5 ± 1ps and this value does not seem to decrease significantly with concentration. Fig. 3 strongly suggests that the motion monitored by the DLS data is much more sensitive to the environment viscosity than the motion monitored by the Raman line and that the ratio of the zero-concentration limit values of τ as measured by the two techniques is

approximately 2.

Fig. 4 illustrates the application of the SED-equation (2) to the DLS and the RA data at two concentrations in n-heptane (DLS) and and at one concentration in CHCl$_3$ (RA). We can see that the DLS-correlation times at the concentrations of 0.13 and 0.77 mol/l do not scale with η/T in contrast to what we would expect from Eq. (2) and to what has been observed in several cases where a common master-curve comprising variations of concentration and different solvents could be obtained on the SED plot [5]. It appears that the observed rotational correlation time increases above the value τ_r which can be assigned to single molecule reorientation as monitored in the dilute solution or by the Raman correlation times. Note also that whereas both the low-concentration DLS and the Raman points lie reasonably well on straight lines the high-concentration DLS points deviate from a straight line approaching the low-concentration line asymptotically in the high-temperature/low-viscosity region.

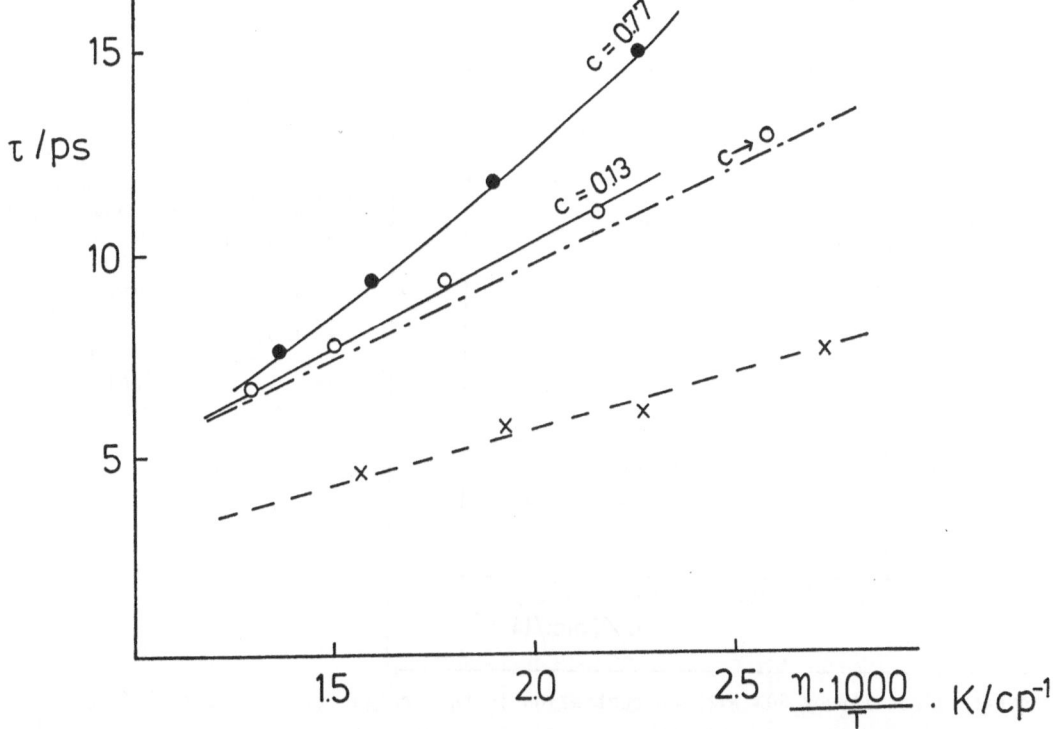

Fig. 4: SED plots of diphenyl. The DLS correlation times were obtained at two concentrations and the dashpoint line illustrates the zero concentration limit. The Raman data are represented by the dashed line. Note the deviation from linearity at 0.77 mol/l. The slope of the Raman results is clearly smaller than the one obtained from DLS. This is shown to be due to the difference of the axis of rotation monitored by the two methods.

Another remarkable feature is that the experimental points of the Raman correla-

tion times lie on a different line than the DLS points with a slope smaller by approximately a factor of two.

DLS correlation times involve, besides the single-molecule rotation correlated rotation of different molecules. The latter is manifested in the concentration dependent correlation times in Fig. 3 and in the difference of the hydrodynamic volumina and activation energies of the dilute and the concentrated solutions.

6.4 DLS and RA results on the terpenyl isomers.

A comparative study of rotational relaxation of these three isomeric molecules in the liquid state is of interest because their symmetry and shape is varied systematically the interaction with the solvent being identical in all three cases.

The DLS intensities obtained at different concentrations and at three temperatures for each sample are illustrated in Fig. 5. The concentration dependence of the intensity is linear for each isomer indicating that intermolecular correlations are not appreciable. On the other hand, the slopes differ markedly. Table 4 illustrates the intensities at one concentration (1% molar), the correlation times at 23°C, and the hydrodynamic volumina obtained from linear SED fits.

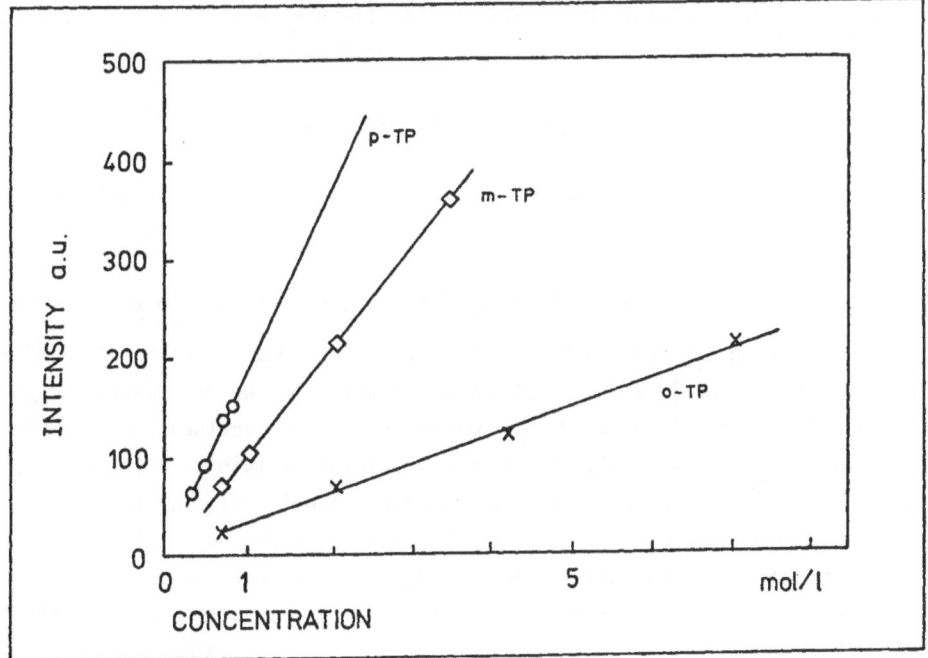

Fig. 5: DLS intensities as a function of concentration of the three terphenyl isomers. Note the strict linearity of the plots and the difference of specific scattered intensity of the three samples.

Table 4: Experimental parameters for the three terphenyl isomers as obtained by DLS in 0.77 mol/l solution in CCl$_4$ and by Raman spectral analysis in the neat liquid. The correlation times are at 23°C.

		I/arb.units	τ/ps	$V_h/\text{Å}^3$	
DLS	p-TP	102	35	280	[22]
(.07 mol/l)	m-T	57	28	223	[22]
	o-TP	17	20	128	[22]
RA	o-TP	--	11	30	[22]
(neat liquid)					

The results illustrated in this table clearly show that the rotational dynamics of terphenyl is, as expected, modified by the change in shape occuring when we go from the p- to the m- and to the o-isomer and/or by the fact that the different shape of the molecules does modify the detected anisotropic scattering.

    a. The DLS intensity is largest in the p-isomer and decreases as the molecules is becoming more compact, apparently because the molecule changes from a quasi rodlike to a more compact shape with a more isotropic polarizability.

    b. The correlation times decrease in the same order.

    c. The hydrodynamic volumina decrease also reflecting the increasing compactness of the molecule in this order.

    d. The hydrodynamic volume. Actually, the result mentioned in section 6.3 according to which the Raman data for DP give a hydrodynamic volume of 38 Å^3 which is smaller by approxi-matly a factor of two as compared to the DLS result, as well as the Raman result in Table 4 for o-TP indicate an interesting feature. In order to understand this we must recall that Raman linewidths pertain to a localized phenyl group vibration and as such reflect the reorientational motion of the phenyl groups. In most of the molecules discussed here the phenyl groups are not free to rotate by a large angle. The observed motion rather reflects the overall tumbling of the molecule and, perhaps, to a lesser extent, restricted libration about the bond to the next ring.

In order to visualize the rotation monitored by the two techniques DLS and RA, we must distinguish between the **molecular polarizability ellipsoid** $\underset{\sim}{A}$ on the one hand and the

Raman ellipsoid for the totally symmetric vibration of each phenyl group R̰ **on the** other. For molecules like DP or p-TP the former is triaxial while the latter is biaxial as indicated in Fig. 6. However, to some extent we can approximate the polarizability ellipsoid to a prolate biaxial one with its high polarizability axis, which under

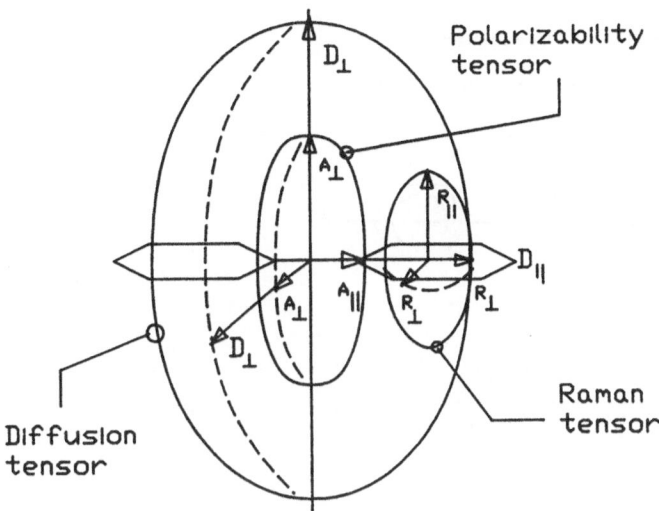

Fig. 6: Relative orientation of the three ellipsoids corresponding to the diffusion tensor $\underset{\sim}{D}$, the polarizability tensor $\underset{\sim}{A}$ of the whole molecule and the Raman tensor $\underset{\sim}{R}$ of the A_{1g} vibration of the phenyl group in diphenyl. The figure shows that while the high polarizability axis $A_\parallel$ is collinear to the $D_\parallel$ axis the symmetry axis $R_\parallel$ of the Raman tensor is perpendicular to it.

this assumption will be its symmetry axis, in the direction roughly parallel to the symmetry axis of the diffusion tensor. The Raman ellipsoid on the other hand has its symmetry axis normal to the phenyl group plane i.e. normal to the symmetry axis of the diffusion and the polarizability tensors. The diffusion ellipsoid $\underset{\sim}{D}$ is triaxial, but can degenerate to a biaxial depending on the relative orientation of the phenyl groups. If we assume biaxial symmetry for the diffusion tensor it will be oblate with the symmetry axis as indicated in the Fig. 6. Thus, overall rotations about the long molecular axis, which is also the symmetry axis of both, the polarizability and the diffusion tensors, will be fast and associated to a small V_h. Rotations about the small axes of the diffusion tensor will be slower and associated to higher values of V_h. Overall spinning or tumbling motions, i.e. reorientations about the ⊥ or about the ∥ axis of the diffusion tensor can be, in principle, detected by DLS and by Raman scattering respectively. This is the reason why DLS and Raman scattering come up with so different values for τ_r and hence for V_h. These results clearly show that the

Table 5: Experimental hydrodynamic volumina V_h^{exp} calculated from DL/RA correlation times and theoretical values for slip boundary conditions $V_h^{\parallel}$, $V_h^{\perp}$ for rotation about the parallel and the perpendicular axis of an equivalent diffusion ellipsoid for some phenyl and diphenyl molecules.

Substance (technique)	Structure	V_h^{exp}	$V_h^{\parallel}$	$V_h^{\perp}$	
DP (DLS)		72	37	73	
DP (Raman)		38	"	"	
p-TP (DLS)		280	69	277	
DPE (DLS)		72	< 10	50	[4]
DPM (DLS)		66	< 10	≈ 50	[5]
DPS (DLS)		67	< 10	≈ 50	[23]
PS (DLS)		44	≈ 20	≈ 40	[23]
o-cresol		60	≈ 20	≈ 70	[24]

combination of DLS and RA methods can allow us to obtain a fairly accurate picture of the three-dimensional asymmetric molecular rotation in liquids. Despite this favorable situation a caveat must be posted at this point. Due to the approximate character of the above symmetry considerations the rotation about either of the axes of the diffusion tensor may leak into the Raman or the DLS spectrum respectively and thus render the calculations less simple. However, depending on the particular molecular geometry, this may, in favorable cases, occur with a small weight.

Table 5 illustrates the above situation on a number of mono- and diphenyl compounds of a rather simple geometry. We can see that for the simple phenyl compounds like phenyl silane (PS) and o-cresol we find with DLS approximately a τ_r of 44 to 60 $Å^3$ in reasonable agreement with the molecular dimensions. The diphenyl compounds like DPE, DPM, DP, DPS which are very similar in stucture and dimensions display a larger V_h varying from 66 to 72 $Å^3$. Finally, the only linear terphenyl compound has a V_h of 280 $Å^3$. These values can be compared with the calculated values in columns 4 and 5 of the hydrodynamic volumina for the rotation about the parallel and the perpendicular axis respectively. We see that indeed **DLS predominantly probes rotation about the perpendicular and RA about the parallel axis**. In order to complete the picture more high resolution Raman data are required on the other substances of the table.

7. Summary

Using depolarized light scattering, Raman bandwidth analysis and fluorescence anisotropy decay one obtains different components of molecular rotation. This depends upon the detailed geometry of the diffusion tensor in relation to the pertinent tensors/vectors for each of these techniques. A simultaneous application of these methods is feasible for a large class of molecules and promises to make some interesting details of molecular reorientation accessible.

Acknowledgements

This contribution is a short summary of several theses and papers in preparation in our laboratory. The details and the primary data are contained in the references [3], [15], [16], [17], [22], [23], [26].

Part of the work was supported by a grant of the "Minister für Wissenschaft und Forschung NRW". The financial support of the "Fonds der Chemischen Industrie" is gratefully acknowledged.

References:

[1] Hwang, L.P. and Freed, J.H., J.Chem.Phys. 63, 118 (1975)

[2] Keyes, T. and Ladanyi, B.M., J.Chem.Phys. 70, 5261 (1979)

[3] Eimer, W., Thesis Universitat Bielefeld 1987

[4] Fytas, G., Lilge, D. and Dorfmuller, Th., J.Chem.Soc. Faraday Trans. 2, 80 (1984), 283

[5] Lilge, D., Eimer, W. and Dorfmuller, Th., J.Chem.Phys. 86 (1987), 391

[6] Weber, G., Ann.Rev.Phys. Chem. 1974

[7] Mildbred, K., Thesis, Universitat Bielefeld

[8] Perrin, F. Journal de Phys. et le Radium VII, 1 (1936)

[9] Hu, Ch. and Zwanzig, R. J.Chem.Phys. 74 (1974), 4353

[10] Patkowski, A., Schneider, G. and Eimer, W. to be published

[11] Aragon, S.R. and Pecora, R., J.Chem.Phys. 82, 5346 (1985)

[12] Riseman, J. and Kirkwood, J.G., J.Chem.Phys. 12, 512 (1950)

[13] Youngren, G.H. and Acrivos, J., J.Chem.Phys. 63 (1975), 3846

[14] Assink, R.A., J.Chem.Phys. 63 (1975) 5045

[15] Nikomanis, F., Samios, D. and Dorfmuller, Th., to be published

[16] Uphoff, R., Thesis Universitat Bielefeld 1987

[17] Uphoff, R. and Dorfmuller, Th., to be published

[18] Hudson, B.S., Kohler, B.G., and Schulten, K., in "Excited States" ed. E.C. Lim, 1982

[19] Rulliere, C. and Declamy, A., Chem.Phys.Lett. 135, 213 (1987)

[20] Szabo, A., J.Chem.Phys. 81, 150 (1984)

[21] Keyes, T. and Kivelson, D., J.Chem.Phys. 56 (1972), 1057

[22] Samios, D. and Dorfmuller, Th., to be published

[23] Lempart, A., Diplom Thesis, Universitat Bielefeld 1987

[24] Wang, C.H., Ma, R.J., Fytas, G. and Dorfmuller, Th., J.Chem.Phys. 78 (1983), 5863

[25] Philips, L.A., Web, S.P. and Clark, H.H., J.Chem.Phys. 83 (1985) 5810

[26] Potthast, R., Thesis Universitat Bielefeld 1986

MOLECULAR ROTATIONAL DYNAMICS IN ISOTROPIC AND ORIENTED FLUIDS STUDIED BY ESR[*]

Jack H. Freed

Baker Laboratory of Chemistry
Cornell University
Ithaca, NY 14853 / USA

Table of Contents

[*]Supported by NSF Grants DMR 86-04200, CHE 83-19826, and NIH Grant GM 25862.

1. Introduction

ESR (like NMR) has for many years been utilized successfully in the study of rotational dynamics in liquids [1-4]. Our aim in this chapter is to present a reasonably up-to-date prospectus on the subject. I have chosen to do this with a variety of recent examples, largely from our laboratory, since I am most familiar with them.

Thus, in Section II, I describe the results of pressure and temperature-dependent spin-relaxation studies which led us to propose a new "quasi-hydrodynamic" model of rotational reorientation: the expanded-volume model. In Section III, ESR observations of deviations from Debye spectral densities at higher frequencies are described, and interpretations in terms of localized dynamic cooperativity are suggested. These studies are based upon the traditional measurements of T_2's or linewidths of motionally-narrowd ESR spectra. In Section IV we introduce experiments that require a more elaborate analysis than those required for the T_1 and T_2 of motionally-narrowed spectra. These are slow-motional ESR studies of rotational dynamics. They are representative of a subject I like to refer to as "Beyond T_1 and T_2", for which considerably more sophisticated analyses may be required, but more microscopic details about the motions may be forthcoming. After first illustrating the special features of rotational motions in liquid crystals in Section V, I review in Section VI our theoretical approach for analyzing slow (and fast) motional spectra in both isotropic and ordered fluids. In the first part I summarize the formulation of the theory in terms of the stochastic-Liouville equation; then I describe the powerful modern computational algorithm that we employ to solve the lineshape problem; this is followed by some of our current ideas on modeling of rotational dynamics and dynamic cooperativity.

The precise nature of the changes in molecular dynamics at a phase transition is a fascinating one, particularly at a second-order phase transition where the macroscopic equilibrium and transport properties appear to diverge. I summarize in Section VII our recent studies on spin-relaxation at the nematic to smectic A phase transition, which includes our model proposed to explain the observed critical divergences. This model emphasizes the special importance of rotational-translational couplings in highly ordered phases such as smectic phases. This suggests that a complete understanding of rotational motions requires that we also understand the translational motions in such phases. Thus, in Section VIII, I summarize our recent efforts to study the translational diffusion of ESR probes in ordered phases.

To illustrate the importance of rotational dynamics in fields other than chemical physics, I describe in Section IX our recent studies on rotational motions in model membrane systems. These are of biophysical interest, but we show how the methods and applications described in the previous Sections can be employed effectively for these more complex "ordered fluids". In particular I discuss the dynamic molecular structure and phase transitions in oriented lipid multilayers and lipid-macromolecule interactions.

Finally, In Section X, I describe new electron-spin-echo techniques that are being developed to deal more effectively with the "beyond T_1 and T_2" regime of study, and which have the promise of revolutionizing how ESR is utilized in the study of rotational motions.

II. Quasi-Hydrodynamic Models of Rotational Reorientation

In an extensive study of the pressure and temperature dependence of the electron-spin relaxation of the small nitroxide spin probe PD-Tempone in toluene solvent [5], we found significant deviations from Stokes-Einstein (SE) behavior for the rotational correlation time τ_R, where τ_R ranged over more than two orders of magnitude (cf. Fig. 1). (We shall mean by the SE relation that $\tau_R = \nu_e \eta / k_B T$ where ν_e is an "effective molecular volume", η is the solvent viscosity and T the temperature.) These deviations were found

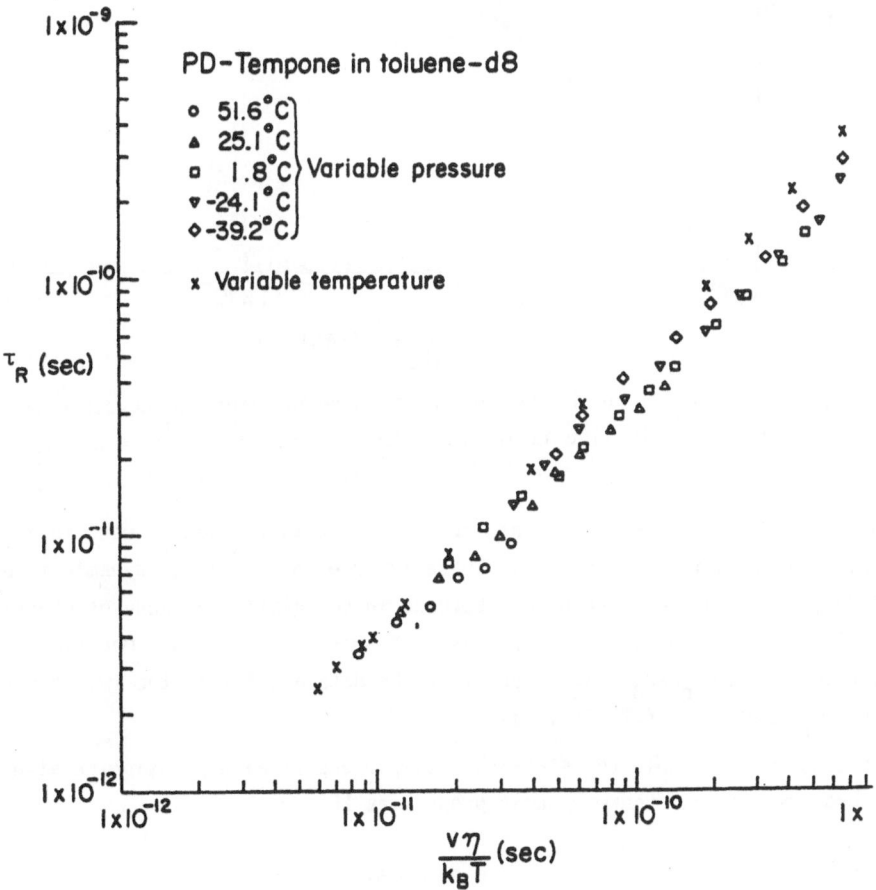

Fig. 1a: τ_R vs. $\nu_e \eta / k_B T$ for PD-Tempone in toluene-d8. Variable pressure and temperature results. T varied from -40°C to +50°C. P varied from 1 bar to 5.5 kbar. (From Ref. [5].)

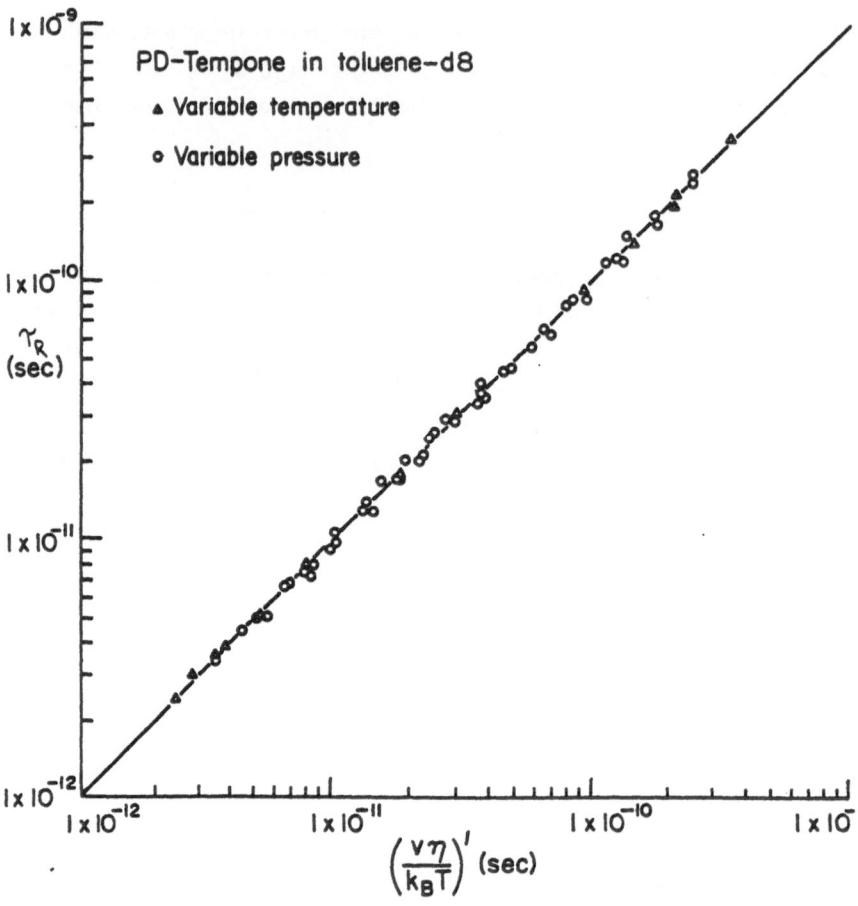

<u>Fig. 1b</u>: τ_R vs. $(\nu_e\eta/k_BT)'$ for PD-Tempone in toluene-d8. Variable pressure and temperature results. The identity line is included for comparison. $(\frac{\nu_e\eta}{k_BT})' = d + a_1P + a_2P^2 + b_2T^2 + cPT$ where coefficients were fit by least squares. (From Ref. 5).

to be inconsistent with hydrodynamic and molecular models previously used to interpret rotational relaxation in liquids [6-8]. These include the quasihydrodynamic free-space model of Dote et al. which has had some success in correlating a range of experimental results [7] as well as the empirical approach of Alms et al. [8] who modified the SE expression to be $\tau_R = \nu_e\eta/k_BT + \tau_R^0$ such that the data are fit to the two adjustable parameters ν_e and τ_R^0 (cf. Table 1).

Instead, we were able to establish a simple empirical expression relating τ_R to basic hydrodynamic and thermodynamic properties [5]:

$$\tau_R = Cn\beta_T(\rho - \tilde\rho)/T \tag{1}$$

where β_T is the isothermal compressibility of the solvent, and ρ is the solvent density. There are just two empirical constants: C and the reference density $\tilde\rho$, ($= 0.819$ gm/cm^3) needed to accurately fit the numerous data points obtained under a

Table I: Least-squares results for $\tau_R(T)$ vs. $\eta(T)/T$ at constant pressure. The approach of Alms et al. [9] implies that $\tau_R{}^0$ should be the same and positive for all isobars (from Ref. [5]).

Pressure (kbar)	Slope (10^{-7} sK/P)[a]	Intercept (10^{-12} s)[a]
0.001	2.47 ± 0.06	-0.0 ± 0.3
0.5	2.89 ± 0.09	-1.6 ± 0.6
1	2.94 ± 0.05	-3.3 ± 0.5
1.5	2.54 ± 0.07	-2.9 ± 0.9
2	2.52 ± 0.05	-4.3 ± 0.9
2.5	2.36 ± 0.09	-4.4 ± 2.5
3	2.6 ± 0.2	-12 ± 7
3.5	2.7 ± 0.2	-21 ± 12
4	2.7 ± 0.2	-30 ± 17
4.5	2.8 ± 0.2	-49 ± 26

a) Uncertainties represent average deviations.

wide range of T and P (cf. Fig. 2). (The accuracy was as good as shown in Fig. 1b which was just an expansion in powers of P and T and required five parameters.) This empirical equation could be interpreted in terms of an "expanded volume" model. That is, $\tilde{V} = \tilde{\rho}^{-1}$ is a solvent reference volume, the "expanded volume", such that as the solvent volume $V \to \tilde{V}$ (where $V = \rho^{-1}$), then $\tau_R \to 0$. Actually this is an ideal reference state, not realized in real systems, because Eq. (1) relates to purely viscous motion, and as $V \to \tilde{V}$, the liquid is becoming more gaslike, so inertial effects would take over. Over the range of our experiments, it was demonstrated that PD-Tempone exhibits purely viscous behavior, so no inertial effects were present to complicate the viscous dynamics. The expanded volume expression takes into account in a natural way the concept of slip of the rotating molecule in the solvent, which is often introduced in an ad-hoc manner [6,7]. In another point of view we can rewrite Eq. (1) based on recognizing $\beta_T \propto c^{-2}$, where c is the sound velocity in the solvent. Then the ratio $(\tilde{V} - V)/c^2$ would appear in Eq. (1), and it is a measure of the importance of the anisotropic intermolecular interactions acting on the solute (proportional to $\tilde{V} - V$) relative to the total intermolecular interactions between molecules in the liquid (proportional to c^2). This point of view is consistent with earlier theoretical analyses of Kivelson and co-workers [10].

We believe this result is potentially important in improving our understanding of molecular rotational dynamics in liquids. Further experiments are most certainly

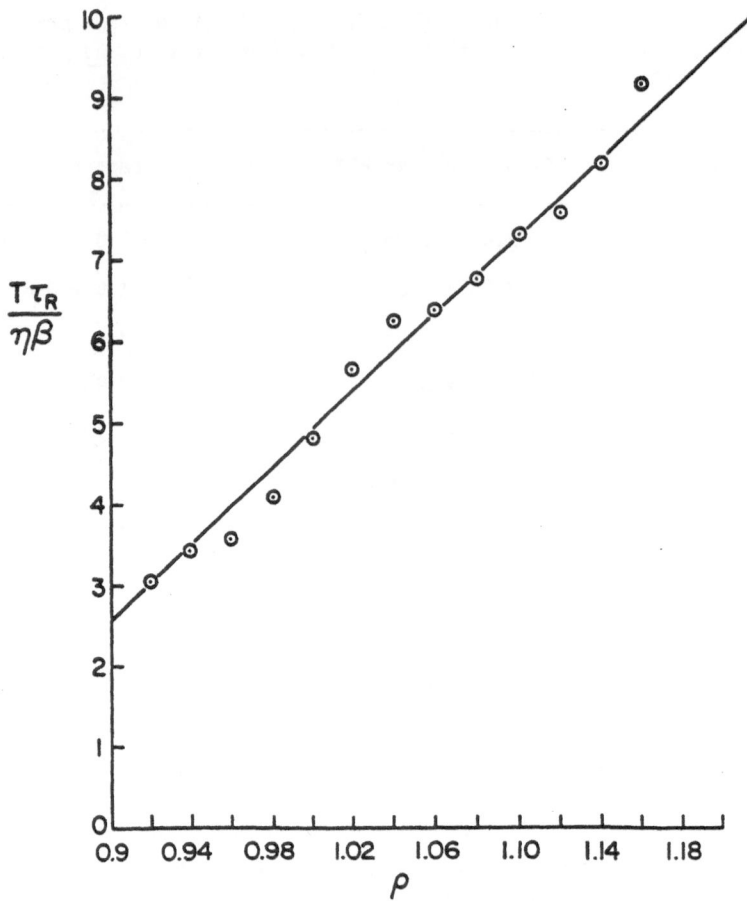

Fig. 2: Plot of average value of $\tau_R T/\eta\beta_T$ for each constant density group of data from Fig. 1 vs. density, ρ, (from Ref. [5]).

required. One should test the applicability of Eq. (1) over a wide range of solvents and solutes by means of pressure-dependent studies. Furthermore, the role of ν_p/ν_s (i.e. the ratio of probe-molecule volume to solvent-molecule volume), which should be important, needs to be studied in some detail. This ratio plays an important role in the theory of Dote et al. [7]. In our expanded volume model we expect that the reference volume $\tilde{V} = \tilde{V}(\nu_p/\nu_s)$, (i.e. it is a function of ν_p/ν_s) and that it probably also depends on the shape (i.e. deviation from sphericity) of the probe molecule. As $\nu_p/\nu_s \to \infty$, we would expect $\tilde{V}/V \to \infty$ consistent with the approach to a Stokes-Einstein limit of Eq. (1). Based upon the partial success of the theory of Dote et al. to a range of data [7], and to our success using Eq. (1) to the very precise and detailed data obtained as a function of P and T, we believe that a "quasi-hydrodynamic" model for rotational relaxation in liquids can be developed, which can fairly accurately fit a wide range of data.

When one compares our results [5] to previous pressure-dependend ESR studies of VO(acac)$_2$ (i.e. Vanadyl acac) in toluene and other solvents [11] and to NMR studies on neat liquids [12], one notes that for the former, the solvent molecules were small compared to the solute molecules, which favors Brownian rotational diffusion, while for the latter, where solvent and solute are the same, there are inertial effects. Our results of PD-Tempone in toluene probably represent an intermediate case, where the solute molecule is only a little larger than the solvent molecule, with somewhat different shape and intermolecular interaction with the solvent, and yet one is still in a regime of T and ρ where inertial effects are negligible. This regime would thus be a very favorable one for further study on molecular dynamics.

III. Non-Debye Spectral Densities

Our group has in recent years obtained extensive data on the frequency-dependence of the spectral densities due to rotational motion of probes such as PD-Tempone and

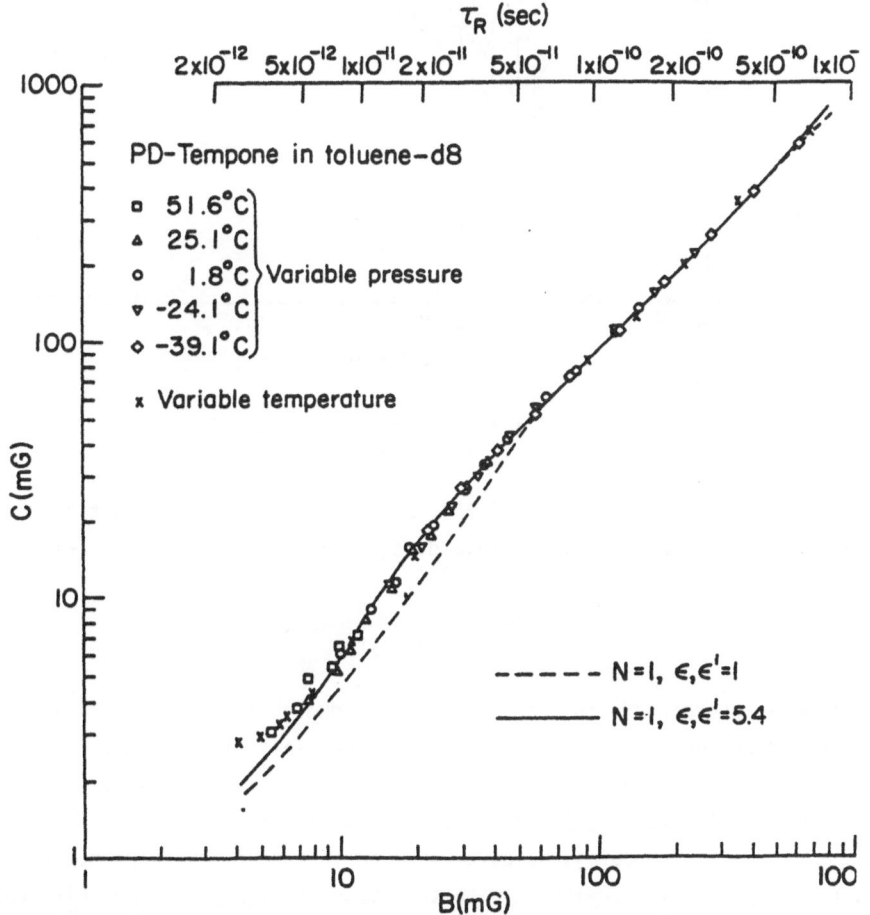

Fig. 3: A comparison of experimental and calculated values of linewidth coefficients C vs. B for PD-Tempone in toluene-d8. Variable pressure and temperature results. (From Ref. [5].)

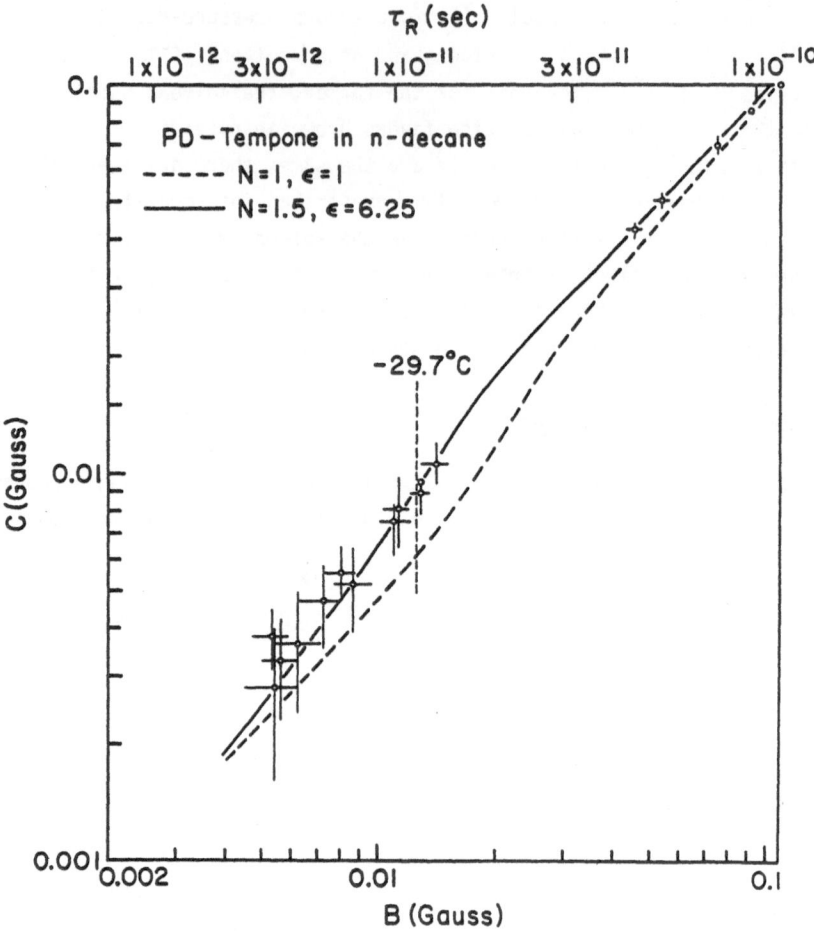

τ_R (sec)

Fig. 4: Comparison of experimental and calculated values of C vs. B for PD-Tempone in n-decane. (From Ref. [13].)

peroxylamine disulfonate (PADS) [5,13-15]. We find, quite generally, that we must modify the Debye spectral density with a dimensionless factor $\varepsilon \neq 1$ (cf. Figs. 3 & 4):

$$j(\omega) = \frac{\tau_R}{1 + \varepsilon\omega^2 \tau_R^2} . \qquad (2)$$

We have interpreted this ε-correction as due to the effects of fluctuating torques acting on the probe, which relax on a time scale of the order of τ_R itself. More phenomenologically, an $\varepsilon \neq 1$ can be attributed to viscoelastic effects in the liquid. Thus, the fluctuating torque model is a particular model for "explaining" the viscoelesticity in these cases. We have found an inverse correlation between ε and solvent polarity that is consistent with this model (cf. Fig. 5) [9]. Also Patron et al [16] have reported results which show $\varepsilon > 1$ for the larger probe $VO(acac)_2$ in long chain hydrocarbons, which would be consistent with our model.

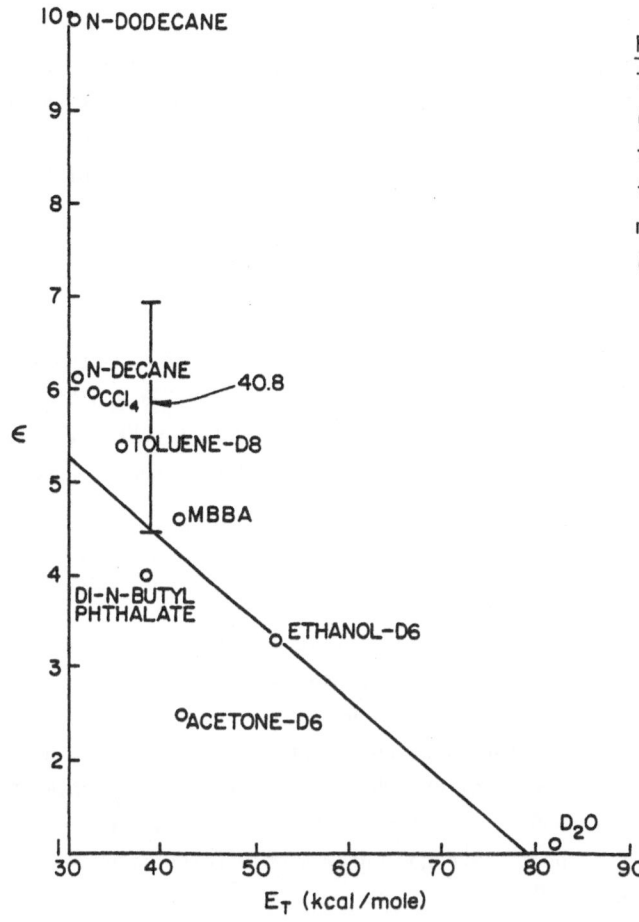

Fig. 5: ε vs. E_T for PD-Tempone in several solvents. (The line is drawn to guide the eye.) E_T, the molar transition energy, is a measure of solvent polarity. (From Ref. [13].)

One should note that it is very difficult to study the ε-correction of Eq. (2) by NMR, because the lower frequencies lead to the extreme narrowing condition for liquids. Blicharska et al. [17] have succeeded, in a study of CH_3OH in $(CD_3)_2SO$ to reach the $\omega\tau_R \propto 1$ regime. They observed definite anomalies, which we find to be consistent with an $\varepsilon > 1$.

Further work on varying probe size and solvent to distinguish between the relative importance of solvent polarity and relative solute size would be valuable, as would frequency-dependent experiments to properly study Eq. (2) and its variations.

In our attempts at interpreting Eq. (2) we have also considered another model which leads to non-Debye spectral densities [13]. It is referred to as the slowly relaxing local structure (SRLS) model, originally introduced [18] in studies on liquid-crystalline solvents. In this model, the slowly fluctuating components of the anisotropic intermolecular potential are regarded as a local structure, which presists for a mean time τ_X, and with respect to which the probe rotates, since $\tau_X > \tau_R$. Then, on this longer time scale τ_X, the local structure relaxes (or else the probe diffuses - or jumps - away).

Our results [5,13] were generally better represented by the form of Eq. (2) with a constant ε than with SRLS, although our analyses could not be regarded as conclusive. Recently van der Drift and Smidt [19] working with a specially constructed cw ELDOR spectrometer obtained results that offered a systematic test to the ε-model of Eq. (2). They found that their data could not be fit with just a constant ε, but instead required a significant SRLS component even for PD-Tempone in toluene. Also, in analyzing their data, they could obtain good agreement between τ_R and τ_J (the angular-velocity correlation-function obtained from the spin-rotational relaxation terms) according to the Hubbard-Einstein relation: $\tau_R \tau_J = I/6kT$ (where I is the moment-of-inertia of the probe). The apparent breakdown of this relation in many ESR and NMR studies has been a matter of significant theoretical concern [14,20]. Further work with modern ESR techniques, such as the spin-echo methods described in Section X should be very rewarding on these matters.

IV. Slow Motional ESR Studies of Rotational Dynamics

Over a number of years we have developed highly sophisticated methods for simulating ESR (and NMR) slow motional spectra [14,15,18,21]. Such spectra are important in the study of molecular rotational dynamics by ESR in viscous media such as in liquid crystals, in spin-labeling applications in biophysics, and in polymer physics. By the slow-motion regime, we mean that the rotational motions which can average the orientation-dependent spin Hamiltonian $H(\Omega)$ (see Section VI) are too slow, i.e. we are in the regime where $\tau_R^2 \overline{H^2(\Omega)} > 1$. These spectra are no longer simple Lorentzians but are, in general, more complex in shape.

Such spectra provide considerably more information about the microscopic models of rotational dynamics than motionally narrowed spectra. Thus, for example, jump models of rotational reorientation lead to slow motional spectra which are distinguishable from Brownian reorientational models. PD-Tempone, in particular, showed deviations from Brownian rotation that appeared to be fit by a model of moderate jumps [14,15,21] (but see Section X). We have, however, suggested that a more fundamental analysis of the motional dynamics may be required as the experimental results associated with slow tumbling become more precise [14]. That is, fluctuating torques or else SRLS may be important, and the slow-motional spectra could be providing unique information on these microscopic details. VO^{2+} complexes with a vanadyl nuclear spin of $I = 7/2$ were found to be more sentitive to motional model than are nitroxides with $I = 1$ [22], (cf. Fig. 6). In particular, the slow tumbling lineshapes seem to be strongly dependent upon the nature of the ligands and of the solvent (cf. Fig. 7). Furthermore, because the vanadyl magnetic tensor components are about an order of magnitude greater than those for nitroxides, the vanadyl spectra exhibit slow motional (hence model-dependent) effects for $\tau_R > 10^{-10}$ sec. (at X-band), i.e., an order of magnitude sooner. Thus a greater range of liquids may be studied in the slow-motional region (before they freeze) by use of vanadyl probes.

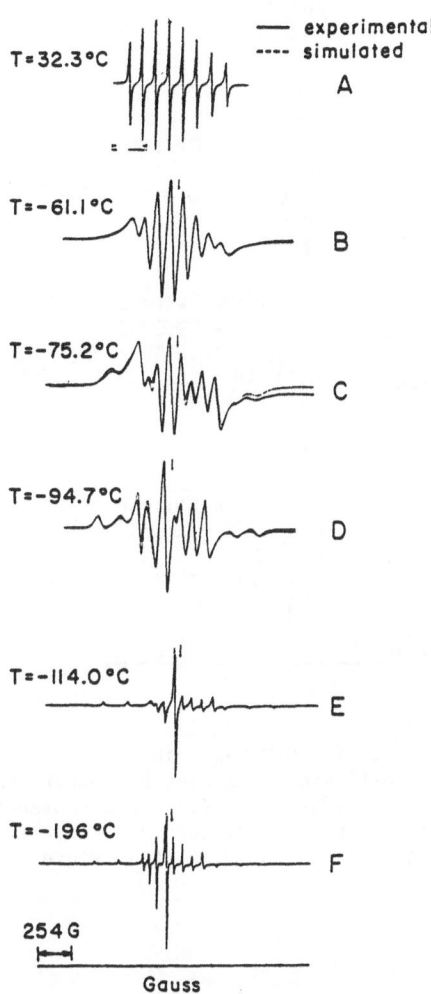

Fig. 6: Comparison of experimental and simulated spectra from the rapid motional to the rigid limit for $VO(acac_2(pn))$ in toluene. All simulations use a Brownian rotational diffusion model. (A) $\tau_R = 2.06 \times 10^{-11}s$, (B) $2.63 \times 10^{-10}s$, (C) $5.00 \times 10^{-10}s$, (D) $2.25 \times 10^{-10}s$, (E) $5.0 \times 10^{-8}s$, (F) rigid limit. (From Ref. [22].)

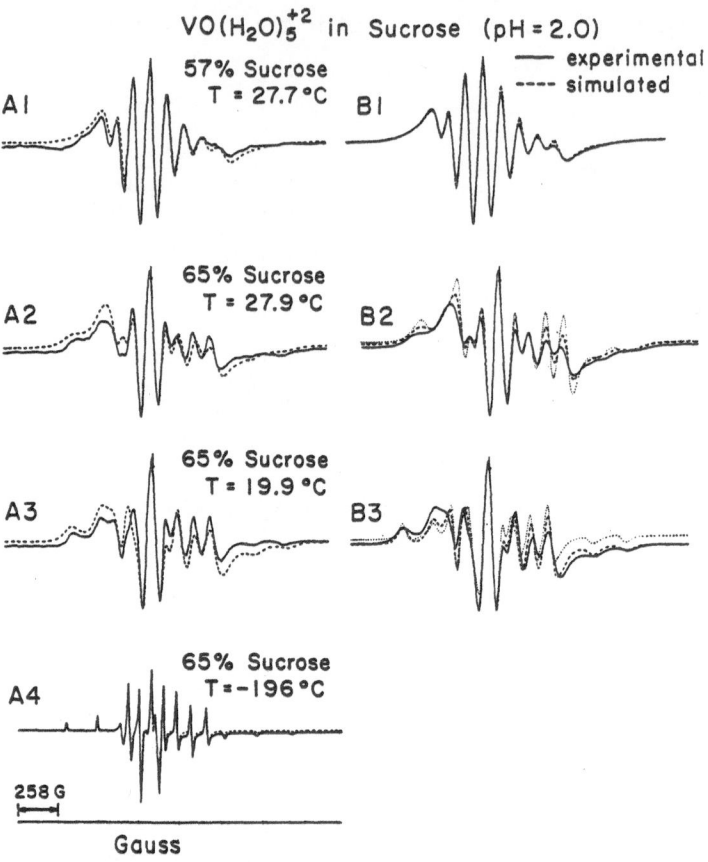

<u>Fig. 7a</u>: Model dependence of $VO(H_2O)_5^{2+}$ in sucrose. (Series A) Comparison of experiment with moderate jump diffusion. (Series B) Comparison of moderate jump diffusion (solid lines) with its free (dashed lines) and Brownian (dotted lines) diffusion equivalent. (Moderate jump gave best agreement in all cases.) (A1) $\tau_R^J = 3.4 \times 10^{-10}s$, (A2) $6.0 \times 10^{-10}s$, (A3) $9.0 \times 10^{-10}s$, (A4) rigid limit.

V. Liquid Crystals

The important new feature of molecular rotational dynamics in liquid crystals is that it must occur relative to a mean orienting potential. This is readily included in the theoretical analysis as described in Section VI.

In our studies on liquid-crystalline solvents we found it necessary to introduce the slowly-realxing local structure (SRLS) model in addition to the effects of the mean orienting potential [18,23,24]. In this SRLS model (as noted above) the slowly fluctuating components of the anisotropic intermolecular potential are regarded as a local structure, which persists for a mean time τ_x, and with respect to which the probe rotates, since $\tau_x \gg \tau_R$. Then, on this longer time-scale τ_x, the local structure relaxes (or else the probe diffuses or jumps away). This seems a reasonable model for

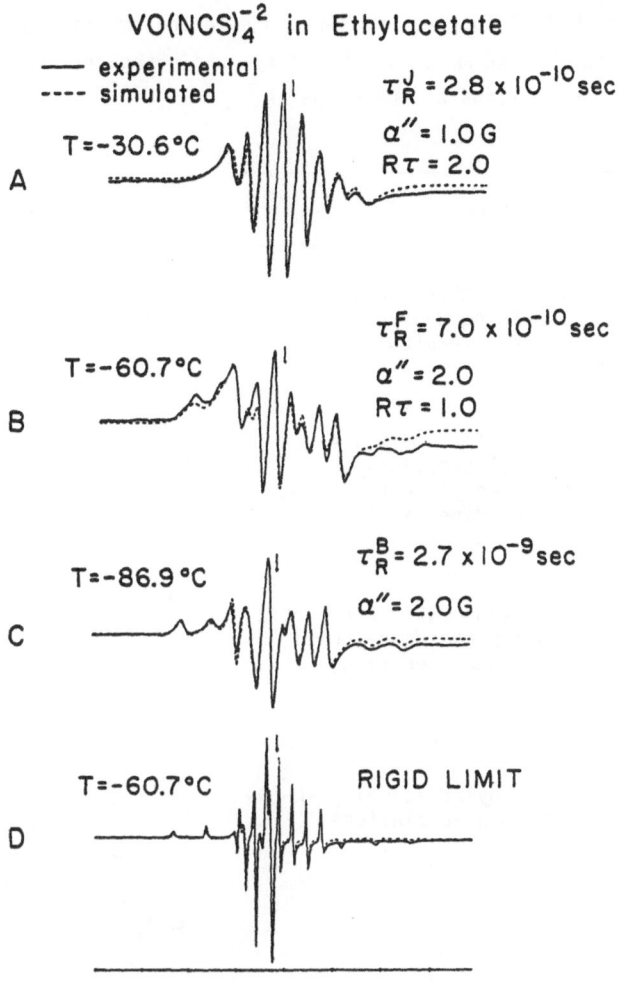

Fig. 7b: Model dependence of $VO(NCS)_4^{2-}$ in ethyl acetate. Note that A is approximately fit with moderate jump, B with free diffusion, and C with Brownian diffusion. (From Ref. [22].)

probes that are smaller than the liquid crystalline molecules, since they reorient in times much shorter than the surrounding solvent molecules. Again the evidence comes from our detailed linewidth studies of smaller probes such as PD-Tempone [18,23,24]. In particular, in a pressure-dependent study of PD-Tempone in the nematic Phase V solvent, we concluded that the SRLS contribution to the linewidths might be comparable to the normal reorientational contribution from which τ_R is estimated [23]. Thus, τ_R's calculated from $j(\omega)$ might be significantly larger than the true value, but the data just from the linewidth coefficients are insufficient to be definitive on this latter point. In our studies of smectics we find even more substantial linewidth anomalies attributable to the SRLS model [24]. This is reasonable in the context of our model (discussed below) for the smaller probe in the smectic phases; viz. these probes are now located in the more flexible alkyl-chain regions, experiencing the slow (cooperative) fluctuations of these chains.

Acronym	Name	Structure
PD-Tempone	2,2', 6,6'-tetramethyl-4-piperidine N-oxyde (perdeuterated)	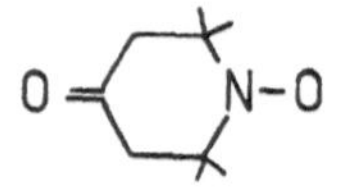
P	2,2', 6,6'-tetramethyl-4-(butyloxyl)-benzylamino-piperidine 1-oxyl (perdeuterated piperidine ring)	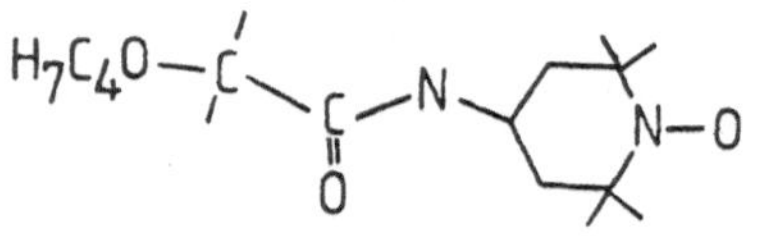
MOTA	4-methylamino 2,2', 6,6'-tetra-methyl-piperidine-1-oxyl (perdeuterated ring)	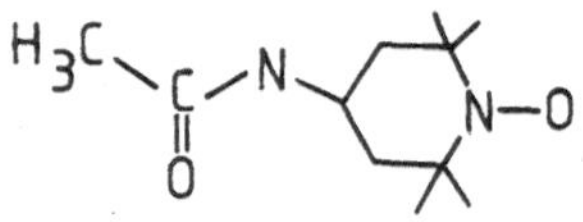
CSL	3,3'-dimethylorazolidinyl-N-oxy 2'3-5a-cholestane	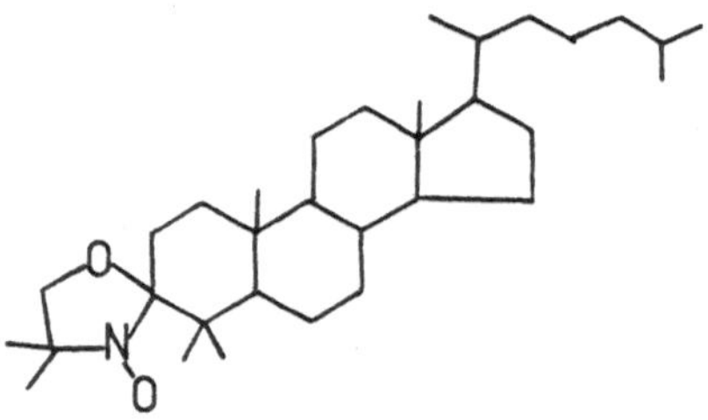

Fig. 8: Structures of spin probes in our studies.

We show in Figs. 8 and 9 some spin probes used in our studies and some liquid crystals we have used. In Fig. 10 we illustrate how ESR lineshapes are sensitive to both the orienting potential and the rotational dynamics relative to this potential for the P-probe, which is comparable in size and shape to a liquid crystal molecule [25]. The variations in ordering and rotational rates for the isotropic, nematic, smectic A and smectic B phases are illustrated in Fig. 11. This probe experiences an orienting alignment similar to that of liquid crystalline molecules (except that there is some flexibility for internal rotational motion of the nitroxide moiety). Shown in the inset to Fig. 11 is the related data [24] for the small PD-Tempone probe. The unusual observation here is the very low activation energy for rotation of this probe in the two smectic phases even though it is substantially higher in the isotropic and nematic phases. This is taken as significant evidence to support the model [24] that this probe

Acronym	Name	Formula
6OCB	4-cyano 4'-n-hexyloxybiphenyl	$NC-\phi-\phi-OC_6H_{13}$
8OCB	4-cyano 4'-n-octyloxybiphenyl	$NC-\phi-\phi-OC_8H_{17}$
40,6	N-(p-butoxybenzylidene)- p-n-hexylaniline	$H_9C_4O-\phi-CHN-\phi-C_9H_{13}$
40,8	N-(p-butoxybenzylidene)- p-n-octylaniline	$H_9C_4O-\phi-CHN-\phi-C_8H_{17}$
8CB	4-cyano 4'-n-octylbiphenyl	$NC-\phi-\phi-C_8H_{17}$
S2	Eutectic mixture of: 50% 4-cyano 4'-n-octylbiphenyl 39% 4-cyano 4'-n-decylbiphenyl 11% 4-cyano 4'-n-decyloxybiphenyl	 $NC-\phi-\phi-C_8H_{17}$ $Nc-\phi-\phi-C_{10}H_{21}$ $NC-\phi-\phi-OC_{10}H_{21}$

Transition Temperatures of some Liquid Crystals:

a. 27% 6OCB - 73% 8OCB : K (24^0) N (31^0) S_A (45^0) N (79^0) I

b. 40,6 : K (18^0) S_B (48^0) S_A (55^0) N (78^0) I

c. 8CB : K (21^0) S_A (34^0) N (41^0)

d. S2 : K (-10^0) S_A (48^0) N (49^0) I

Fig. 9: Some liquid crystals used in our studies.

is expelled into the chain region of the smectic layers, since it is known that the activation energy for reorientation for PD-Tempone in neat aliphatic hydrocarbons is just 2-4 kcals [13].

Additional strong evidence for this "expulsion effect" in the smectic phases is shown in Figs. 12 and 13. In Fig. 12 we show the ordering of PD-Tempone (in 40,8) in the different phases [24]. In the S_A phase this probe experiences a reduction in its ordering as the temperature decreases which is contrary to the increasing ordering of the liquid crystal molecules (cf. Fig. 11). This is readily interpreted as due to the probe being increasingly expelled into the less ordered aliphatic chain region. In Fig. 13 we show a plot of the PD-Tempone hf splitting for different solvents arranged according to their relative polarity [24]. Aliphatic hydrocarbons, being the most non-planar, yield the lowest hf splitting. Indeed PD-Tempone in 40,6 and 40,8 shows a decrease in its hf splitting as the temperature is lowered into the smectic phases in a manner consistent with its being dissolved more into the aliphatic hydrocarbon region.

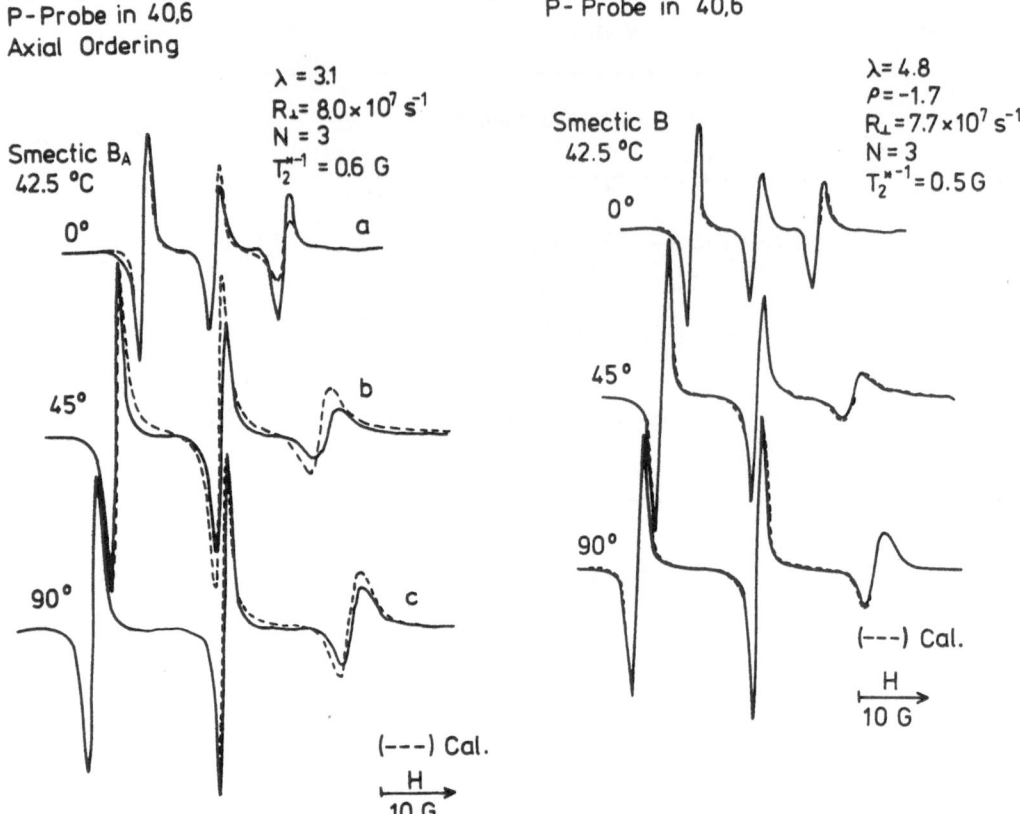

Fig. 10: (___) Experimental and (---) calculated ESR spectra of P probe dissolved
in 40,6 oriented between plates and in the smectic B_A phase of 42.5°C. The angle Θ
between the magnetic field B and the plate normal is denoted in the Figures. The
ordering parameters and rotational rates (given by $R_\perp$, the perpendicular component
of the rotational diffusion tensor, and $N = R_{\parallel}/R_\perp$ the rotational asymmetry) are
on the figures. (a) corresponds to cylindri-cally symmetric ordering given by
$\lambda(5/4\pi)^{1/2} \equiv \epsilon_0^2/kT$, while in (b) an asymmetry term of $\rho(5/4\pi)^{1/2} \equiv \epsilon_2^2/kT$ [cf.
Eq. (7)] is allowed. (From Ref. [25].)

VI. Theoretical Approach

A) Stochastic Liouville Equation

The basis for our analysis in the above studies is the stochastic Liouville equation
(SLE). One starts with the spin density-matrix equation of motion for ρ :

$$\frac{\partial\rho(t)}{\partial t} = -i[H(t),\rho]$$ (3)

where H(t) is the time-varying spin-Hamiltonian due to fluctuations of Euler angles
Ω of the molecule with respect to the lab frame. A classical stationary Markovian

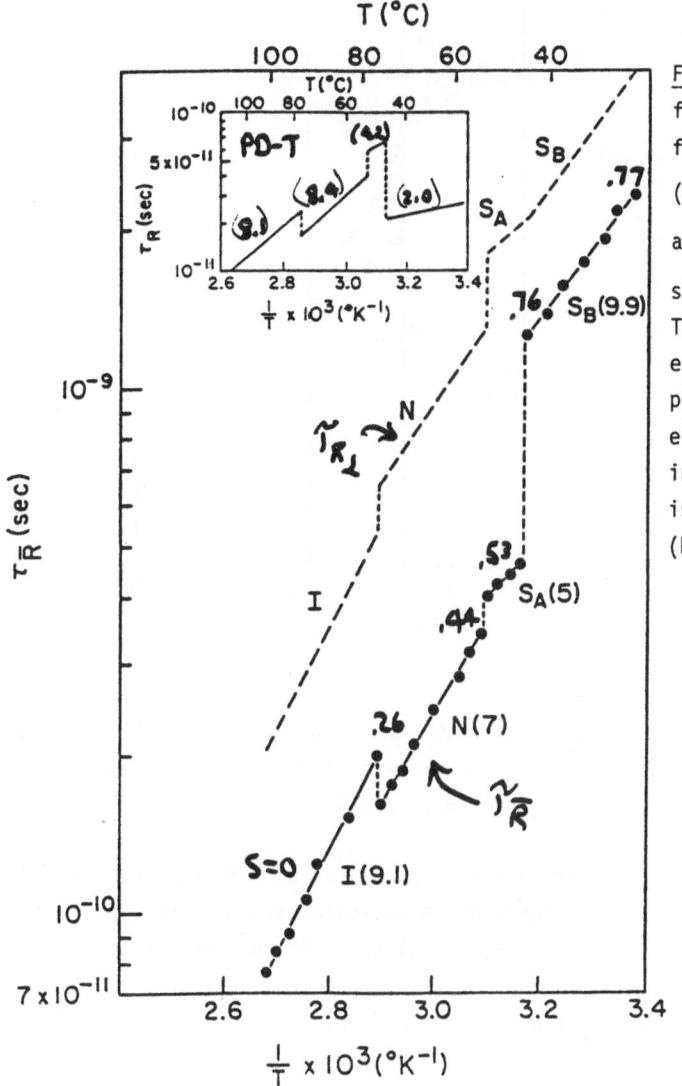

Fig. 11: Rotational rates for $\tau_{\bar{R}}$ and $\tau_{R_\perp}$ vs. 1/T for the P-probe $^\perp$ in 40,6. (Here $\tau_{\bar{R}} \equiv 6^{-1}(R_{||}\, R_\perp)^{-1/2}$ and $\tau_{R_\perp} = 6_\perp^{-1} R^{-1}$). Insert shows τ_R vs. 1/T for PD-Tempone in 40,6. Activation energies in the respective phases are shown in parenthesis, and the range of ordering for P-probe in each phase is also given. (From Ref. [25].)

description for the orientational probability distribution $P(\Omega, t)$ is assumed. That is we let

$$\frac{\partial P(\Omega, t)}{\partial t} = -\Gamma_\Omega P(\Omega, t) \tag{4}$$

where Γ_Ω is the time-independent "diffusion" operator. Then one can show that the composite set of spin and space variables obeys the SLE:

$$\frac{\partial \rho(\Omega, t)}{\partial t} = -i([H(\Omega), \rho] - \Gamma_\Omega \rho(\Omega, t) \tag{5}$$

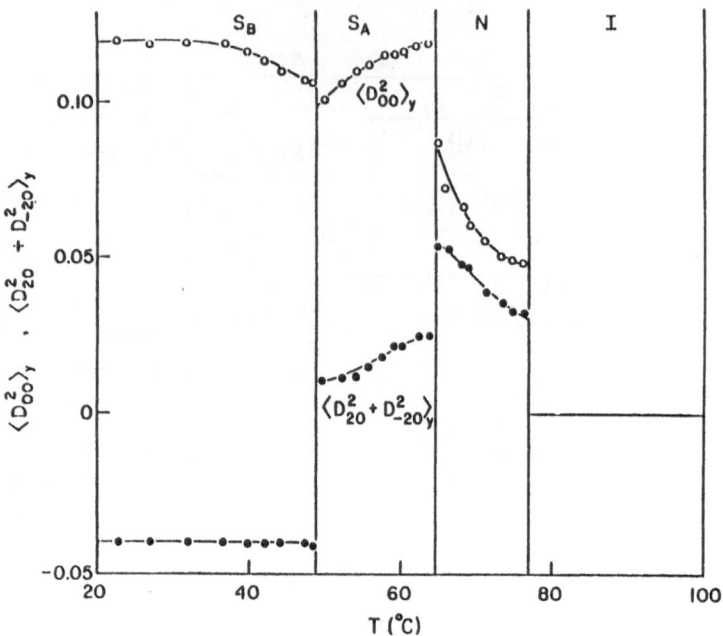

<u>Fig. 12:</u> Order parameters for PD-Tempone in 40,8. Here $<D^2_{00}> = (5/4\pi)^{1/2} <Y^2_0>$ and and $<D^2_{00} + D^2_{0-2}> = (5/4\pi)^{1/2} <Y^2_2 + Y^2_{-2}>$. The break in ordering parameters at N-S_A transition is an experimental artifact. (From Ref. [24].)

where now $\rho(\Omega,t)$ is simultaneously the quantum-mechanical spin-density matrix and the classical probability distribution function in molecular orientation. To simplify the analysis one typically chooses [18,21,28] a Smoluchowski equation for Γ_Ω :

$$\Gamma_\Omega = -\underset{\sim}{M} \cdot \left[\left(\underset{\approx}{R} \, M \, \frac{U(\Omega)}{kT} \right) + \underset{\approx}{R} \cdot \underset{\sim}{M} \right] . \qquad (6)$$

Here $\underset{\sim}{M}$ is the vector space for infinitesimal rotation, $\underset{\approx}{R}$ is the rotational diffusion tensor, and $U(\Omega)$ is the equilibrium potential for the orientation of the probe, which may be expanded in spherical Harmonics, $Y^L_K(\Omega)$ according to:

$$U(\Omega) = \sum_{\substack{even \\ L}}^{\infty} \left\{ \epsilon^L_0 \, Y^L_0 \, (\Omega) + \sum_{K>0}^{L} \epsilon^L_{K\pm} \left[Y^L_K(\Omega) \pm Y^L_{-K}(\Omega) \right] \right\} . \qquad (7)$$

Typically only the $L = 2$ terms are kept (but sometimes terms through $L = 4$ are included [29], cf. Fig. 14). One can even allow for dynamic cooperativity by letting

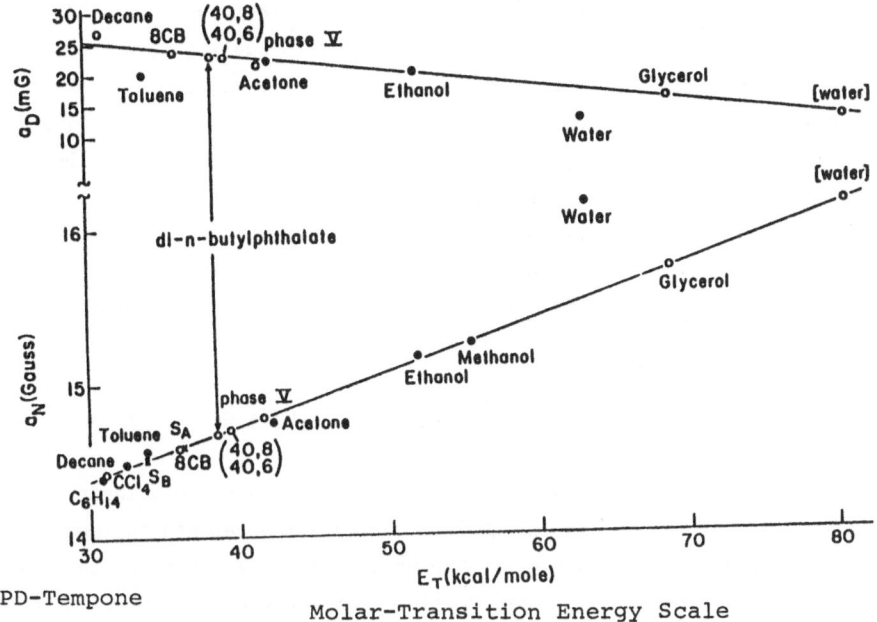

Fig. 13: Variation of hf splittings a_N and a_D for PD-Tempone with E_T (the molar transition energy) in different solvents. (From Ref. [24].)

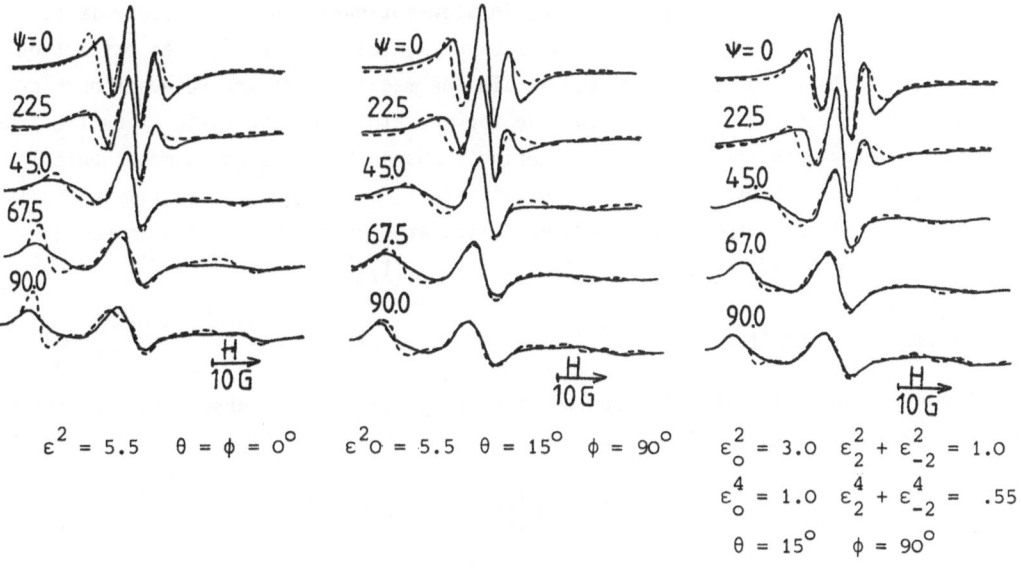

$$\varepsilon^2 = 5.5 \quad \theta = \phi = 0^\circ$$

$$\varepsilon^2_0 = 5.5 \quad \theta = 15^\circ \quad \phi = 90^\circ$$

$$\varepsilon^2_0 = 3.0 \quad \varepsilon^2_2 + \varepsilon^2_{-2} = 1.0$$
$$\varepsilon^4_0 = 1.0 \quad \varepsilon^4_2 + \varepsilon^4_{-2} = .55$$
$$\theta = 15^\circ \quad \phi = 90^\circ$$

Fig. 14: Experimental spectra of homeotropically aligned CSL in S2 at -8°C for tilt angle ψ between the liquid crystal director and the external magnetic field H_0 (solid lines). Dashed lines denote simulated spectra with anisotropic diffusion in a high ordering potential. $(0,\theta,\phi)$ denote Euler angles between the magnetic frame and the ordering frame. The ε^L_K denote coefficients in the expansion of the ordering potential in spherical harmonics. (From Ref. [29].)

$U(\Omega) \to U(t)$ such that the probe reorients in the instantaneous potential field of its surroundings. This approach is used for SRLS and for coupling to hydrodynamic and critical modes [30,31]. In smectic phases the orientational potential felt by a molecular probe should depend upon the probe location within the smectic bilayer in order to be consistent with the observation of reduced ordering for some probes as they are expelled into the alkyl chain region. (Related NMR observations have also been made [32-34]). Other evidence exists that translational diffusion perpendicular to the smectic layers is highly hindered in the smectic phase [33]. Thus the probe will experience a <u>coupled orientation-position</u> potential as it diffuses in the spatially non-uniform smectic. Moro and Nordio have studied this [35] using as their combined potential:

$$U/kT = \{A + B\cos(2\pi z/d)\} \ Y_0^2(\beta) + C\cos(2\pi z/d) \qquad (8)$$

which is of the form utilized by McMillan [36] in his mean-field theory of the smectic phase transition. (Here β is the angle of orientation between the nematic director and the principal axis of ordering of the probe, while z measures the position in a smectic layer of thickness d). They consider small probes of dimension $d_0 \ll d$, so that they undergo many reorientations in the time they translate across a smectic layer. This model then has all the ingredients of a SRLS model. Indeed their detailed calculations showed that "the relaxation mechanism due to this order modulation is similar to those often referred to as 'slowly relaxing local structures'... " [18].

In summary, the fundamental problem in slow-motional ESR spectroscopy is to compare solutions of the SLE (Eq. (5)) with experimental spectra so as to extract out the correct stochastic operator Γ and obtain the magnitude of the relevant physical parameters. We may refer to this as solving for the "inverse stochastic Liouville Transform" (by analogy with the "inverse scattering transform" in the quantum mechanical theory of scattering). In practice this is not possible, so one constructs simple models for Γ_Ω as illustrated above, and then one calculates predicted spectra to compare with experiment as illustrated (cf. Figs. 6, 7, 10, 11).

B. Computational Algorithm

It follows from Eq. (5) that in the linear response regime, the absorption intensity $I(\Delta\omega)$ is given by [21,26,27]:

$$I(\Delta\omega) = \frac{1}{\pi} \ \text{Re} < v \left| [i(\Delta\omega \underset{\sim}{1} - \underset{\sim}{L}) + \underset{\sim}{\Gamma}]^{-1} \right| v > \qquad (9)$$

where $\Delta\omega$ is "the sweep variable", $\underset{\sim}{L}$ is the Liouville operator associated with the spin Hamiltonian $H(\Omega)$ of the spin probe, and $\underset{\sim}{\Gamma}$ is the diffusion operator for the reorientation that modulates the magnetic interactions. Also $|v>$ is the so-called "starting vector" constructed from the spin transition moment averaged over the equilibrium ensemble. The vectors and operators are defined in the direct product space of the ESR transitions and of the functions of the Euler angles Ω. (In ESR we usually have $\Delta\omega = \omega - \omega_0$ where ω_0 is the Larmour frequency at the center of the spectrum and ω is the angular frequency of the applied radiation field.) We may rewrite Eq. (9) as:

$$I(\Delta\omega) = \frac{1}{\pi} \, \text{Re} \, <v|u(\Delta\omega)> \tag{10}$$

where $|u(\Delta\omega)$ is the solution of the equation

$$\underset{\sim}{A'}|u> = |v> \tag{11}$$

the matrix $\underset{\sim}{A'}$ is defined as $\underset{\sim}{A'} = i\Delta\omega \underset{\sim}{1} + \underset{\sim}{A}$ where $\underset{\sim}{A} = \underset{\sim}{\Gamma} - i \underset{\sim}{L}$. Eq. (9) can be solved either by inversion of $\underset{\sim}{A'}(\Delta\omega)$ for a range of values of $\Delta\omega$, or alternatively by diagonalizing $\underset{\sim}{A}$ only once [21].

The matrix $\underset{\sim}{A}$ is in general very large and sparse. The conventional methods [21] for solving Eq. (11) by inversion or by diagonalizing $\underset{\sim}{A}$ prove to be too cumbersome. One soon runs out of memory even on mainframe computers, and the solution requires prohibitive amounts of computer time. To remedy this situation the Lanczos algorithm has been developed for complex-symmetric matrices, since $\underset{\sim}{A}$ is typically of this form [37,38]. It is an efficient method for tri-diagonalizing $\underset{\sim}{A}$ and is particularly suited to the solution of large-sparse matrices. We have shown that it can lead to at least order of magnitude reductions in computation time, and it yields solutions of Eq. (9) to a high degree of accuracy [37,39]. This Lanczos algorithm is also appropriate for the general class of Fokker-Planck equations (including models of rotational motion) which can be represented by a complex symmetric Fokker-Planck operator $\underset{\sim}{A}$ (after symmetrization). For such cases, Eq. (9) would be associated with the spectral density , which is the Fourier transform of the time correlation function of a dynamical variable $v(t)$ [37,38]. In fact, more generally, it is possible to establish the close connection between the Lanczos algorithm based upon a scheme of projection operators in Hilbert space, and the Mori projection scheme in statistical mechanics [38,40].

The LA tri-diagonalization proceeds by recursive steps or projections. If we let N be the dimension of the matrix, and n_s , the number of recursive steps needed to converge to an accurate spectrum, then we find $n_s \ll N$. This inequality becomes more dramatic the more complicated the problem. In this sense, the Lanczos projections rapidly seek out, from an initial finite subspace of dimension N , a smaller subspace spanned by the Lanczos vectors. That is the LA constructs subspaces that progressively approximate the "optimal reduced space" for the problem. These subspaces, spanned by the Lanczos vectors are related to Krylov subspaces [41,42] and are generated from the sequence $\underset{\sim}{A}^{k-1}|v>$ for $k = 1,...,n$. Thus, the choice of $|v>$ as the "starting vector" biases the projections in favor of this "optimal reduced space". It is easy to show that this Krylov sub-space can only contain eigenvectors of $\underset{\sim}{A}$ with a non-zero component along $|v\rangle$. In general, the time required for the LA tri-diagonalization goes approximately as $n_s N(2n_E + 21)$, where n_E is the average number of non-zero matrix elements in a row of $\underset{\sim}{A}$ [37].

Very recently [43] we have learned how to blend the LA with the conjugate gradient method to "turbo-charge" the Lanczos Algorithm. This more powerful version supplies objective criteria for truncation of basis sets and recursive steps.

We briefly summarize the Lanczos algorithm. We first identify the starting vector $|v\rangle$ as the first Lanczos vector $|\Phi_1\rangle$. Then a Schmidt orthogonalization on the Krylov sequence $\underset{\sim}{A}^{k-1}|v\rangle$ for $k = 1,\ldots,n$ allows one to iteratively generate the set of orthonormal Lanczos vectors $|\Phi_k\rangle$ according to

$$\beta_{k+1}|\Phi_{k+1}\rangle = (1-P_k)\,\underset{\sim}{A}|\Phi_k\rangle \tag{12}$$

where β_{k+1} is the normalizing coefficient such that

$$\langle\Phi_{k+1}|\Phi_{k+1}\rangle = 1 \tag{13}$$

and $\underset{\sim}{P}_k$ is the projection operator on the Krylov subspace spanned by $|\Phi_j\rangle$ $j = 1,\ldots,n$ given by:

$$P_k = \sum_{j=1}^{k} |\Phi_j\rangle\langle\Phi_j| \quad . \tag{14}$$

Eq. (11) leads to a three-term recursive relation for generating the $|\Phi_j\rangle$:

$$\beta_{k+1}|\Phi_{k+1}\rangle = (A-\alpha_k)|\Phi_k\rangle - \beta_k|\Phi_{k-1}\rangle \tag{15}$$

where

$$\alpha_k = \langle\Phi_k|\underset{\sim}{A}|\Phi_k\rangle \tag{16}$$

and

$$\beta_k = \langle\Phi_k|\underset{\sim}{A}|\Phi_{k-1}\rangle \quad . \tag{17}$$

It may easily be shown that $\underset{\sim}{A}$ has a tridiagonal representation, $\underset{\sim}{T}_n$ in the basis of Lanczos vectors $|\Phi_j\rangle$ such that

$$\langle\Phi_k|\underset{\sim}{A}|\Phi_j\rangle = 0 \quad \text{if} \quad k \neq j, \ j \mp 1 \tag{18}$$

while Eqs. (16) and (17) give the non-zero matrix elements. That is, given the vectors $|\Phi_k\rangle$ in terms of their components $x_{j,k}$ in the original basis set, $|f_j\rangle$, $j = 1,\ldots,N$

$$|\Phi_k> = \sum_j x_{j,k}|f_j> \tag{19a}$$

$$x_{j,k} = <f_j|\Phi_k> \tag{19b}$$

then the column vectors $\underset{\sim}{x}_k$ form the orthogonal matrix $\underset{\sim}{Q}_n$ such that $\underset{\sim}{Q}_n^{tr}\underset{\sim}{Q}_n = \underset{\sim}{1}_n$ and

$$\underset{\sim}{T}_n = \underset{\sim}{Q}_n^{tr}\underset{\sim}{A}_N \underset{\sim}{Q}_n . \tag{20}$$

We have described the conventional Lanczos algorithm for real symmetric (or Hermitian) matrixes $\underset{\sim}{A}$ such that Eq. (13) involves the usual norms in Hilbert space.

For our present applications to ESR (and Fokker-Planck equations) for which $\underset{\sim}{A}$ is complex symmetric (or else can be transformed to complex symmetric form [37,38,40]) Moro and Freed [37,38] showed that one must introduce the Euclidean pseudo-norm. That is, first consider the general non-Hermitian case. One must introduce a biorthonormal set of functions Φ_j and $\Phi^{j'}$ such that

$$<\Phi^{j'}|\Phi_j> = \delta_{j,j'} \tag{21}$$

or alternatively (letting $\underset{\sim}{x}^{j'}$ and $\underset{\sim}{x}_j$ be their column vector representations):

$$\underset{\sim}{x}^{j'} \cdot \underset{\sim}{x}_j = \delta_{j,j'} . \tag{22}$$

However, for the case of (non-defective [42]) complex symmetric matrices $\underset{\sim}{A}$, it is possible to let

$$\underset{\sim}{x}^j = \underset{\sim}{x}_j^* \tag{23}$$

such that Eq. (22) becomes:

$$\underset{\sim}{x}_j^{tr} \cdot \underset{\sim}{x}_j = \delta_{j,j'} \tag{24}$$

and then the recursion method of Eqs. (12) - (18) remains applicable with Eq. (24) defining the Euclidean pseudo-norm, whereby the bra vectors are defined without the usual complex conjugation in a Hilbert space.

Finally we note that the complex symmetric tridiagonal matrix $\underset{\sim}{T}_n$ can easily be diagonalized by one of several methods [42,44]. This is not necessary for cw-ESR spectra (or for simple spectral densities from Fokker-Planck equations), since one can use a continued-fraction solution, but it is needed for 2D-ESE spectra (cf. Sect. X).

The computational algorithms for simulating magnetic resonance spectra in terms of molecular dynamics as well as for spectral densities for molecular dynamics have recently been reviewed [83].

C. Modeling of Rotational Dynamics

In recent years [45] we have developed a useful method of stochastic modeling, which has been summarized elsewhere [46]. It is a generalization of the stochastic-Liouville method whereby the basic physics of the "relevant" degrees of freedom and their coup-lings may be introduced in a transparent manner, in conjunction with the stochastic features of the bath variables. The resulting incomplete stochastic Liouville equation is then subjected to the constraints required for detailed balance in order to correct-ly include the "back-reaction" of the bath on the "relevant" degrees of freedom. The resulting augmented stochastic Liouville equation may then be efficiently solved by the Lanczos algorithm. By means of this approach one can develop expressions for fluc-tuating torque, SRLS, and related models. Partially based on such ideas, we have devel-oped soluble models for coupled reorientation of rod-like liquid crystal molecules in order to improve on the hydrodynamic model used for describing spin relaxation by di-rector fluctuations [47]. Also, we have developed an approach. the dynamic cluster model (DCM), to incorporate the dynamic localized cooperativity in the rotational relaxation of liquid crystal molecules in nematics.

D. Modeling of Dynamic Cooperativity

Our past ESR analyses in terms of local cooperativity were based upon the slowly relax-ing local structure (SRLS) model, which has now been improved by using our augmented Fokker-Planck approach. But in the highly ordered phases typical of liquid crystals, the cooperativity in reorientation may well be too great to be modeled in such a simple fashion. With the recent theoretical and computational advances, it could be feasible to analyze more realistic models of cooperative dynamics for the longer time scales (i.e. $\gg 10^{-12}$ sec) that are important for our magnetic resonance studies [48]. In this sense, such methods could be a useful alternative to full molecular dynamics calcula-tions [49,50]. The particular model we have been studying is a "dynamic cluster model" (DCM) [48]. There have already been treatments of the role of short range order on the equilibrium properties of liquid crystalline phases using various forms of Bethe's cluster method [51]. In this method a central molecule is surrounded by γ nearest neigh-bors which form the outer shell. The central molecule interacts with this cluster through pairwise potentials $U(\Omega_0, \Omega_i)$ $i = 1,2,\ldots,\gamma$, with the Ω_i being the Euler angles for the i-th molecule. While the γ-neighbors do not interact with each other, they all feel an orienting potential $V(\Omega_i)$ representing the mean potential of the fluid. Then a self-consistency relation is introduced so that the central molecule orders exactly in the same way as do the outer shell molecules. This leads to non-trivial integral equations for solving for $V(\Omega_i)$. Ypma and Vertogen [51] obtain best results for this model for $\gamma = 3$ or 4.

The DCM we have been studying is just the dynamical version of this. We model the $\gamma + 1$ particle system by the appropriate set of coupled Smoluchowski equations for the joint probability distribution $P(\Omega_0...\Omega_\gamma,t)$. These coupled Smoluchowski equations are then solved for correlation functions such as

$$\langle Y_0^2[\Omega_0(t)]Y_0^2[\Omega_0(0)]\rangle - \langle Y_0^2[\Omega_0(0)]\rangle^2$$

by means of the Lanczos algorithim. If we regard the central molecule as a probe, then in a sense, it is not necessary to achieve self-consistency. However, the matter of self-consistency is conveniently dealt with by solving numerically the Ypma-Vertogen integro-differential equations. These equations then give us the self-consistent values of $V(\Omega_i)$ for a given value of $U(\Omega_0,\Omega_i)$. In our solution of the several-body diffusion equation $\partial P/\partial t = \Gamma P$, (actually we solve the symmetrized form), we use the analogy to the problem of the quantum mechanics of a many-electron atom. Let us call the operator Γ for the dynamic cluster problem $\Gamma_{cluster}$, while the standard problem of the diffusion of a single particle in the mean-field of its surrounding is represented by the operator Γ_{MF} (and is typically given in the form of Eq. (6)). We first solve the conventional mean field problem for $P^1(\Omega_1,t)$ in the usual manner (i.e. Eq. (4)) to yield the one-particle mean-field eigenvalues and eigenvectors appropriate for a given meanfield potential. Then we select the mean-field solution corresponding to order parameter $S = \langle Y_0^2 \rangle$ which is (nearly) equal to that for the cluster problem to be solved. That is, we regard these Γ_{MF} as the diffusional analogue of (Hartree) SCF theory, i.e. an approximation to the best one-particle solution. Then we solve for $\Gamma_{cluster}$ by methods analogous to configuration-interaction starting with these one-particle solutions of Γ_{MF}. That is, the basis states are products of the one-particle states from Γ_{MF}, and we then diagonalize Γ in this basis. The problem with this method, as in any configuration interaction, is whether we have chosen enough excited "configurations" for the calculation to converge.

Utilizing approximate basis sets (suitable for computation on a PDP11 mini-computer), we have obtained initial results for the correlation function for $Y_0^2(\Omega_1)$, and we have compared them to the mean-field results as illustrated in Fig. 15. Interestingly enough, we find that for the nematic phase, the results for the full DCM lie rather close in shape to those for the simple MF problem. However, this is not so for the isotropic phase, especially for the phase transition region for $\gamma = 3$ or 4 (where γ is the number of cluster particles).

114

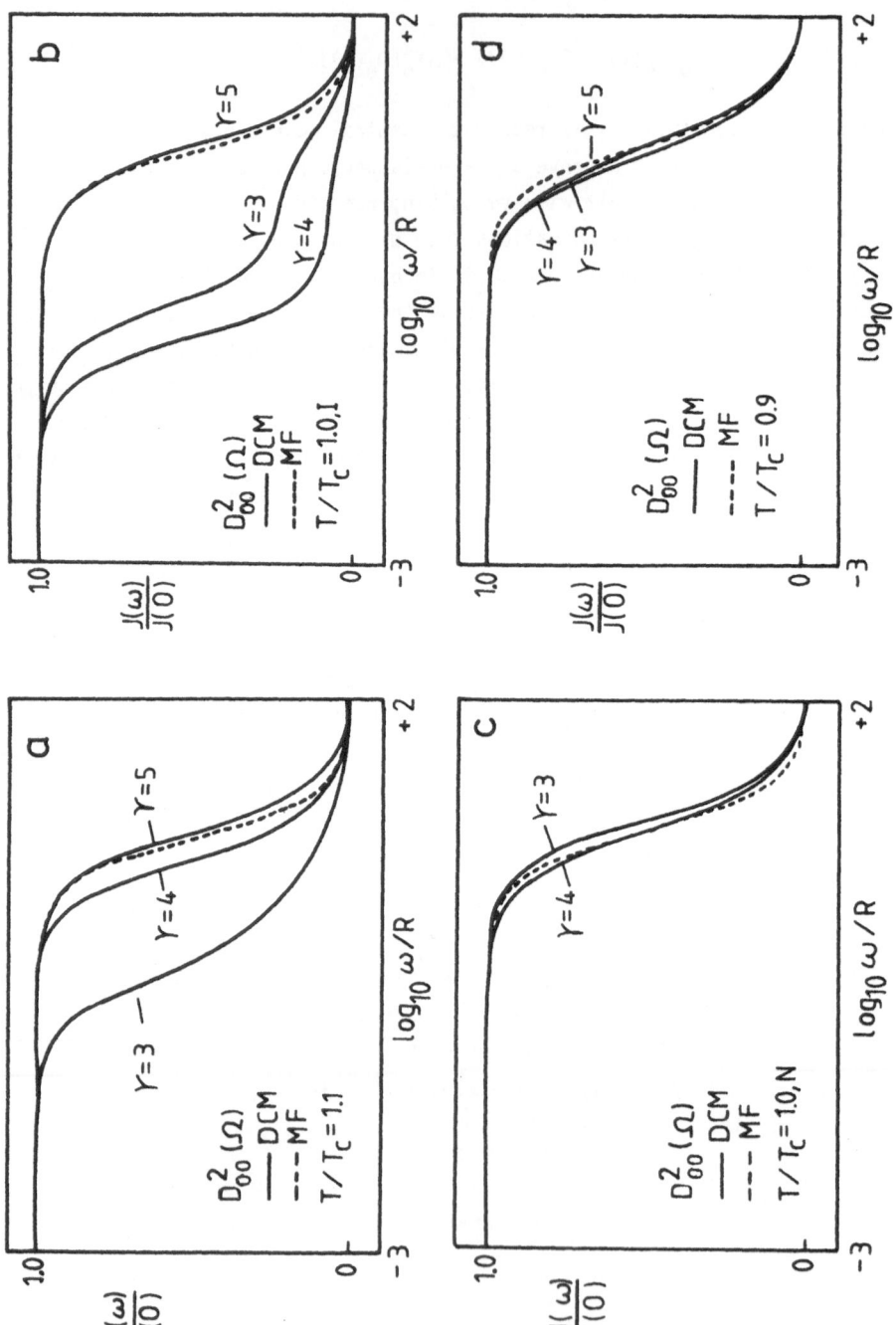

Fig. 15: Dimensionless spectral densities $J(\omega)/J(0)$ vs. dimensionless frequency ω/R plotted logarithmically for the dynamic cluster model; a) $T/T_c = 1.1$, i.e. in the isotropic phase above the liquid crystal phase; b) $T/T_c = 1.0$, in the isotropic phase at the phase transition; c) $T/T_c = 0.9$, deep in the liquid crystal phase at the phase transition; d) $T/T_c = 0.9$, deep in the liquid crystal phase. Dashed line is the mean field result and solid lines are for $Y = 3, 4,$ and 5 cluster particles.

VII. Molecular Dynamics at the Nematic to Smectic A-Phase Transition

A) Experiments

We studied in considerable detail the N/S_A phase transition with several nitroxide probes in two different liquid crystalline solvents: 40,6 and a mixture of 80CB-60CB, which exhibits a re-entrant nematic phase [31,52-54]. Our X-band ESR spectrometer is designed for milliKelvin temperature control. What we measure are the ESR linewidths δM of the three hyperfine lines (due to ^{14}N), and we study B and C, which are appropriate linear combinations of them (i.e. $\delta M = A + BM + CM^2$, where M is the z-component of ^{14}N spin). These show divergences at the various phase transitions [31,42,54]. We summarize in Tables 2a and 2b critical exponents and their magnitudes, which are obtained by non-linear least squares fits to the data after one subtracts off the background widths due to spin relaxation associated with the overall reorientation. For the smaller probes PDT and MOTA (cf. Fig. 8), we observe a universal result in the nematic phase, just above the SmA phase, viz. a critical exponent of $\gamma = -1/3$, within experimental error.

However, the larger spin probe P-probe showed no discernable critical effect at this transition in the nematic phase. On the other hand, it did show a critical exponent of $-.22 \pm 0.03$ on the smectic phase side of this transition in 40,6 solvent (the only solvent studied with this probe); whereas PDT and MOTA show no clear critical effect, only a weak hint of a pre-transitional anomaly on the smectic phase side.

The re-entrant nematic-smectic $(RN-S_A)$ transition in 60CB-80CB was also studied, and it showed a critical exponent of nearly -1/3 for PDT but only -0.13 for MOTA on the smectic side of this phase transition. Again, there is only a weak hint of a pre-transitional anomaly on the re-entrant nematic side.

Critical-type divergences are also observed on both sides of the I-N transition. The I-N phase transition is characterized (on either side) by spin relaxation parameters which diverge with exponent close to -1/2 as is consistent with a Landau - de Gennes mean-field theory of fluctuations in the orientational order parameter [55]. This results in a slowly-fluctuating orientational potential at the site of the probe molecule, which is able to modulate the rotational reorientational of the probe, thereby leading to the observed critical-type of effect on the spin relaxation [30,55]. Our virtually universal observations of -1/2 for the critical exponent on either side of the I-N transition is in support of the reliability of the technique.

While the weak first order I-N phase transition is generally well-characterized by mean-field theory, this is not so for the $N-S_A$ transition, which is most likely second-order for 40,6 [56] and to which scaling lawas analogous to the λ-transition in He have been applied [57,58]. In the dynamic scaling approach of Brochard [58] and of Jähnig and Brochard [59], the coherence length ξ characterizing fluctuations in the smectic order parameter $\psi(\vec{r},t)$, which is complex, is predicted to diverge as $(T-T_c)^{-0.66}$

Table 2a: CRITICAL EFFECTS: 4O,6

	PDTempone		MOTA		P-Probe	
	B	C	B	C	B	C
ISOTROPIC PHASE (Nr. Nematic)						
k:	7.6 ± 1.7	17.4 ± 2.1	58 ± 5	68 ± 6	161 ± 49	66 ± 37
γ:	-0.49 ± 0.11	-0.45 ± .09	-0.47 ± .02	-0.50 ± .02	-0.49 ± .12	-0.50 ± .15
NEMATIC PHASE (Nr. Isotropic)						
k:	4.46 ± .05	9.7 ± 0.2	43.0 ± 0.15	44.0 ± 0.2	210 ± 1	125 ± 1.5
γ:	-0.48 ± .01	-0.54 ± .02	-0.48 ± 0.01	-0.50 ± 0.01	-0.51 ± 0.1	-0.49 ± .02
NEMATIC PHASE (Nr. Smectic A)						
k:	4.1 ± 1.7	6.0 ± 0.8	8.0 ± 0.2	5.2 ± 0.4	No Critical Effects Observed	
γ:	-0.33 ± .07	-0.38± 0.06	-0.32 ± 0.01	-0.33 ± 0.02		
SMECTIC PHASE (Nr. Nematic)						
k:	No Critical Effects Observed		No Critical Effects Observed		713 ± 3	702.8 ± .7
γ:					-0.21 ± .03	-0.23 ± .02

Above Based on Fits to Form

$B,C = k(T-T*)^{\gamma}$ (in mG.)

Errors shown are from the non-linear least squares fits

(From Ref. 52)

Table 2b: CRITICAL EFFECTS: 6OCB - 8OCB

	PDT			MOTA	
	B	**C**		**B**	**C**

ISOTROPIC PHASE (Nr. Nematic)

k: 14   ± 2   46   ± 2   42   ± 5   39 ± 16

γ: -0.43 ± .09   -0.40 ± .03   -0.48 ± 0.08   -0.48 ± 0.04

NEMATIC PHASE (Nr. Isotropic)

k: 16.2 ± .03   29.9 ± .03

γ: -0.50 ± .01   -0.56 ± .01   No Critical Effects Observed

NEMATIC PHASE (Nr. Smectic A)

k: 17   ± 3   12.5 ± 1.2   75   ±   8   77 ± 6

γ: -0.30 ± .02   -0.36 ± .03   -0.36 ± .02   -0.35 ± .02

SMECTIC PHASE (Nr. Re-entr. Nematic)

k: 19   ± 3   32   ± 3   207 ± 10   262 ±   33

γ: -0.33 ± .02   -0.38 ± .02   -0.13 ± 0.01   -0.13 ± 0.02

Above Based on Fits to form
$$B, C = k(T-T^*)^{\gamma} \text{ (in mG)}$$
(From Ref. 52)

[66]. The true story is actually more complicated, involving separate critical expo-
nents for $\xi_{||}$ and $\xi_{\perp}$, the coherence lengths parallel and perpendicular, respect-
ively to the nematic director, and these exponents appear to vary from one liquid-
crystal to another [56,60].

B) Proposed Model

We have proposed the following model to explain our principal results [31]. As discussed above, the probe has a preference to be located in the lower density regions
of the smectic layer, i.e. the alkyl chain region [24]. As the smectic phase is approached from higher temperature, and smectic layering forms as a pre-transitional
phenomenon (i.e. cybotactic clusters), there is "expulsion" of the probe to the lower
density regions of the transitory smectic layer. Molecular parameters which affect
spin relaxation (e.g. the nematic ordering parameters S_p and/or τ_R) are affected
by this expulsion effect. The onset of smectic layers near the transition is described
by density fluctuations: $\rho(\vec{r},t)$ which also affects the translational motion of the
probe. Since the critical fluctuations in $\rho(\vec{r},t)$ occur on a much longer time-scale
separation of the two types of motions which simplifies the analysis. Thus, as cybotactic clusters form and break up in different regions, molecular dynamics and therefore the spin relaxation of the probe is modulated.

In our formal approach, we first expand the relevant relaxation parameter Q
(= e.g. S_p or τ_R) as a Taylor's series in the deviation of the density from its
mean value ρ_0, i.e. $\Delta\rho(\vec{r},t) \equiv \rho(\vec{r},t) - \rho_0$. That is

$$Q(\vec{r}_B,t) = Q_0 + Q_1\Delta\rho(\vec{r}_B,t) + Q_2[\Delta\rho(\vec{r}_B,t)]^2 + \ldots \qquad (25)$$

where the subscript B refers to the location of the probe. The translational diffusion of the probe is taken to obey a Smoluchowski equation with a time-dependent
potential [30]:

$$\frac{\partial P(\vec{r}_B,t)}{\partial t} = -\nabla \cdot \underset{\sim}{D} \cdot \{\nabla + [\nabla U(\vec{r}_B,t)]/kT\}P(\vec{r}_B,t). \qquad (26)$$

In Eq. (26) $P(\vec{r}_B,t)$ is the probability density of finding the probe at $\vec{r}_B$ at
time t, $\underset{\sim}{D}$ is the translational diffusion tensor with components $D_{||}$ and $D_{\perp}$,
while the potential of mean force on the probe is a functional of the density fluctuations, i.e.

$$U(\vec{r}_B,t) = U[\Delta\rho(\vec{r}_B,t)] . \qquad (27)$$

Now $\Delta\rho(\vec{r})$ is related to the complex order parameter $\psi(\vec{r})$ in the usual
manner [79,80]:

$$\Delta\rho(\vec{r}) = \frac{\rho_0}{\sqrt{2}} \text{Re}\left[\psi(\vec{r})e^{iq_s z}\right] = \frac{\rho_0}{\sqrt{2}} \left|\psi(\vec{r})\right| \text{Re}(e^{iq_s[z-u(\vec{r})]}). \qquad (28)$$

Here $q_s = \frac{2\pi}{d}$, where d is the smectic layer spacing, and the phase, $q_s u(\vec{r})$ of $\psi(\vec{r})$ locates the smectic layer's density maxima and minima in each part of the sample. If for simplicity we follow dynamic scaling according to Jähnig and Brochard [58,59], we obtain for the $\vec{q}$-th Fourier component of $\psi(\vec{r})$ the time correlation function:

$$<\psi^*(\vec{q},t)\psi(\vec{q},0)> = <|\psi_{\vec{q}}|^2> e^{-\Gamma_{\vec{q}} t} \qquad (29)$$

$$<|\psi_{\vec{q}}|^2> = \frac{k_B T}{2A} \left[(1 + q_\perp^2 \xi_\perp^2 + q_{||}^2 \xi_{||}^2)V\right]^{-1}. \qquad (30)$$

Here $q_{||}$ and $q_\perp$ are the components of $\vec{q}$ parallel and perpendicular (to the nematic director), V is the sample volume, and A is the coefficient in the term quadratic in $\psi(\vec{r})$ in the Landau expansion of the smectic free-energy. (A goes to zero almost as ξ^{-2}). The damping $\Gamma_{\vec{q}}$ of the $\vec{q}$-th mode is given by:

$$\Gamma_{\vec{q}}^{-1} = \frac{\tau_m}{(1 + q_\perp^2 \xi_\perp^2 + q_{||}^2 \xi_{||}^2)^x} \qquad (31)$$

with τ_M a characteristic relaxation time for the cybotactic clusters and $x = 3/4$. The relaxation time τ_m is expected to diverge as $\xi^{3/2} = (T-T_c)^{-1}$.

In the spirit of a Landau expansion we consider only the lowest order terms in $\Delta\rho(\vec{r}_B)$ to represent the time-dependent fluctuations in Q. That is:

$$<\Delta Q(\vec{r}_B,t)\Delta Q(\vec{r}_B,0)> = Q_1^2 C(t) + \text{H.O.T} \qquad (32)$$

where

$$C(t) \equiv <\Delta\rho(\vec{r}_B,t)\Delta\rho(\vec{r}_B,0)>/\rho_0^2. \qquad (33)$$

The method of approach for calculating $C(t)$ including the critical hydrodynamics of the phase transition and the translational diffusion of the probe is based upon methods we have previously developed [30]. The dominant contribution to the ESR linewidths should be from terms involving $J(\omega) \approx J(0)$ yielding (in the limit $q_s \to 0$):

$$J(0) = \frac{Mk_B T}{8\pi} \frac{z\tau_m}{\xi} (1 + z^{1/2})^{-1}$$

$$\xrightarrow{z \cong 1} \frac{Mk_B T}{16\pi} \left(\frac{\tau_m}{\xi}\right) \propto \xi^{1/2} \propto (T-T_c)^{-1/3}$$

$$\xrightarrow{z \ll 1} \frac{Mk_B T}{8\pi D} \xi \propto \xi \propto (T-T_c)^{-2/3} \qquad (34)$$

where $z \equiv (1 + D\tau_m/\xi^2)^{-1}$ measures the relative importance of translational diffusion over the coherence length ξ vs. relaxation of the order parameter in providing averaging the fluctuations in Q. Here $z \cong 1$ corresponds to relaxation dominated by director fluctuations, while $z \ll 1$ corresponds to relaxation dominated by molecular translational diffusion. The limiting form for $z \cong 1$ predicts the experimentally observed critical divergence of -1/3.

In the second case we keep $q_s = \frac{2\pi}{d}$ but introduce other simplifications. Then for $J(0)$ we obtain

$$J(0) = \frac{Mk_BT}{16\pi} \; \frac{\tau}{\xi} \; \frac{\sqrt{1+c} - 1}{c} \; . \tag{35}$$

Here $c \equiv q_s^2 D_\parallel \tau_m$ measures the relative importance of averaging out the effects of density fluctuations $\Delta\rho(\vec{r})$ in a $\underline{single}$ smectic-like layer through diffusion of the probe in the direction normal to the layer vs. the relaxation of the smectic layers. As $c \to 0$, corresponding to probe diffusion being unimportant, one obtains essentially the previous result for $z \approx 1$. For $c \gg 1$, $J(0) \propto \tau_m^{1/2}/\xi_\parallel \propto \xi^{-1/4}$, and it does $\underline{not}$ diverge, but rather goes to zero.

Based upon measurements of $\xi_\parallel$ [56] and D [52,61,62] we estimate $\xi_\parallel \sim 10^{-5}$ cm for $T - T_c = 0.1°K$ and $D \sim 10^{-6} cm^2$/sec. Also we estimate $\tau_m \sim 10^{-5}$ to 10^{-6} sec at $T - T_c = 0.1°C$ [31], so $D_\parallel\tau_m/\xi^2 \sim 10^{-2}$ to 10^{-1} (for $T - T_c \sim 0.1°C$), while $D_\parallel\tau_m q_s^2 \sim 10$ to 10^2. Thus, while it may be reasonable to ignore the averaging effects of translational diffusion over the distance of ξ_1, this is questionable for diffusional averaging over a single smectic layer of thickness d .

However, the above simple model, i.e. Eq. (32), implicitly ignores the potential of mean force $U(\vec{r}_B,t)$ in Eq. (26). If U is a very sensitive functional of $\Delta\rho(\vec{r})$ (cf. Eq. (27)), then as $\Delta\rho(\vec{r})$ diverges as the critical point is reached, Eq. (26) would predict virtually no diffusion parallel to the normal to the smectic phases in the cybotactic clusters. Instead, the probe would reside entirely in the alkyl chain regions in such clusters, i.e. the "expulsion effect" referred to above. Thus, the modulation of the parameter Q would be primarily determined by the formation and break-up of the cybotactic clusters, with the probe rapidly adjusting its location within the layers accordingly. This effect would be measured by the correlation function: $\langle\psi(r_B,t)\psi(r_B,0)\rangle$ and we would obtain Eq. (34) for the (simplified) spectral densities.

Our model for cooperative molecular dynamics and critical effects at the N-S phase transition is still a rather simple one, and we have not yet explicitly considered the role of a finite $U[\Delta\rho(\vec{r}_B,t)]$ despite its presumed importance. On the other hand, we recall the Moro-Nordio theory [35] for spin-relaxation within the smectic phase. These two theoretical analyses are really related to one another. The more general approach would yield smectic-like fluctuations experienced by the spin probe near the

N-S phase transition, but a well-defined smectic-like potential deep in the smectic phase. It may be written in terms of a potential with the following form:

$$U[\beta, \Delta\rho(\vec{r}_B)] = AY_0^2(\beta) + B'[\Delta\rho(\vec{r}_B)]Y_0^2(\beta) + C'[\Delta\rho(\vec{r}_B)] \ . \tag{36}$$

If we assume that $B'[\Delta\rho(\vec{r}_B)]$ and $C'[\Delta\rho(\vec{r}_B)]$ are linearly proportional to $\Delta\rho(\vec{r}_B)$, then deep in the smectic we can let $B' \propto \frac{\rho_0}{2} |\psi_0| \cos \frac{2\pi}{d} z$ so that $B' = B \cos \frac{2\pi}{d} z$ (and similarly for C'). This is just the form Moro and Nordio use (cf. Eq. (8)). This suggests that near the phase transition we should use as an explicit form for $U[\Delta\rho(\vec{r}_B), t]$:

$$U = AY_0^2(\beta) + B'\Delta\rho(\vec{r}_B)Y_0^2(\beta) + C''[\Delta\rho(\vec{r}_B)] \tag{36}$$

in conjunction with the Fourier analysis of $\Delta\rho(\vec{r}_B)$ outlined above.

Such a model would more explicitly describe how the probe ordering and combined rotational-translational dynamics is modulated by the smectic-like pre-transitional fluctuations. It may also enable one to correlate observations at the N-S phase transition with SRLS type of behavior deep in the smectic phase.

VIII. Translational Diffusion

Our primary motivation here is to obtain the translational diffusion coefficients for the probe in liquid crystals in conjunction with studies of cooperative rotational dynamics and phase transition studies. The mechanism of Heisenberg Spin exchange can be used to study translational diffusion over molecular dimensions, while the new ESR imaging technique measures diffusion over macroscopic dimensions and can detect anisotropies in the diffusion coefficient.

A) Heisenberg Spin-Exchange

Our results [52,61] with PDT (cf. Fig. 16) show trends that are consistent with the model we have proposed [24] that in mono-layer smectics like 40,6 the PDT probes are gradually "expelled" from the central aromatic core regions to the aliphatic chains as one proceeds to lower temperature phases; but for cyanobiphenyls the probe expulsion takes place prior to the formation of the smectic mesophase, as is evidenced by translational diffusion coefficients, which show no discontinuities (in magnitude or activation energy) at the phase transitions in the latter case. Also the low activation energy of about 2.5 kcal/mole for 60CB-80CB and the decreasing activation energy of 4.7 to 1 kcal/mole for 40,6 are more characteristic of diffusion through aliphatic chains. This emphasizes the important role such studies can play in developing an understanding of motional dynamics.

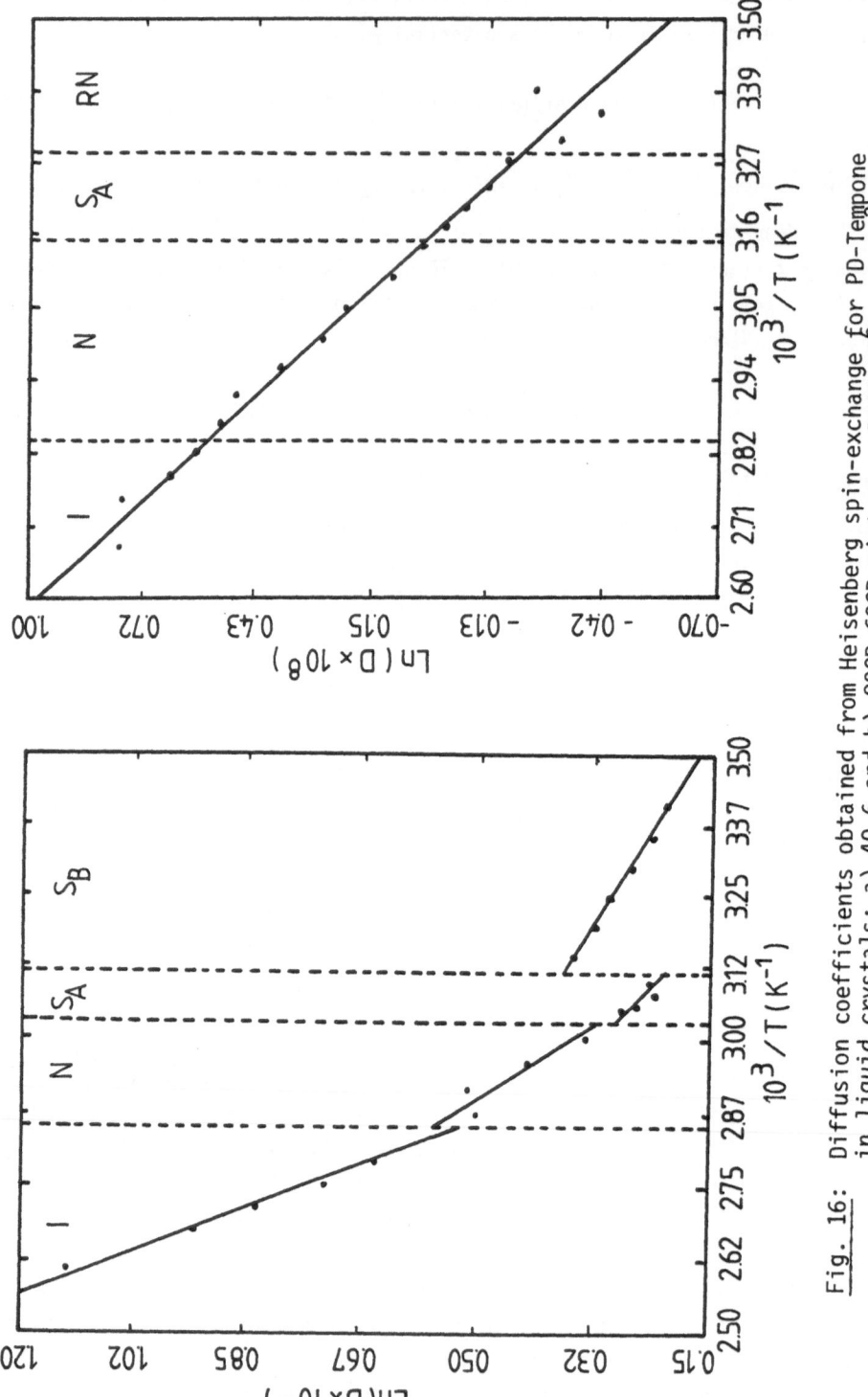

Fig. 16: Diffusion coefficients obtained from Heisenberg spin-exchange for PD-Tempone in liquid crystals: a) 40,6 and b) 8₂CB-60CB mixture. $\ln(D \times 10^6)$ vs. $10^3/T$ is plotted, where D is in units of $cm^2 s^{-1}$. (From Ref. [63].)

B) ESR Imaging and Macroscopic Diffusion

The relaxation techniques for measuring translational diffusion as discussed above for HE are based on bimolecular collisions of spin probes. As such, they measure diffusion over dimensions of (several) molecular length(s). However, they do not provide direct means for measuring diffusional anisotropy. Also, the analysis leading to the diffusion coefficient depends upon the choice of molecular model. It is therefore useful to have measurements of diffusion that do not depend upon one's choice of model and that can be used to study anisotropiy in diffusion. Such studies could then be compared with the sub-microscopic ones to understand much better the nature of the molecular model.

We have developed a technique employing ESR imaging, which we have found to be convenient for measurement of D in the range of $10^{-8} < D < 10^{-5}$ cm^2/sec, and we have succeeded in measuring the anisotropy of D in the liquid crystal 5,4 where $D \gtrsim 5 \times 10^{-7}$ cm^2/sec [32]. In the nematic phase we find $D_{||}/D_{\perp} = 0.71 \pm 0.1$ which is consistent with a pre-transitional smectic-like effect [33] that could be expected, since 5,4 does exhibit a N-S transition.

Utilization of such an imaging technique to study the anisotropic translational diffusion properties of the various smaller and larger spin probes in the different liquid crystalline phases would be useful for comparing with studies of sub-microscopic rotational dynamics (and sub-microscopic translational dynamics). Also one might hope to observe pre-transitional anomalies at the liquid crystalline phase transitions (e.g. a decrease in $D_{||}$ as the $N-S_A$ phase transition is approached from the nematic phase, cf. Ref. [31]).

IX. Rotational Dynamics in Model Membranes

A) Dynamic Molecular Structure and Phase Transitions in Lipid Multilayers

The use of defect-free oriented samples enabled us to clearly observe lipid phase transitions through the appearance of composite spectra in the transition (two-phase) region [63]. From ESR observations on low-water content DPPC and DMPC (cf. Fig. 17) three phase transitions were found over a temperature range below $180^{\circ}C$: two were assigned to the main transition and to the isotropic transition by reference to the transition temperatures in the literature. The remaining one, at $100-110^{\circ}C$, was characterized as a "chain-orientational" transition.

The ordering and the rotational diffusion tensor of the various spin labels could be determined accurately as a function of temperature, % H_2O, and phase (e.g. Table III). CSL, 5PC, and 16PC exhibit in all phases decreasing order parameters S, according to CSL > 5PC > 16PC and increasing motional rate (measured by $\bar{R}$ the mean rotational diffusion coefficient) again according to CSL < 5PC < 16PC, while the anisotropy in rotational motion obeys CSL < 5PC < 16PC, consistent with the well-known concept of the increased flexibility as one proceeds down the chain [64]. However, we have been

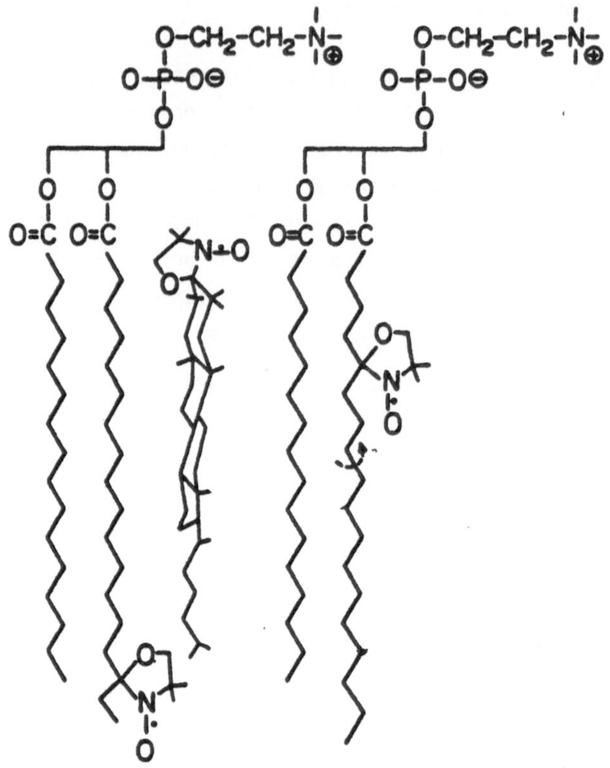

Fig. 17: Schematic structures of CSL, SPC, 16PC (from Ref. [63]).

able to quantify this flexibility gradient in terms of its reduced ordering and its symmetry, as well as the increased motional rate [63].

Using these results we can characterize the main (or gel-to-Lα(1)) transition as primarily a "chain-diffusional" transition, while, as noted above, the new high-temperature one is characterized as a "chain-orientational" transition: the ordering parameter S experiences a more significant relative reduction at the second transition compared to that at the main transition, whereas the diffusion coefficient $R_\perp$ for the chain probes (i.e. 5PC and 16PC) experiences a more significant relative increase at the main transition. Thus, a relatively smaller reduction in molecular ordering more effectively "unfreezes" the chain motions at the main transition compared to the second one. The relative increase in $R_\perp$ for CSL at the two transitions is, however, comparable suggesting that while local chain motion increases more significantly at the main transition, the overall molecular motions exhibit comparable relative changes at both phase transitions. Also, whereas $R_\perp$ shows substantial change at the phase transitions, $R_\parallel = NR_\perp$, which measures the motion about the long chain axis, is much less affected. This undoubtedly reflects the existence of significant motion of this type in the gel phase, which may be due to its relatively unhindered nature. Finally, we find that at

Table III: Parameters for Molecular Ordering and Anisotropic Rotation of CSL in DPPC[d]

t, °C	phase	$(D^2_{00})^d$	$(D^2_{02} + D^2_{0-2})^d$	$R_\perp$, c s⁻¹	$R_\parallel$, c s⁻¹	N	E_a, kcal/mol	τ_2^{*-1}, G
			(A) Hydrated to 3 wt %					
40	I	0.90	−0.01	2.9×10^5	4.4×10^7			
50		0.90	−0.01	4.0×10^5	6.0×10^7	150	7.7	1.5
60		0.90	−0.01	6.2×10^5	9.3×10^8			
70		0.90	−0.01	8.3×10^6	1.2×10^8			
80	II	0.76	−0.03	6.8×10^6	3.4×10^8			
85		0.73	−0.03	8.0×10^6	4.0×10^8	50	(8.7)	1.2
90		0.67	−0.03	9.6×10^8	4.8×10^9			
110	III	0.28	0.08	1.1×10^8	1.7×10^9			
120		0.21	0.06	1.5×10^8	2.4×10^9	16	9.2	1.0
130		0.14	0.03	2.0×10^8	3.2×10^9			
140	IV	0.13	0.03	2.5×10^8	4.0×10^9			
160		0	0	5.8×10^8	2.9×10^9	5	5.0	1.0
170		0	0	6.7×10^8	3.4×10^9			
180		0	0	7.5×10^8	3.8×10^9			
			(B) Hydrated to 7 wt %					
40	I	0.88	−0.005	3.0×10^5	4.5×10^7			
50		0.88	−0.005	5.0×10^5	7.5×10^7	150	(8.7)	1.5
60		0.88	−0.005	7.0×10^6	10.5×10^8			
70	II	0.78	−0.007	9.0×10^7	4.5×10^8			
80		0.74	−0.009	1.2×10^7	6.0×10^8	50	(7.0)	1.2
90		0.65	−0.015	1.6×10^7	8.0×10^8			

[a] Estimated errors: ±2% in (D^2_{00}), ±30% in $(D^2_{02}+D^2_{0-2})$, ±10% in $R_\perp$, ±20% in $R_\parallel$, ±20% in N, ±20% in E_a, and ±01. G in τ_2^{*-1}. Note $R_\parallel = NR_\perp$. [b] I, biaxial gel phase; II and III, liquid-crystalline phases; IV, isotropic phase. [c] Correlation times: $\tau_\perp = 1/6R_\perp$, $\tau_\parallel = 1/6R_\parallel$, and $\bar{\tau} = 1/6\bar{R} = 1/6(R_\parallel R_\perp)^{1/2}$. [d] The relationship between (D^2_{00}) and λ and ρ is given by the following expression:

$$(D^2_{00}) = \int_{\phi'} \int_{\theta'} P(\theta',\phi')\,1/2(3\cos^2\theta' - 1)\sin\theta'\,d\theta'\,d\phi'$$

and $(D^2_{02} + D^2_{0-2}) = \int_{\phi'}\int_{\theta'} P(\theta',\phi')(6^{1/2}/2)\sin^2\theta'\cos2\phi'\sin\theta'\,d\theta'\,d\phi'$, where θ' denotes the angle between the principal axis z' of the ordering tensor and the principal axis z" of the director frame. $P(\theta',\phi')\sin\theta'\,d\theta'\,d\phi'$ is the distribution of z' relative to z" given by $P(\theta',\phi') \propto \exp[- 1/2(3\cos^2\theta' - 1) + (6^{1/2}/2)\rho \sin^2\theta' \cos2\phi']$. (From Ref. [65].)

both phase transitions there is a more significant relative reduction in ordering at the end of the chain but a smaller increase in fluidity (as measured by $R_\perp$). Thus while there is greater "melting" of orientational order at the end of the chain, the end-chain motions are not as significantly tied to the ordering.

B) Lipid-Gramicidin Interactions

Given that the ESR spectra from dispersions is somewhat ambiguous to interpret, we adapted our alignment methods of sample preparation to prepare very well-aligned uniform samples containing the stable polypeptide gramicidin A (DA). This polypeptide is frequently used to mimic the effects of protein on phospholipid bilayers [65]. Its advantages are its known chemical structure and its known helical conformations, its considerable stability, and its ready availability. Chapman and co-workers have found that the dimeric GA is incorporated into the lipid bilayer, and they regard it as a model for the interactions of the polypeptide segments of transmembrane proteins within the hydrocarbon regions of the lipid bilayers [65].

Our principal findings are the following [66]: 1) In the gel phase we observed distinct two-component spectra which could be assigned to highly oriented bulk lipids and to a disordered component, and the latter was fit by a model of "molecular disorder" such that the ordering of these molecules is greatly reduced, but its rotational-motional properties are not appreciably changed (cf. Figs. 18A and 19A). The disordered region at lowest concentration of GA is estimated to consist of about 30-40 lipid pairs, or about five times the number required to coat the GA dimer. This corresponds to a disordered region in the bilayer extending radially about 3 lipid molecules. This effect of disordering is significantly reduced by increasing the wt.% of water, but it appears to be independent of temperature. 2) In the liquid crystalline phase, heterogeneity is not distinguished from the ESR spectrum. Instead, the primary effect of GA is to significantly reduce the observed ordering of all the lipids, with only a very small decrease in motional rates (cf. Figs. 18B, 19B). However, in the high-temperature weakly-ordered phase, addition of GA actually leads to significant increase in ordering. This increase in ordering is also observed in high-water-content dispersions in the liquid crystalline phase. There is no hint of features usually assigned to "immobilized" species in any of the spectra obtained from well-aligned samples. However, such features are present in dispersion samples of 4 M% GA prepared from the same materials as the well-aligned ones. If we associate these spectral features with "trapped lipids" due to aggregation of GA, then it follows that microscopically well-aligned samples do not allow for such aggregation.

We conclude from these findings that the principal lipid-GA interaction is that of a boundary effect such that the GA induces disorder in the low-temperature and low water content lipids, but it induces order for high-temperature and high water content (i.e., less ordered) lipids. It has only slight effects on lipid fluidity, in general reducing only slightly the rates of rotational reorientation.

127

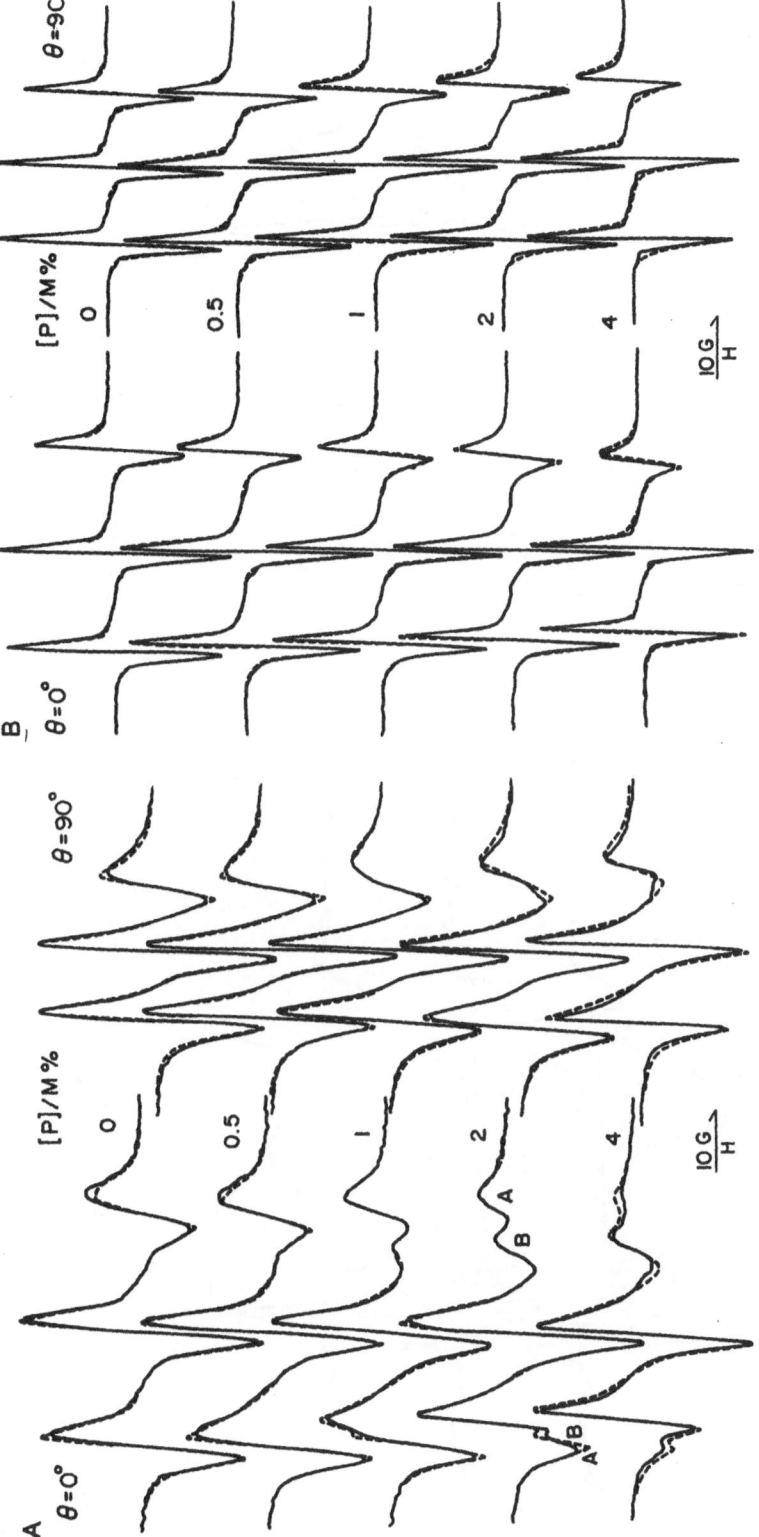

Fig. 18: ESR spectra from 16PC in DPPC hydrated to 7 wt. % for concentrations of GA ranging from 0 to 4 M%. Spectra are shown for θ = 0° and 90°: (A) 50°C corresponding to phase I; (B) 80°C corresponding to phase II. Dashed spectra are simulations. In (A) the ordered and disordered components are labeled A and B respectively. (From Ref. [66].)

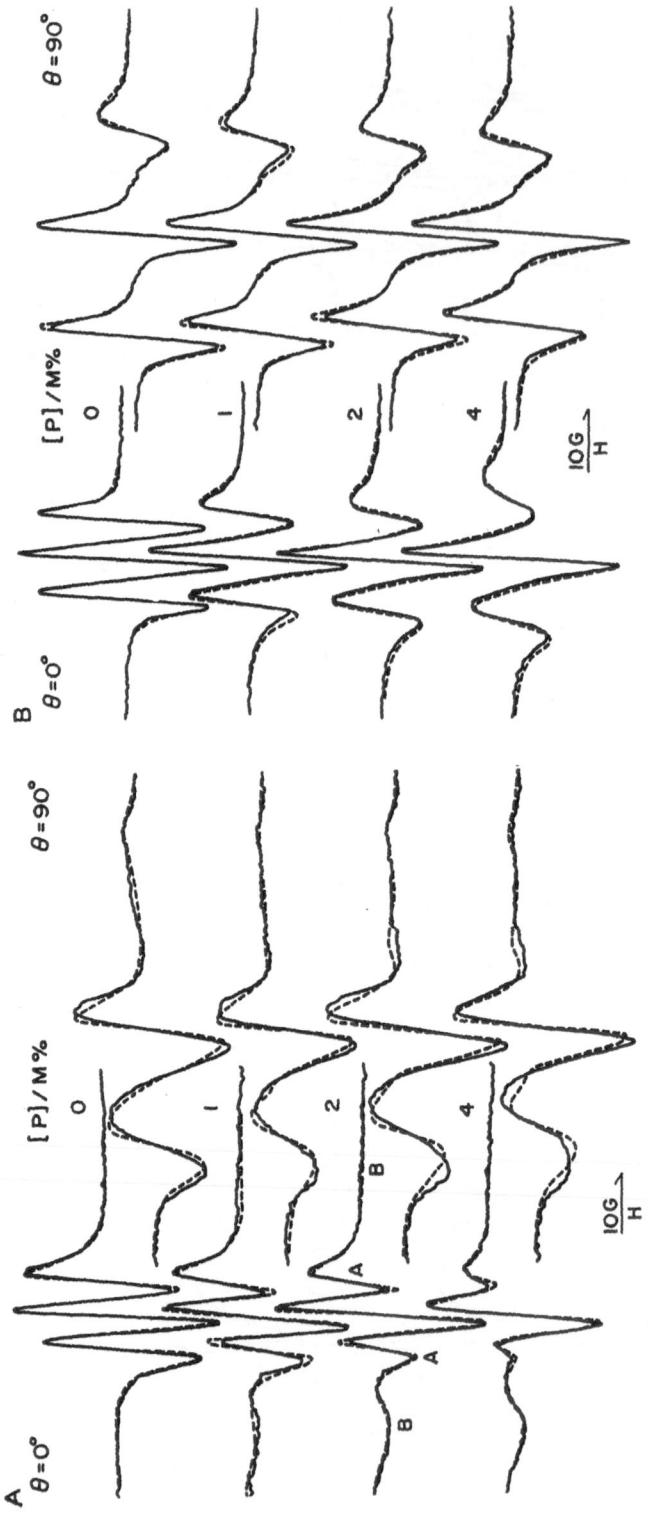

Fig. 19: Same as Fig. 18 but for CSL in DPPC. (From Ref. [66].)

We believe that these various effects can be explained as the consequence of two competing features of the lipid-GA interaction: a "disordering feature" and a "hardening feature". The former induces a disordering of the lipids in their vicinity, while the latter makes them more solid-like, as exemplified by somewhat reduced fluidity and by increasing the order. Furthermore, we require that disordering is dominant under conditions of low fluidity, while hardening is dominant when there is high fluidity. The notion of two apparently opposite effects of the macromolecule on the ordering of lipids has been incorporated into a simple model by Jähnig [67]. He proposed that the ordering at the boundary of a protein should be lower than that for the "ordered" phase but greater than that for the "fluid" phase. Our low water content results are consistent with this model when we apply it to the $L\alpha(1)$ to high-temperature-liquid-crystal phase transition.

The heterogeneity induced by GA at very low concentrations is a distinctly different phenomenon from the one usually assigned to "immobilized" or "trapped" lipids. The clear discrimination of the heterogeneity in the bilayer induced by the GA, and the determination of the molecular properties of this heterogeneity should be significant in understanding the polypeptide - lipid interaction.

X. Electron-Spin Echoes and Rotational Relaxation

It is our expectation that over the next several years the modern electron-spin echo (ESE) technique will become increasingly important in the study of spin relaxation and rotational dynamics. The possibilities have been greatly enhanced by instrumental developments in several laboratories around the world including our own [4, 68-73].

The principal motiovations for applying ESE techniques to spin relaxation studies include: (1) the ability to separate homogeneous from inhomogeneous contributions to the linewidths (or T_2) as well as to profit from the resulting increase in resolution; (2) the ease of simultaneously performing T_1 measurements; (3) the possibility that special ESE techniques could provide information on motional dynamics in addition to that from cw-studies; and (4) the possibility of extending the range of study to slower motions.

(A) ESE and Slow Motional Theory

Wihile ESE work on nitroxides in liquids showed good agreement with the motionally-narrowed line widths extracted by cw-techniques, we were especially motivated by the initial observation [74] that in the slow motional regime, for $\tau_R \sim 10^{-7}$ to 10^{-6} sec, the phase memory time, T_M was found to be proportional to $(\tau_R)^\alpha$ with $\alpha \sim 1$ (cf. Fig. 20). Simple arguments suggest this was to be expected. That is, in the slow motional regime, reorientational jumps should lead to spectral diffusion wherein each jump takes place between sites of different resonance field. This would be an uncertainty-in-lifetime broadening that is analogous to the slow exchange limit in the classic two-site case, and it contributes to T_M. The broadening would then be given by τ_R^{-1}, the jump frequency. This result suggested that studies of the ESE T_M in slow-motional spectra would supply complementary information on motional dynamics to cw-lineshape studies.

A rigorous theoretical basis for the analysis of slow-motional ESE was developed in order to interpret such experiments with confidence [75]. It is based on the same stochastic Liouville equation (SLE) which is used to predict cw-lineshapes (cf. Sect. VI). This emphasizes that echoes relate to the same type of motional effects as do the cw-lineshapes.

Our theoretical results on simple $90^o - \tau - 180^o - \tau$ echoes demonstrated that [75]: 1) the ESE decay envelopes show a short-time behavior with an $e^{-c\tau^3}$ dependence on τ and a long-time behavior of $e^{-\tau/T_M^\infty}$. 2) The asymptotic phase-memory decay constant T_M^∞ shows a significant dependence on models, and this was traced to the different

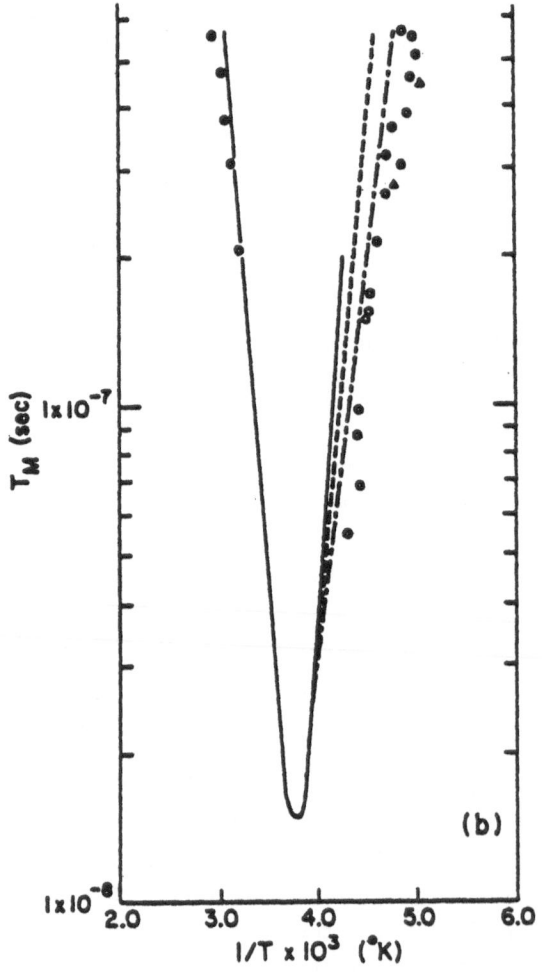

Fig. 20: Graph of T_M (or T_2) vs. T^{-1} for PD-Tempone in 85% glycerol/15% water. The circles show T_M data. The triangles show the T_M from 2D ESE. The different lines represent T_M in the central spectral regime calculated for the models of jump diffusion (solid line), free diffusion (dashed line), and Brownian diffusion (dashed-dotted line). The calculations employed the values of τ_R extrapolated from the motional narrowing regime. (From Ref. [76a].)

mechanisms of spectral diffusion induced by these models. 3) The T_M obtained from selective echoes on different parts of the (nitroxide) spectrum show significant differences. 4) The short-time behavior yields a $c \propto \tau_R^{-1}$, and c is independent of diffusion model.

We also showed that in an ideal two-pulse ESE experiment, the decay envelope is described by [75]:

$$S(\tau) = \text{Re} \sum_{1,j} a_{1,j} \exp [\Lambda_1 + \Lambda_j^*] \tau \qquad (38)$$

which is the real part of a sum of complex exponential functions. The relevant parameters are determined from the SLE, and the Λ_j are the eigenvalues of the stochastic Liouville operator in the rotating frame. Their imaginary parts (i.e. $\text{Im}\Lambda_j = \omega_j$) represent the resonance frequencies of the associated "dynamic spin packets", while the real parts (i.e. $\text{Re}\Lambda_j = T_{2,j}^{-1}$) represent their natural or homogeneous widths and are associated with the observed T_2. The relative contribution of the various spin packets depends upon the coefficients $a_{1,j}$, also provided by the theory. One finds from the simulations that the nitroxide slow-motional cw-spectrum will typically consist of 50-200 such spin packets which make significant contributions. These spin packets overlap with one another (especially when the inhomogeneous width is included), and the typical broad envelope, i.e., the cw-spectrum, is observed. Thus if the $\text{Re}\Lambda_j$ are different in magnitude from one another, one may expect to observe a T_2 that varies across the spectrum. We have developed a two-dimensional ESE technique specifically designed to study variations of the natural width across the spectrum.

A) Two-Dimensional ESE

In our original experiments [76,77], we utilized a standard $90°-\tau-180°$ two-pulse echo sequence and monitored the echo height at 2τ while slowly sweeping through the ESR spectrum by sweeping the dc-magnetic field. This generates an "echo-induced ESR spectrum" for each value of τ. The resonant microwave field H_1 is kept small enough that it does not introduce any distortion into the echo-induced field-swept ESR spectrum (i.e., $\gamma_e H_1 \ll \Delta\omega_s$, where $\Delta\omega_s$ is a measure of the width of spectral detail in the cw-spectrum). After collecting a family of such spectra, a Fourier transform with respect to τ is performed at each field value. The resulting 2D spectrum yields the inhomogeneously broadened absorption-like echo-induced ESR spectrum in one dimension and the homogeneous lineshape in the other dimension. The two-dimensional spectrum is given by:

$$S(\omega,\omega_0) \propto \sum_j a_{j,j} \frac{T_{2,j}}{1+\omega^2 T_{2,j}^2} \exp[-2\tau_d/T_{2,j}] \exp[-(\omega_0-\omega_j)^2/\Delta^2] \qquad (39)$$

where $\omega_0 = \gamma_e H_0$ and a Gaussian inhomogeneous broadening of width Δ has been assumed, and we have included the effect of the finite dead-time τ_d. Note that for

$\omega = 0$ we almost recover the expression for an ESR-like absorption spectrum with Gaussian inhomogeneous broadening. Along the ω-axis one observes a blend of Lorentzian line-shapes from the various "dynamic spin packets". In general, we find that while T_2 varies across the spectrum, the observed 2D-ESE line-shape at each position ω_0 is close to a simple Lorentzian in ω. Examples of such 2D-ESE spectra appear in Fig. 21.

It is preferable, however, to study the normalized contours to obtain useful information. This representation is developed by dividing every slice of the spectrum along the "width" (or ω)-axis by its corresponding amplitude at 0 MHz and then generating contour lines at every 10% change in height. The resulting map reveals the homogeneous line-shape as a function of field location unaffected by differences in signal height. One finds that these contours are very sensitive not only to the rate of reorientation but also to the model of molecular reorientation, (e.g. whether it is by jumps, free diffusion, of Brownian motion) with different characteristic <u>patterns</u> for each! We show in Fig. 22 an actual experimental demonstration of the sensitivity to motional anisotropy by comparing the results for tempone, which tumbles nearly isotropically vs. those for CSL, whose motion is anisotropic. While the T_2's are comparable, the shapes are significantly different, emphasizing the large anisotropy for CSL.

It should also be emphasized that Fig. 22 shows patterns that are consistent with a Brownian reorientation model, since jump models predict parallel contours with no features [76a].

We have already applied the 2D-ESE technique to oriented lyotropic liquid crystalline samples [76b] (cf. Fig. 21). In Fig. 23, we show a sequence of experimental contours from oriented multilayers of low water-content DPPC-doped with CSL for different temperatures and angle of tilt θ , and in Fig. 24 we show typical simulations which relate to these results showing specific sensitivities to the orienting potential as well as details of the dynamics.

We wish to emphasize the importance of the latter. Our studies with the CSL spin-label in the oriented samples (cf. Section IX) have shown that even in the slow-motional region, where cw-spectral simulations are only slightly sensitive to motion, it is very difficult to obtain a unique set of parameters characterizing the system under study. The 2D-ESE results are much more sensitive to these matters as illustrated in the simulations of Figs. 25. In Fig. 25a we show a cw-ESR simulation for high ordering ($S = 0.87$) and very slow motion $R \approx 10^4$ sec^{-1}. We superimpose the results for isotropic ($N = 1$) and very anisotropic ($N = 100$) motions to demonstrate that they are almost indistinguishable. However, in Fig. 25b we show the 2S-ESE contours and 0 MHz slices for the same parameters. They clearly differ both in magnitude and shape and are very easily distinguishable!

As described above, the 2D-ESE spectrum is obtained by monitoring the height of a spin echo from a two-pulse sequence, as the magnetic field is scanned.

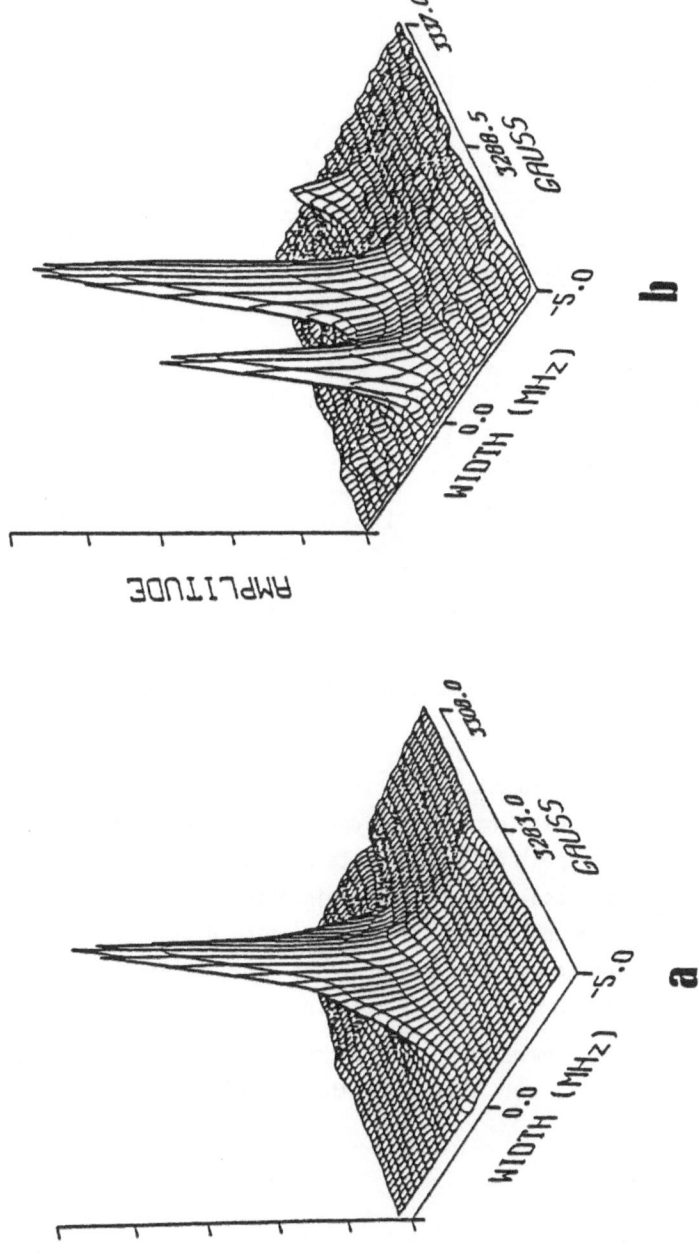

Fig. 21: Experimental 2D-ESE spectra from CSL spin probe in oriented multilayers of low water content DPPC: a) θ = 0° T = -20°C b) θ = 90° T = -20°C where θ denotes the orientation of the director with respect to the external and applied magnetic field H. (From Ref. [76b].)

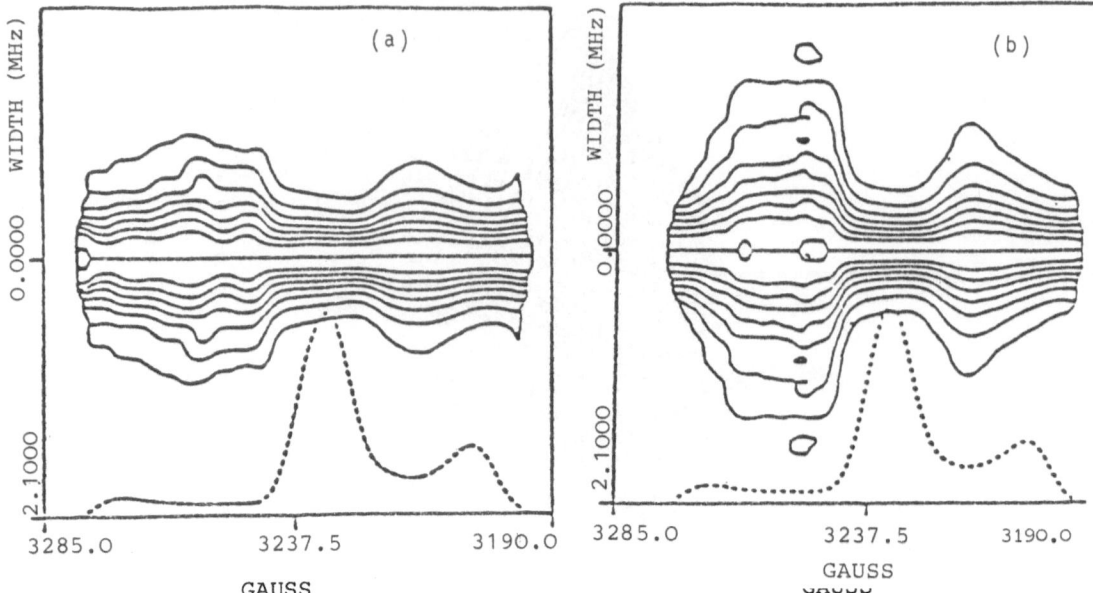

Fig. 22: Normalized contours from spectra of two different nitroxides. (a) shows the spectrum of Tempone in 85% glycerol/H_2O at -75°C. (b) shows the spectrum of CSL in n-butylbenzene at -135°C. The T_M's for these spectra, under these conditions, are approximately the same. (This result was obtained by G.L. Millhauser in these laboratories.)

Two difficulties arise from this approach. First, to avoid so-called FFT window effects, it is necessary to collect data over a considerable time range. This means that a considerable amount of time is spent collecting data when the signal-to-noise ratio is low, and, hence, the spectral resolution is low. This effect is more pronounced in curves with rapidly relaxing components, and so, the resulting distortions are not uniform across the spectrum. The second difficulty arises from the spectrometer dead time, which tends to filter out the more rapidly relaxing components.

To remedy these problems we have found that a linear prediction method (LPSVD [77,78]) for processing the data is very useful for these two dimensional spectra [79] (cf. Fig. 26).

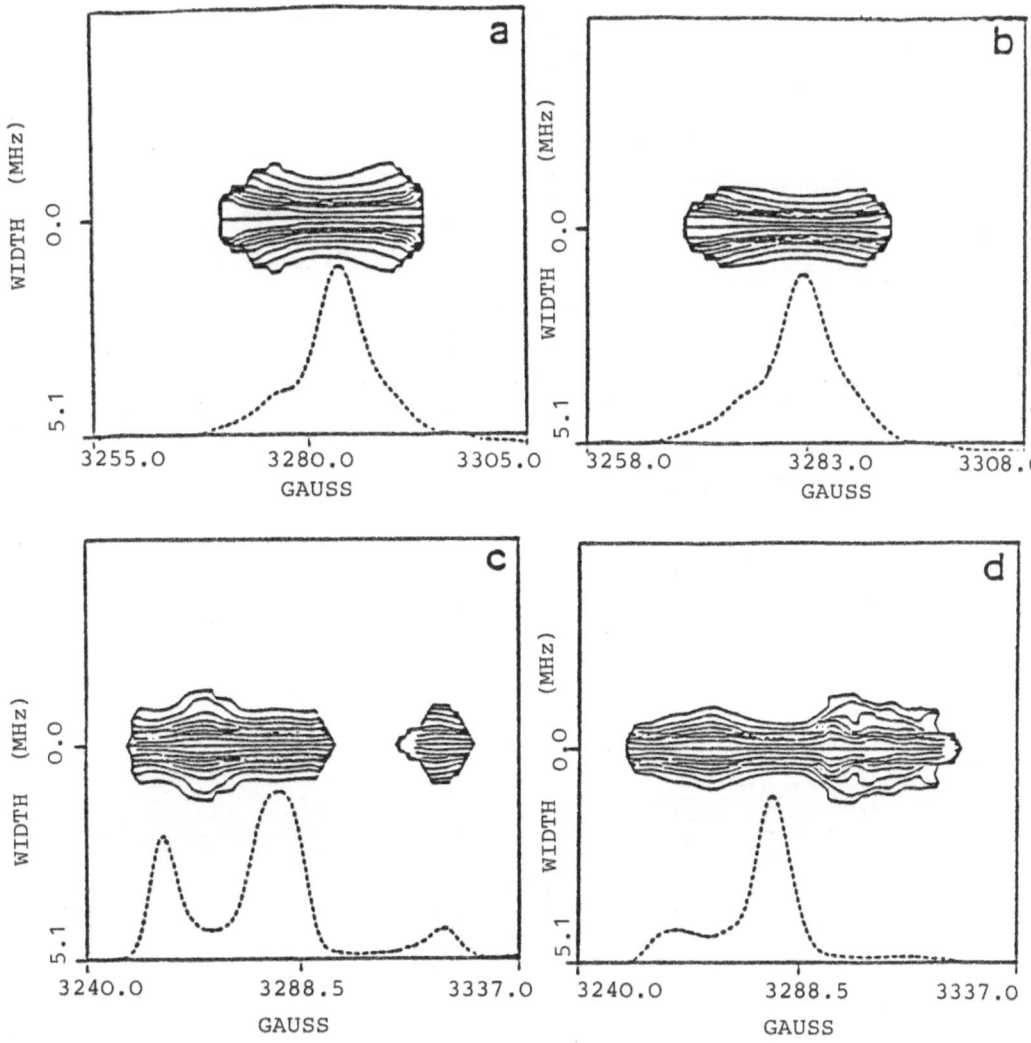

Fig. 23: Normalized contours of experimental spectra from oriented multilayers of low water content DPPC doped with CSL spin probe.

(a) $\theta = 0^0$, $T = 0^0C$; (b) $\theta = 0^0$, $T = -20^0C$; (c) $\theta = 90^0$, $T = -20^0C$;

(d) $\theta = 45^0$, $T = -20^0C$. (From Ref. [76b].)

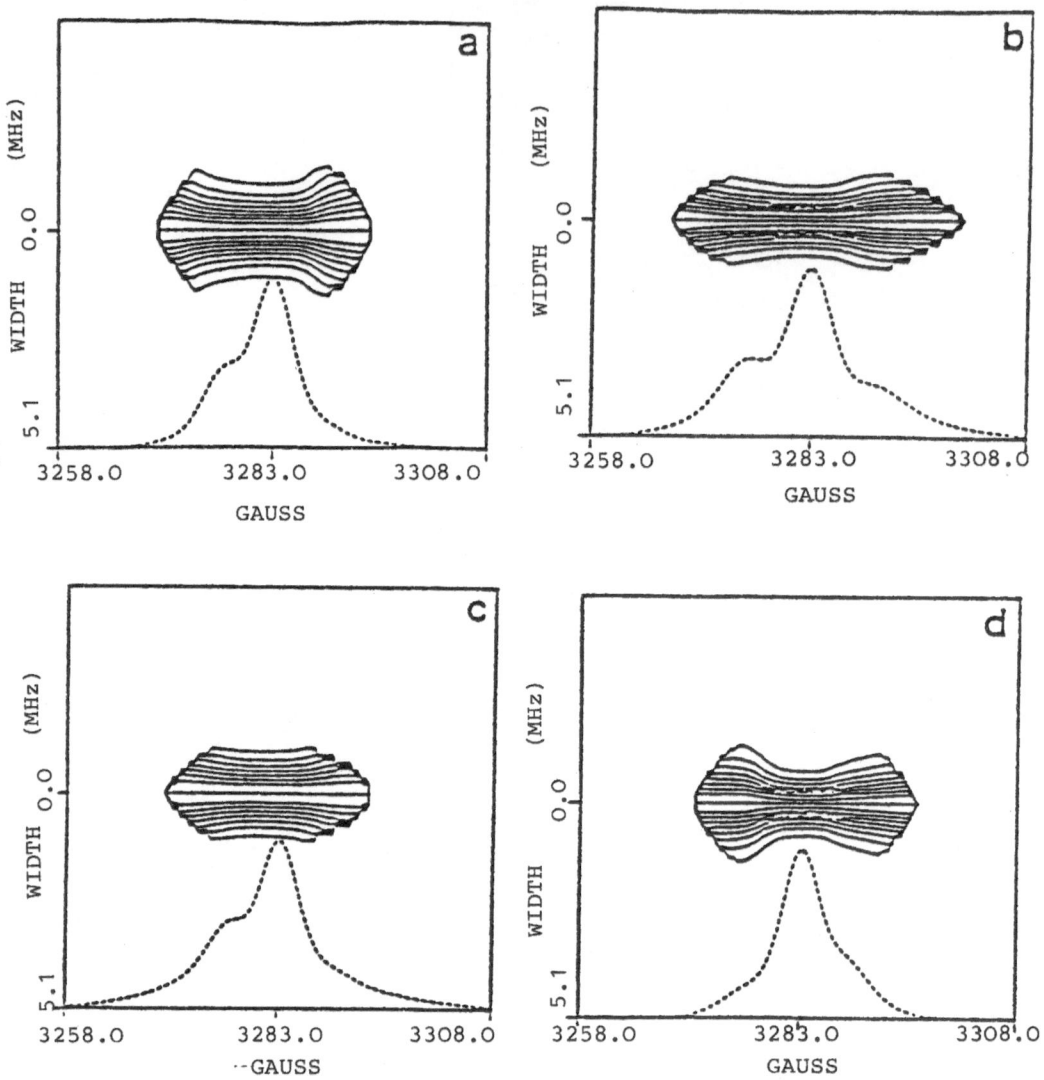

Fig. 24: Normalized contours of simulated DPPC/CSL oriented spectra $(\theta = 0^0)$ to illustrate the sensitivity to motion and ordering. (a) $R_{||} = 4 \times 10^4 \, s^{-1}$, $R_\perp = 0.8 \times 10^4 \, s^{-1}$, or $N = 5$, $\lambda_0^2 = 8.0$; (b) $R_{||} = 2 \times 10^4 \, s^{-1}$, $R_\perp = 0.5 \times 10^4 \, s^{-1}$, $N = 4$, $\lambda_0^2 = 8.0$; (c) $R_{||} = R_\perp = 1 \times 10^4 \, s^{-1}$, $\lambda_0^2 = 4.0$; (d) $R_{||} = R_\perp = 1 \times 10^4 \, s^{-1}$, $\lambda_0^2 = 12.0$. The effects of an intrinsic T_2^{ss} (0.7 µs) due to "solid-state" contributions and inhomogeneous broadening (3.2 gauss) have been included. (From Ref. [76b].)

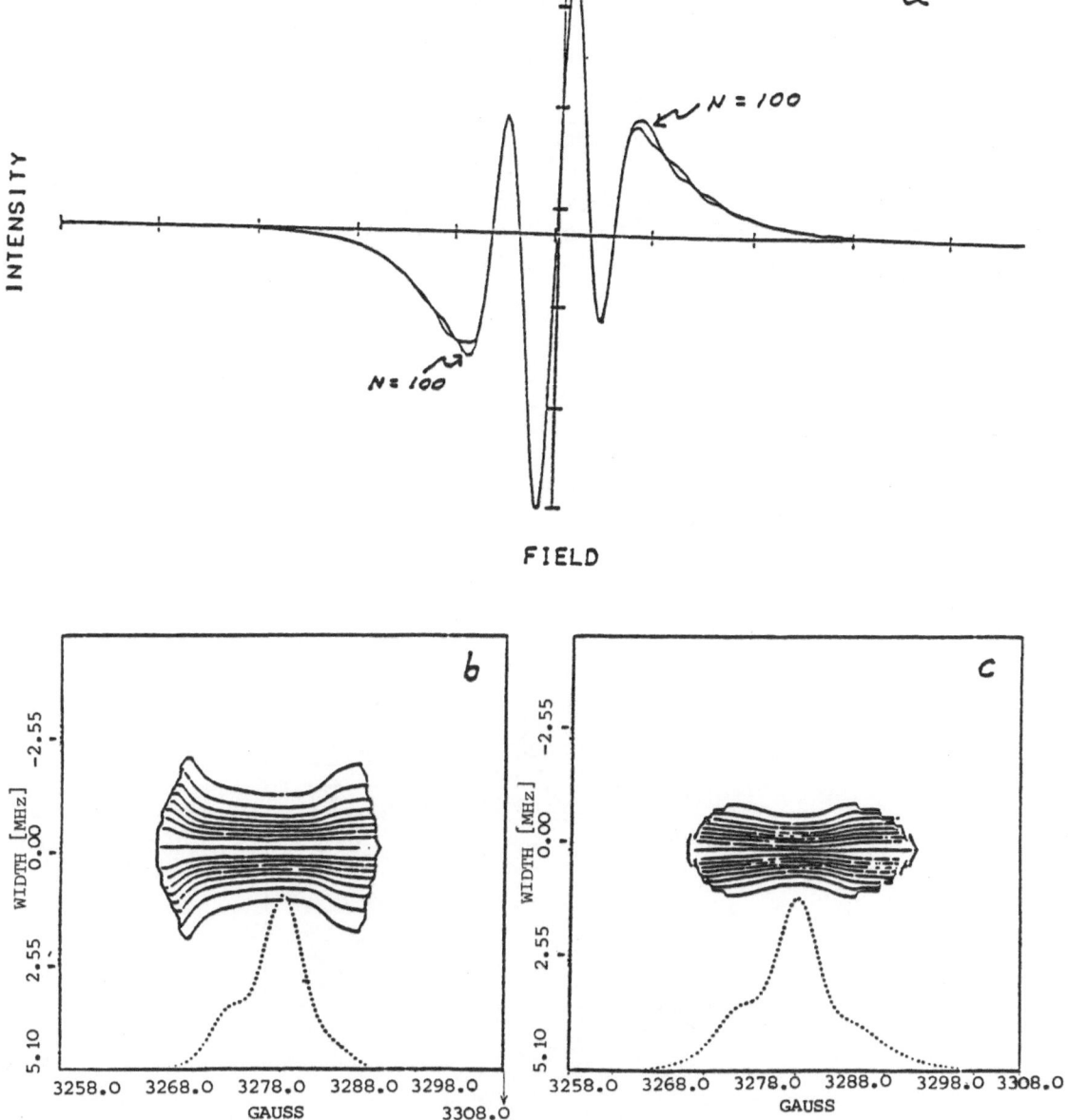

Fig. 25: A comparison of the relative sensitivity of cw vs. 2D ESE to motional anisotropy. (a) Two superimposed spectral simulations where one spectrum has $R_\perp = 10^4 s^{-1}$ and $N = 1$, and the other has the same $R_\perp$ but with $N = 100$. The markers on the x-axis are 9.77 Gauss apart. The normalized contours are simulated from the same parameters with (b) $N = 100$ and (c) $N = 1$. Case of high ordering, $S = 0.87$ and $\theta = 0^0$.

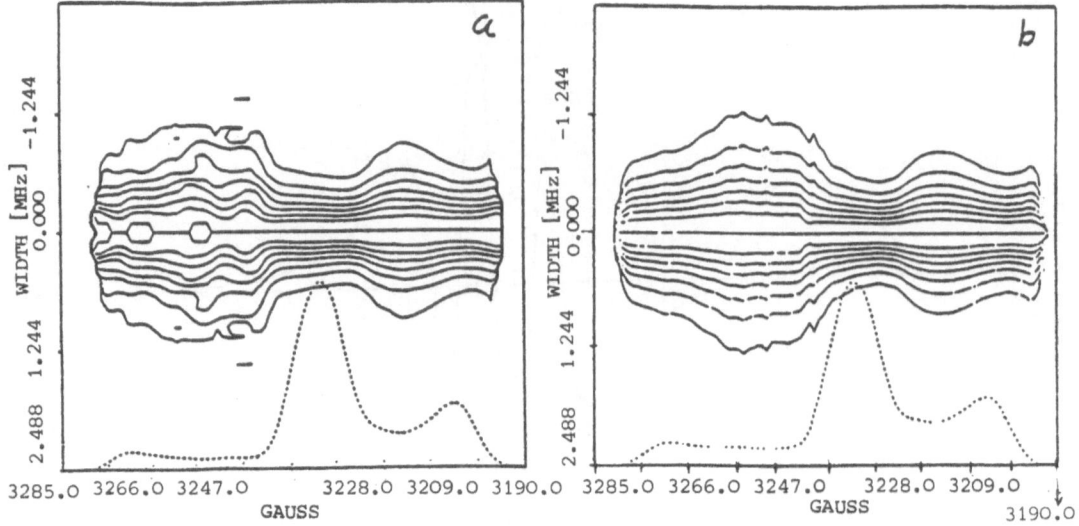

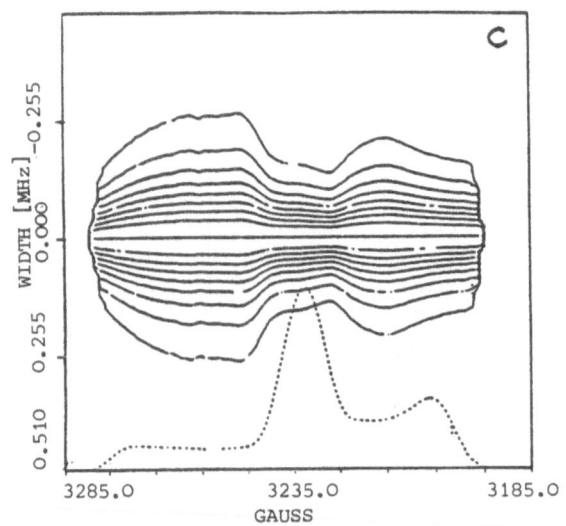

Fig. 26: Normalized contours showing the resolution enhancement obtained from the LPSVD treatment. (a) is from data of Tempone in 85% glycerol/H_2O at -75^{0}C treated by conventional FFT. (b) is from the same data set, but treated with LPSVD. (c) is a different data set that was collected from the same system in a manner that maximizes the efficiency of the LPSVD algorithm. (From Ref. [79].)

C. Newer Techniques

The 2D ESE technique has now been applied to give a map of the rate of magnetization transfer out of each spectral position [80,81]. It is produced by analogy to the "T_2 maps" described above, but is based on the stimulated echo, and it requires one to subtract out the simple T_1 decay. It can produce dramatic indication of the motional model without even requiring sophisticated analysis. We illustrate this in Fig. 27 for the case of physisorbed NO_2 (on a crushed vycor surface), since it dramatically demonstrates anisotropic reorientation (about an axis parallel to the line through the two oxygen atoms labelled as the y-axis. (The study of slow motions on surfaces by a variety of these new methods has recently been reviewed [82].)

Most recently Fourier-Transform (FT) ESR techniques have been developed [71-73], which hold the promise of revolutionizing the use of ESR to study motional dynamics [72,82]. It should permit one-to-two orders of magnitude reduction in experimental times for the 2D-ESE techniques described above, as well as new magnetization transer experiments, in which one could directly correlate the transition rate from one molecular orientation to another by an FT-2D exchange-transfer technique [72].

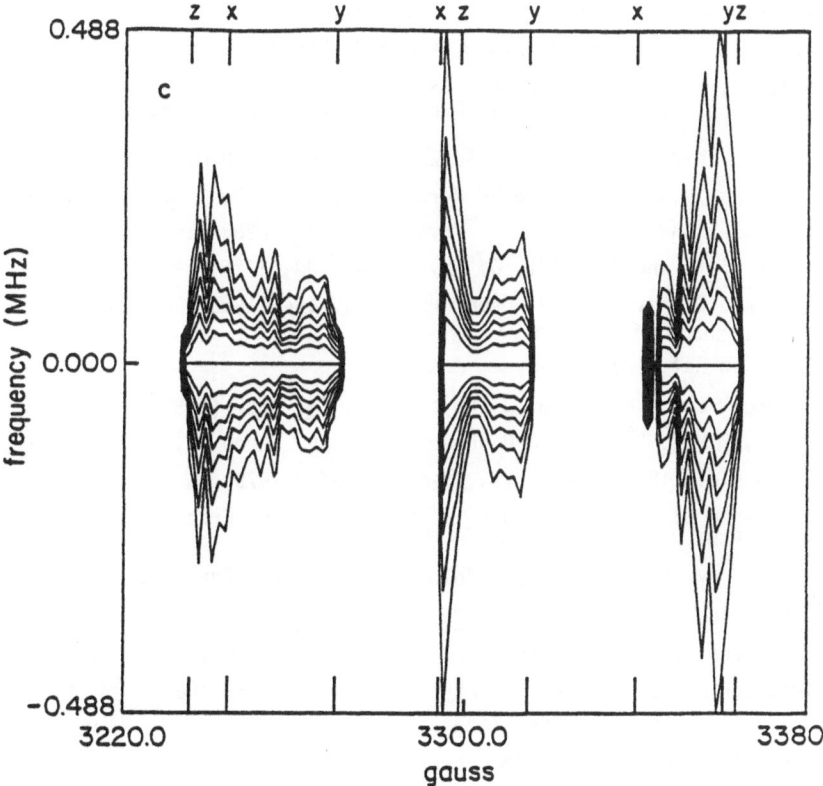

Fig. 27: 2D-ESE contours from the Stimulated Echo sequence for NO_2/Vycor at 35°K showing rates of magnetization transfer. The exponential decay term in T_1 has been subtracted out (from Ref. [80]).

References

[1] L.T. Muus and P.W. Atkins, Eds., Electron-Spin Relaxation in Liquids, Plenum, NY (1972)

[2] L.J. Berliner, Ed., Spin Labeling: Theory and Applications, Academic NY (1976)

[3] M. Dorio and J.H. Freed, Eds., Multiple Electron Resonance Spectroscopy, Plenum, NY (1979)

[4] L. Kevan and R.N. Schwartz, Eds., Time Domain Electron-Spin Resonance, Wiley, NY (1979)

[5] S.A. Zager and J.H. Freed, J. Chem. Phys. 77, 3360 (1982)

[6] D. Kivelson and P.A. Madden, Ann. Rev. Phys. Chem. 31, 523 (1980)

[7] J.L. Dote, D. Kivelson, and R.N. Schwartz, J. Phys. Chem. 85, 2169 (1981)

[8] W.A. Steele, Adv. Chem. Phys. 34, 1 (1976)

[9] G.R. Alms, D.R. Bauer, J.I. Brauman, and R. Pecora, J. Chem. Phys. 58, 5570 (1973); 59, 5310 (1973)

[10] D. Kivelson, M.G. Kivelson, and I. Oppenheim, J. Chem. Phys. 52, 1810 (1970); D. Kivelson, Chem. Soc. Faraday Symp. 11, 7 (1977)

[11] J.S. Hwang, D. Kivelson, and W.Z. Plachy, J. Chem. Phys. 58, 1753 (1973)

[12] D.J. Wilbur and J. Jonas, J. Chem. Phys. 62, 2800 (1975); M. Fury and J. Jonas, ibid. 65, 2206 (1976)

[13] S.A. Zager and J.H. Freed, J. Chem. Phys. 77, 3344 (1982)

[14] J.S. Hwang, R.P. Mason, L.P. Hwang, and J.H. Freed, J. Phys. Chem. 79, 489 (1975)

[15] S.A. Goldman, G.V. Bruno, C.F. Polnaszek, and J.H. Freed, J. Chem. Phys. 56, 712 (1972); 59, 3071 (1973); G.V. Bruno, Ph.D. Thesis, Cornell University (1972)

[16] M. Patron, D. Kivelson, and R.N. Schwartz, J. Phys. Chem. 86, 518 (1982)

[17] B. Blicharska, H.G. Hertz, and H. Versmold, J. Mag. Res. 33, 531 (1980)

[18] C.F. Polnaszek and J.H. Freed, J. Phys. Chem. 79, 2283 (1975)

[19] E. van der Drift and J. Smidt, J. Phys. Chem. 88, 2275 (1984)

[20] R. Wilson and D. Kivelson, J. Chem. Phys. 44, 154 (1966); B. Kowert and D. Kivelson, ibid. 64, 5206 (1976)

[21] J.H. Freed in Ref. [2], Ch. 3.

[22] R.F. Campbell and J.H. Freed, J. Phys. Chem. 84, 2668 (1980)

[23] J.S. Hwang, K.V.S. Rao, and J.H. Freed, J. Phys. Chem. 80, 1490 (1976)

[24] W.J. Lin and J.H. Freed, J. Phys. Chem. 83, 379 (1979)

[25] E. Meirovitch, D. Igner, E. Igner, G. Moro, and J.H. Freed, J. Chem. Phys. 75 3157 (1982)

[26] J.H. Freed, G.V. Bruno, and C.F. Polnaszek, J. Phys. Chem. 75, 3385 (1971)

[27] J.H. Freed, Ann. Rev. Phys. Chem. 23, 265 (1972)

[28] C.F. Polnaszek, G.V. Bruno, and J.H. Freed, J. Chem. Phys. 58, 3185 (1973)

[29] E. Meirovitch and J.H. Freed, J. Phys. Chem. 88, 4995 (1984)

[30] J.H. Freed, J. Chem. Phys. 66, 4183 (1977)

[31] S.A. Zager and J.H. Freed, Chem. Phys. Letts. 109, 270 (1984)

[32] G.J. Krüger, Phys. Rep. 82, 230 (1982)

[33] M.E. Moseley and A. Loewenstein, Mol. Cryst. Liq. Cryst. 90, 117 (1982); 95, 51 (1983)

[34] L.S. Selwyn, R.L. Vold, R.R. Vold, J. Chem. Phys. 80, 5418 (1984)

[35] G. Moro and P.L. Nordio, J. Phys. Chem. 89, 997 (1985)

[36] W.L. McMillan, Phys. Rev. A4, 1238 (1971); 6, 936 (1972)

[37] G. Moro and J.H. Freed, J. Chem. Phys. 74, 3757 (1981)

[38] G. Moro and J.H. Freed, in Large-Scale Eigenvalue Problems, Eds. J. Cullum and R. Willoughby, Vol. 127, Mathematics Studies Series (North-Holland, 1986), 143-160.

[39] G. Moro and J.H. Freed, J. Phys. Chem. 84, 2837 (1980)

[40] G. Moro and J.H. Freed, J. Chem. Phys. 75, 3175 (1981)

[41] G.H. Golub and C.F. Van Loan, Matrix Computations, The John Hopkins University Press, Baltimore, Maryland (1983)

[42] J.K. Cullum and R.A. Willoughby, Lanczos-Algorithms for Large Symmetric Eigenvalue Computations, Vol. I. Birkhauser (Boston, 1985)

[43] K.V. Vasavada, D.J. Schneider, and J.H. Freed, J. Chem. Phys. 86, 647 (1987).

[44] R.G. Gordon and T. Messenger in Ref. [2], Ch. XIII.

[45] A.E. Stillman and J.H. Freed, J. Chem. Phys. 72, 550 (1980)

[46] J.H. Freed in Stochastic Processes - Formalism and Applications, G.S. Agarwal Ed., Lecture Notes in Physics 184, 120 (Springer-Verlag, 1983)

[47] G.P. Zientara and J.H. Freed, J. Chem. Phys. 79, 3077 (1983)

[48] G.P. Zientara and J.H. Freed, Proceedings of the Ninth International Liq. Crystal Conference, Bangalore (1982); G.P. Zientara and J.H. Freed (in preparation)

[49] W.F. van Gunsteren and J.J.C. Berendsen, Mol. Phys. 34, 1311 (1977); P. Rychaert, G. Ciccotti, and J.H.C. Berendsen, J. Comp. Phys. 23, 327 (1977)

[50] G. Zannoni and M. Guerra, Mol. Phys. 44, 849 (1981)

[51] J.F.J. Ypma and G. Vertogen, J. Phys. 37, 557 (1976); Solid State Commun. 18. 475 (1976)

[52] A. Nayeem, Ph.D. Thesis, Cornell University (Oct. 1985)

[53] J.H. Freed (in preparation)

[54] A. Nayeem, V.S.S. Sastry, R. Shankar, and J.H. Freed (in preparation)

[55] K.V.S. Rao, J.S. Hwang, and J.H. Freed, Phys. Rev. Letts. 37, 515 (1976)

[56] a) R.J. Birgeneau, C.w. Garland, G.V. Kasting, and M. Ocko, Phys. Rev. A24, 2624 (1981)
      b) C.W. Garland et al., Phys. Rev. A27, 3234 (1983)

[57] P.G. de Gennes, Solid State Commun. 10, 753 (1972)

[58] F. Brochard, J. Phys. (Paris) 34, 411 (1973)

[59] F. Jähnig and F. Brochard, J. Phys. (Paris) 35, 301 (1974)

[60] A.R. Kortan, H. von Känel, R.J. Birgeneau, and J.D. Litster, J. Physique 45, 529 (1984)

[61] A. Nayeem, S. Rananavare, and J.H. Freed (in preparation)

[62] J.P. Hornak, J. Moscicki, D. Schneider, and J.H. Freed, J. Chem. Phys. 84, 3387 (1986)

[63] H. Tanaka and J.H. Freed, J. Phys. Chem. 88, 6633 (1984)

[64] O.H. Griffith and P. Jost, Ch. 12 and H.M. McConnell, Ch. 13 in Ref. [2].

[65] D. Chapman, J.C. Gomez-Fernandez, F.M. Goni, FEBS Lett. 98, 211 (1979) and references therein.

[66] H. Tanaka and J.H. Freed, J. Phys. Chem. 89, 350 (1985)

[67] F. Jähnig, Mol. Cryst. Liq. Cryst. 61, 157 (1981)

[68] W.B. Mims and J. Peisach in Biological Magnetic Resonance, Vol. 3, Eds. L.J. Berliner and J. Reuben, Plenum, NY (1981), Ch. 5 and references therein

[69] J.F. Norris, M.D. Thurnauer, and M.K. Bowman, Adv. Biol. Med. Phys. 17, 365 (1980)

[70] A.E. Stillman and R.N. Schwartz, J. Phys. Chem. 85, 3031 (1981), and references therein

[71] J.P. Hornak and J.H. Freed, J. Mag. Res. 67, 501 (1986)

[72] J. Gorcester and J.H. Freed, J. Chem. Phys. 85, 5375 (1986).

[73] M. Bowman, Bull. Am. Phys. Soc. 31, 524 (1986)

[74] A.E. Stillman, L.J. Schwartz, and J.H. Freed, J. Chem. Phys. 73, 3502 (1980); 76, 5658 (1982)

[75] L.J. Schwartz, A.E. Stillman, and J.H. Freed, J. Chem. Phys. 77, 5410 (1982)

[76] a) G.L. Millhauser and J.H. Freed, J. Chem. Phys. 81, 37 (1984)

      b) L. Kar, G.L. Millhauser and J.H. Freed, J. Phys. Chem. 88, 3951 (1984)

[77] R. Kumaresan and D.W. Tufts, IEEE Trans. ASSP-30 833 (1982)

[78] H. Barkhuijsen, R. de Beer, W.M.M.J. Bovée and D. van Ormondt, J. Mag. Res. 61, 465 (1985); D. van Ormondt (private communication).

[79] G.L. Millhauser and J.H. Freed, J. Chem. Phys. 85, 63 (1986)

[80] L.J. Schwartz, G.L. Millhauser and J.H. Freed, Chem. Phys. Letts. 127, 60 (1986)

[81] G.L. Millhauser, Ph.D. Thesis, Cornell University (1986)

[82] G.L. Millhauser, J. Gorcester, and J.H. Freed in Electron Magnetic Resonance of the Solid State, J.A. Weil, Ed. (Can. Chem. Soc. Publication, in press).

[83] D.J. Schneider and J.H. Freed in Lasers, Molecules, Methods, J. Hirschfelder, R. Wyatt, R. Coalson, Eds. (Adv. Chem. Phys., Wiley NY, in press).

COMPUTER SIMULATION OF PRETRANSITIONAL PHENOMENA IN HARD-CORE MODELS FOR LIQUID CRYSTALS[*]

D. Frenkel

FOM Institute For Atomic and
Molecular Physics
Amsterdam
The Netherlands

I. Introduction

Van der Waals [1] was the first to suggest that the local structure of dense simple liquids is largely determined by short range repulsive forces. The transformation of this idea from a bold assumption to an almost commonplace statement owes much to the fruitful combination of computer simulations and thermodynamic perturbation theory [2,3,4]. For example, the freezing of atomic liquids can be understood on basis of the fluid-solid transition of hard spheres, first observed by Alder and Wainwright [5]. The liquid-vapor transition does not occur in a pure hard-core system, but can be explained in terms of a weak attractive perturbation added to the repulsive interactions.

It is much less clear to what extent excluded volume effects can help us to understand the thermodynamics and phase transitions of 'complex' liquids, in particular liquid-crystal forming materials. Actually, the question to be addressed is two-fold. Firstly: is it at all possible to explain a particular type of phase transition in a liquid crystal in terms of excluded volume effects alone, or is the presence of other interactions, such as attractive dispersion forces essential? And secondly: even if we find that hard repulsive forces can cause a particular phase transition in a model system, do we have reasons to assume that excluded volume effects are also at the root of such phase transitions in real liquid crystals? To illustrate the latter point, consider the following example: gravitational attraction <u>could</u> cause a liquid-vapor transition in an atomic fluid. Yet, it is clear that gravity is not <u>the</u> explanation for this phenomenon. By analogy, it is conceivable that hard-core repulsions could drive a transition from a nematic to, say, a columnar phase (incidentally, in a simple model system they can), but this does not necessarily imply that repulsive forces are also the dominant factor determining the structure of real discotic liquid crystals.

This paper focuses on the first question which I shall phrase here as: Can hard-

---

[*]Part of the material reviewed in this article appeared in the January 1987 issue of Molecular Physics.

core models exhibit the phase transitions and pre-transitional behavior found in liquid crystals? But I stress that a positive answer to this first question only implies that we now have to face the second, i.e.: Do hard-core models contain the essential physics? At present the answer to this second question is open.

Below I shall present the results of recent computer simulations on non-spherical hard-core models. As we shall see, a surprising number of phenomena observed in real liquid-crystal forming materials are reproduced by these simple models. The remainder of this paper is organized as follows: first I shall discuss the effect of molecular shape on the relative stability of crystalline and liquid crystalline phases. Next I shall present some preliminary results on pretransitional fluctuations in nematogens, in particular collective orientational fluctuations. And finally I shall present some recent results on smectic ordering in hard-core model systems.

However, before continuing I should warn the reader. The discussion in this paper is limited to hard-core models. This bias is intentional. I wish to explore the limits of applicability of the hard-particle concept. But I wish to stress that there exists an extensive literature dealing with simulations of other model systems, such as for instance the 'Lebwohl-Lasher' model which has been studied in detail by Luckhurst and collaborators [6] or the simulation of a realistic model for an existing nematogen (see Ref. [7]). Liquid crystals are complex systems, and only by combining the results obtained by experiment, theory and different computer simulation techniques can we hope to increase our understanding of these fascinating materials.

II. Solid or Nematic?

Most isotropic molecular liquids freeze as the temperature is lowered. But only some form liquid crystals. The obvious question is: what determines whether a molecular liquid forms a liquid crystal before it freezes?

Let us first briefly summarize the essential characteristics of liquid crystals (for details the reader is referred to Ref. [8]). Liquid crystals are partially ordered fluids. The simplest is the nematic phase in which the molecules are translationally disordered but, unlike the isotropic fluid, the molecular orientations are no longer distributed randomly. Rather, the molecular axes tend to align parallel to some, otherwise arbitrary, direction. This alignment axis is usually referred to as the nematic director $\vec{n}$. A quantitative measure for the degree of alignment is the nematic order parameter S. For axially symmetric molecules, S is defined as $<P_2(\cos\theta)>$, where θ is the angle between the molecular symmetry axis and the nematic director. In the isotropic phase, $S = 0$. In a perfectly aligned nematic $S = 1$. The phase transition from isotropic liquid to nematic liquid crystal is weakly first order. By 'weakly first order' I mean that the transition, although first order, is preceded by strong precursor effects as is usually the case in the vicinity of continuous phase transitions. The transition from both the nematic and the isotropic phases to the solid is a normal first order transition.

All molecules that form nematics are non-spherical, but the converse is not true. In other words, a certain degree of non-sphericity is a necessary, but not a sufficient condition for liquid crystal formation. Several other factors also play a role in determining the relative stability of the nematic phase. For instance, attractive intermolecular interactions due to dispersion forces also tend to favor parallel alignment of the molecules. There are a number of models that attribute the orientational ordering in liquid crystals to the effect of dispersion forces [9.10], including the original mean-field model due to Maier and Saupe. Although these models yield fair estimates for a number of properties of nematic liquid crystals, they cannot be used to explain the stability of the nematic phase with respect to the solid. It is quite conceivable that in some cases dispersion forces would stabilize the solid more than the nematic. Another factor that influences the stability of liquid crystals is the presence of flexible tails (often alkane or alkoxy chains) connected to the 'rigid' core of the nematogen. That flexible chains are indeed important for the stability of liquid crystals follows from the experimental observation that essentially all liquid crystal forming molecules ('mesogens) have them [11]. If the tail is shortened too much, the liquid freezes without ever becoming nematic. Intuitively, the role of flexible tails in stabilizing the liquid crystal with respect to the solid is clear: if a flexible tail has to fit into a crystal lattice, it will loose much much of its conformational degrees of freedom with a concomitant loss in entropy. A model that takes into account the effect of the flexible tails on the relative stability of isotropic, nematic and smectic phases was developed by Dowell and Martire [12]. But again, to my knowledge, a quantitative comparison with the solid has not been made.

The question which remains is: which of the factors mentioned above has the most important effect? For someone who is familiar with the theory of simple liquids it is natural to assume that the hard-core repulsions are most important. Of course, the 'natural' assumption is not necessarily the correct one, and the only way to find out is to try. Trying in this case means computer simulation. Still I would like to present some justifications for my assumption that molecular shape effects are of primary importance while dispersion forces and flexibility effects can be considered as 'perturbations'. Onsager [13] has shown that the isotropic-nematic transition can take place in a simple model that considers only the non-spherical excluded volume effects. In contrast, flexible tails alone cannot cause nematic order. Neither can anisotropic dispersion forces. To be more precise: a model which only includes dispersion forces but no hard-core repulsion will collapse. Of course, if one adds a spherical hard-core repulsion, anisotropic dispersion forces can explain the transition to a nematic phase, but in nature a molecule with a spherical hard core will not have anisotropic van der Waals interactions. Hence, in the real world, anisotropy in the dispersion forces implies a non-spherical hard core anyway. Of course, the converse is also true but, at least for sufficiently non-spherical particles, dispersion forces appear relatively unimportant. In summary, non-spherical hard-core repulsion is the most plausible mechanism to explain the formation of a nematic phase without the help of any of the other factors. It seems reasonable then, to treat hard-core repulsion as the 'primary' effect. One may hope that some kind of the thermodynamic perturbation theory can, at a later stage, take the effect of dispersion and flexibility into account. However, I should stress that at this stage we do not yet have any evidence that such a perturbation scheme will work.

Before we turn to the computer simulation let me briefly sketch what we know from experiment about the relation between molecular shape and the tendency to form liquid crystals [14]. First of all, we know that liquid crystals can consist of both rodlike and platelike molecules. Most rodlike mesogens have length-to-breadth ratios between 3 and 4 (it should be noted, however, that this ratio is a rather fuzzy concept for molecules with flexible tails). What we know at present about disclike mesogens (see e.g. 15) suggests that these molecules are even less spherical. It is not straightforward to tell from experiment how the isotropic-nematic and the nematic-solid transition depend on the shape of the repulsive core of the molecules. The problem is that although a wealth of data on the phase transitions of molecules with different shapes are available, we do not know how to disentangle the effects of molecular shape, dispersion forces and flexible tails.

Next, let us consider what we know from theory and earlier computer simulations. For hard spheres we do know the melting point from the computer simulations of Hoover and Ree [16]. But these simulations do not tell us what will happen to the melting point as the molecular shape changes from spherical to rodlike or platelike. The Onsager theory [13] predicts that thin spherocylinders with length L and diameter D undergo a transition from the isotropic to the nematic phase at a density of order $1/(L^2 D)$. At this density the fraction of the volume occupied by the spherocylinders is still vanishingly small (order D/L). The same holds for infinitely thin hard platelets [17]. For neither of these model systems do we know the melting point.

III. Computer Simulations of Nematogens

III.1 General Remarks

To fill this gap in our knowledge about the phase behavior of non-spherical hard-core molecules, we performed Monte Carlo simulations on a simple model system, viz. hard ellipsoids of revolution. The shape of such a spheroid is characterized by a single parameter, x, the ratio of the length of the major axis ($2a$) to the minor axis ($2b$): $x = a/b$. Special cases are: $x = 1$ (hard spheres), $x \to \infty$ (hard needles [13]) and $x \to 0$ (hard platelets [17]). Technical details of the Monte Carlo simulations have been published elsewhere [18]. Here I shall concentrate on the results. Nevertheless, in order to facilitate the 'reading' of the figures, I must explain the reduced units in which the results are expressed. All static properties of a (classical) many body

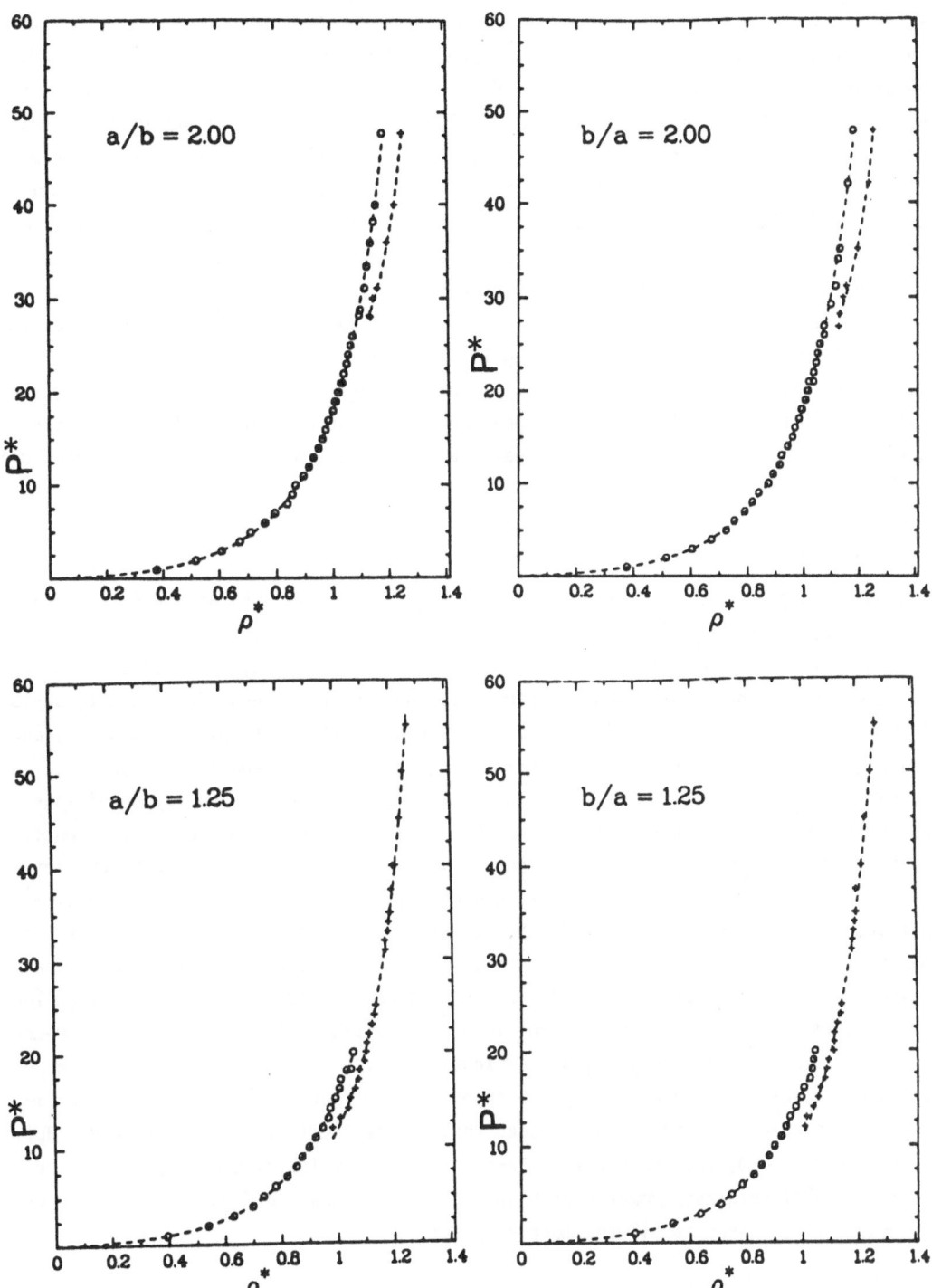

Fig. 1: Equation of state of hard ellipsoids of revolution with length-to-breadth
ratios x = 2, 1,25, 08 and 0.5. The pressure is in units kT/8ab², the density in
units (8ab²)⁻¹. Open circles: (isotropic) fluid branch, pluses: solid branch.

system can be expressed in units which are derived from two basic units: the unit of length and the unit of energy. For time dependent properties, we need one more fundamental unit: usually, the unit of mass. For a hard-core system, the only 'natural' energy scale is the thermal energy kT. Hence, we shall express all energies in units kT. For the unit of length we clearly need a quantity related to the size of the molecules. For ellipsoids it is convenient to use a unit of volume that is proportional to the volume of the particles ($v_0 = (4\pi/3)ab^2$). We use as unit of volume the quantity $8ab^2$. The reason for this choice is that it facilitates comparison of the data for spheroids with the known hard-sphere results. For hard spheres, our unit of volume reduces to σ^3, where σ is the hard sphere diameter. One important consequence of this choice of units is that the reduced density of close regular packing for all spheroids equals $\sqrt{2}$. The obvious choice for the unit of mass is the mass of one molecule. In addition, the dynamics of non-spherical molecules depend on the moment of inertia. In the example to be discussed below, we have assumed that the mass of the molecules is distributed uniformly over their proper volume.

III.2 Equation of State

With this background we can consider the Monte Carlo results. We carried out constant pressure Monte Carlo simulations for hard spheroids with 8 different length-to-breadth ratios x in the range between $1/3$ (oblate) and 3 (prolate). Typical examples of the resulting isotherms are shown in Figure 1. The isotherms shown in Figure 1 consist of two branches: a low density branch corresponding to the isotropic fluid phase, and a high density solid branch. Direct coexistence between solid and liquid cannot be observed in these small systems ($O(10^2)$ particles) with periodic boundary conditions. In Figure 1 no nematic branch is observed because the molecules are not sufficiently anisometric. What is striking about Figure 1 is the strong resemblance between isotherms for particles with reciprocal length-to-breadth ratios. Actually, we find such almost symmetric behavior for all spheroids that we studied. A more quantitative measure of this symmetry is shown in Figure 2. In this figure the percentual difference of the pressure of oblate and prolate ellipsoids is plotted as a function of density. At low densities this difference goes to zero. This is understandable because, in the present units, the second virial coefficients of spheroids with inverse length-to-breadth ratios are equal [19]. However, no such symmetry holds for the higher virial coefficients. In particular, Onsager has argued that in the limit $x \rightarrow \infty$, $B_3/B_2^2 \rightarrow 0$. But in the limit $x \rightarrow 0$, this same ratio tends to a finite value of 0.4447 [17]. The approximate oblate-prolate symmetry at higher densities shown in Figure 2 is therefore not exact, which makes it all the more surprising.

III.3 Orientational Order

Let us next look for orientationally ordered phases. Figure 3 shows the equation of states of spheroids with $a/b > 2.75$ and $a/b < (1/2.75)$ at densities around the

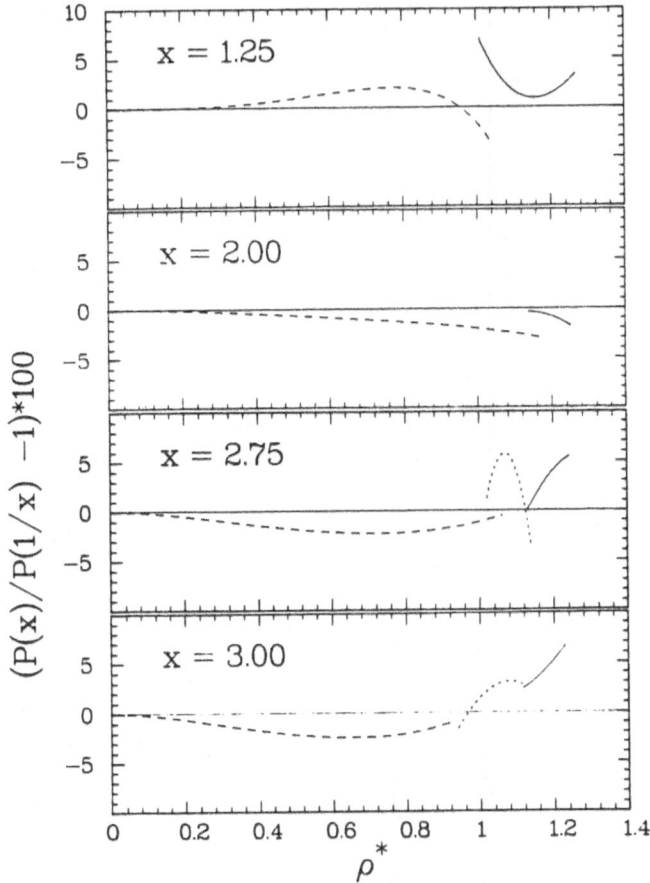

Fig. 2: Deviations from perfect symmetry of the equation of state of hard ellipsoids of revolution with inverse length-to-breadth ratios. The ordinate shows the percentual difference between the pressure of a fluid of hard ellipsoids with a/b = x and b/a = x, both at density ρ. Dashed curve: isotropic branch, dotted curve: nematic branch, drawn curve: solid branch. The curves shown in this figure were obtained from fits to the equation of state date (see Ref. [18]).

melting point. Between the solid and the isotropic fluid branch we observe a third branch which corresponds, as we shall show, to a nematic liquid crystal. There are several methods to detect a nematic phase in a computer simulation. For instance, one may compute the eigenvalues of the collective molecular orientation tensor $\overleftrightarrow{Q}$:

$$Q_{\alpha\beta} = (1/N \Sigma \{3u_\alpha u_\beta - \delta_{\alpha\beta} \}/2. \tag{1}$$

In the isotropic phase, all eigenvalues vanish in the thermodynamic limit. In the nematic phase, the largest eigenvalue equals the nematic order parameter S. This method, although straightforward in principle, requires some care when applied to finite systems. In particular, the largest eigenvalue of $\overleftrightarrow{Q}$ is always positive. In the isotropic phase it is of order $1/N$, where $\sqrt{N}$ is the number of particles in the system. This implies that, for instance, in a system consisting of 100 particles, this 'order parameter' is never less than 0.1. Hence it is not a particularly sensitive probe to monitor the onset of orientational order. A less common but more useful measure for the degree of orientational order is -2 times the smallest (in absolute value)

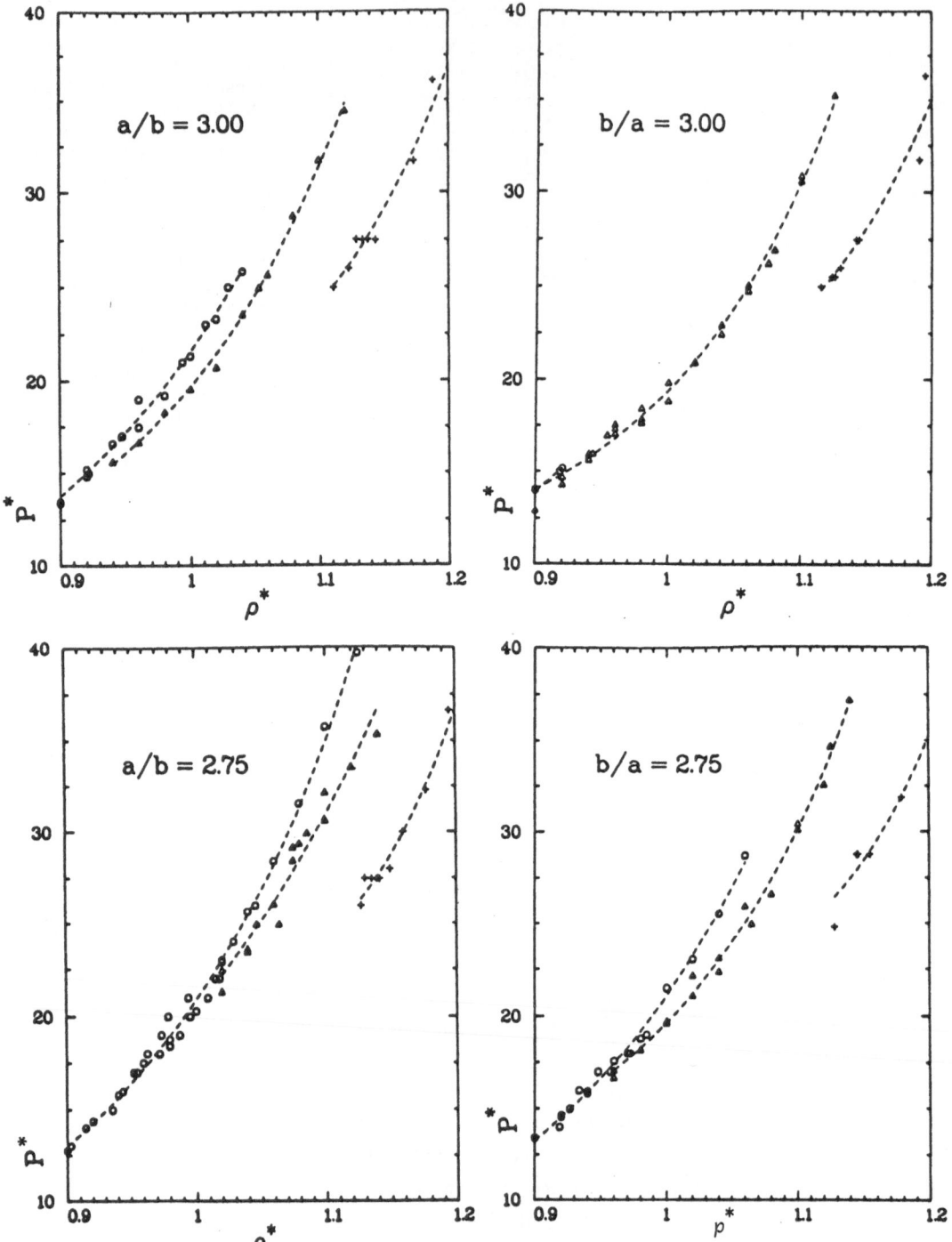

Fig. 3: Equation of state of hard ellipsoids of revolution with length-to-breadth ratios 3, 2.75, 1/2.75 and 1/3, in the density range where a nematic branch is observed. Open circles: (isotropic) fluid branch, triangles: nematic branch, pluses: solid branch. The units are as in Figure 1.

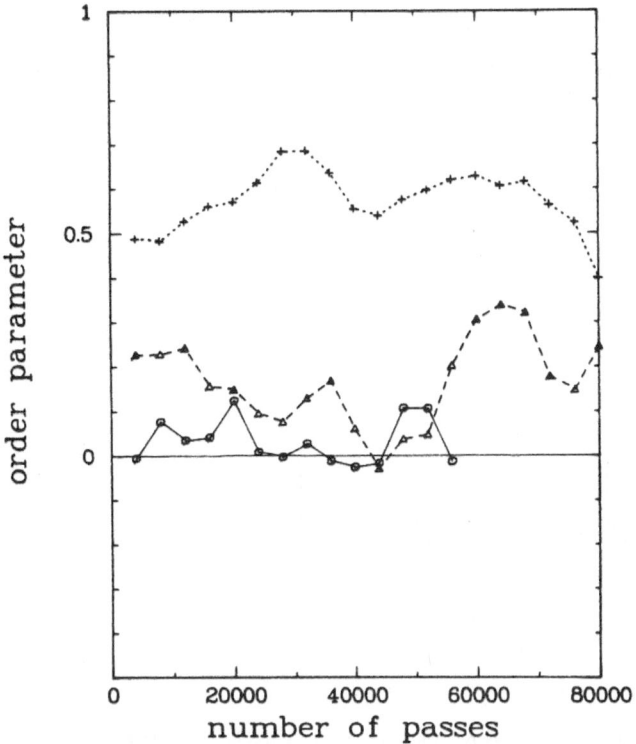

Fig. 4: Slow fluctuations in the nematic order parameter during a Monte Carlo simula-tion of hard ellipsoids of revolution. Pluses: b/a = 3, ρ/ρ_0 = 0.69 (nematic). Open tri-angles: b/a = 3, ρ/ρ_0 = 0.65 (overexpanded nematic). Open squares: a/b = 3, ρ/ρ_0 = 0.636 (isotropic). One Monte Carlo pass corresponds to one trial move for every particle.

eigenvalue of $\overset{\Rightarrow}{Q}$ [17]. This quantity has the desirable property that it is of order 1/N (rather than 1/√N̄) in the isotropic phase and, moreover, that it fluctuates around zero. In a (uniaxial) nematic phase, the latter definition of the order parameter is only equivalent to the previous one in the thermodynamic limit. Figure 4 shows that typical fluctuations in this order parameter in the vicinity of the isotropic-nematic transition in a fluid of hard ellipsoids.

An equally simple method to detect the onset of orientational order is to monitor the behavior of the pair distribution function, in particular the part that depends on the relative orientation of molecules at a distance r. A useful measure of the degree of orientational order is given by $g_2(r)$, defined as the average value of $P_2(\cos\theta)$ for molecules with a center of mass distance r; θ is the angle between the molecular axes. In the isotropic phase $g_2(r)$ decays to zero within a few molecular diameters. In contrast, in the nematic phase $g_2(r)$ becomes long ranged. As $r \to \infty$, $g_2(r) \to S^2$, where S is the nematic order parameter. Of course, in a computer simulation we can never measure correlations at distances greater than half the periodic box diameter, L. But, as Figure 5 shows, even the behaviour of g_2 at r = L/2 is a useful indicator of the onset of orientational order. Figure 5 illustrates another important feature of

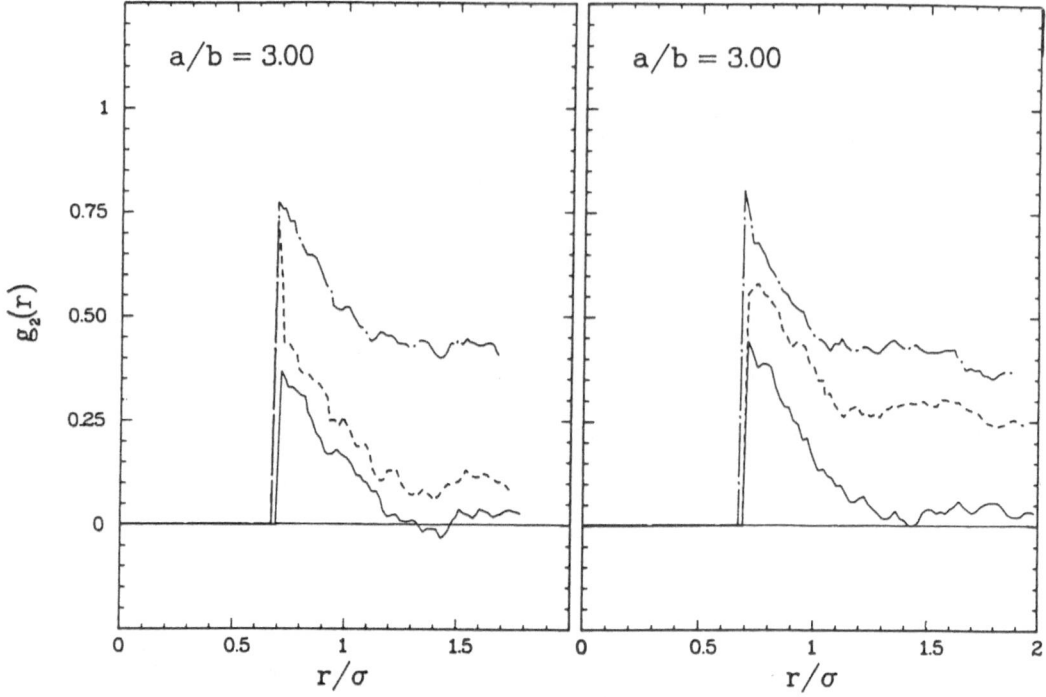

<u>Fig. 5</u>: Example of the behavior of the orientational correlation function $g_2(r)$ = $<P_2(\vec{u}(0)\cdot\vec{u}(r))>$ of hard ellipsoids of revolution with a/b = 3 in the vicinity of the I-N transition. Densities: ρ/ρ_0 = 0.65 (drawn curve), ρ/ρ_0 = 0.707 (dashed curve), and ρ/ρ_0 = 778 (dash-dot). The left-hand side shows the behavior of $g_2(r)$ on compression. The right-hand side shows the corresponding results for expansion.

the isotropic-nematic transition in a system of prolate hard ellipsoids, namely the fact that it exhibits hysteresis. Hysteresis is usually observed at first order phase transitions. However, the isotropic-nematic transition is only weakly first order and, as a consequence, hysteresis may be suppressed by fluctuations. For oblate ellipsoids, hysteresis appears to be weaker than for prolate ellipsoids. Why this should be so is not clear at present.

III.4 Free Energies and Phase Transitions

In order to map the phase behaviour of the hard ellipsoids as a function of their anisometry, the location of all thermodynamic phase transitions must be determined. Without this information we cannot tell whether the nematic branch in Figure 3 corresponds to a thermodynamically stable phase, and if so, over what range of densities. In computer simulations of other model systems, similar questions arise every time a first order phase transition is observed. It is clearly important to establish whether a new phase is thermodynamically stabe or not. Unfortunately, the computational effort involved in locating the coexistence points for first order phase transitions of molecular systems is appreciable. Moreover, the computational techniques that must be used are not widely known. This warrants a brief 'technical' digression on the numerical location of phase coexistence points.

Let us recall that the condition which must be satisfied at a phase transition is that the temperature T, the pressure P and the chemical potential μ of the coexisting phases are equal. From the Monte Carlo simulations described above we know the pressure as a function of density for all phases (for hard-core systems the temperature plays no role; we choose kT as our unit of energy). However, the chemical potential μ cannot be derived directly from the equation of state data. The usual method to determine the chemical potential in real experiments is by thermodynamic integration along a reversible path from a reference state of known chemical potential to the state point under consideration. This approach is also applicable to computer simulations. For the isotropic fluid phase the reference state is the ideal gas. Using:

$$P = -(\partial F/\partial V)_T , \qquad (2)$$

where F is the Helmholtz free energy of the system, we can compute F at a density ρ by integration:

$$F(\rho,T) = F_{id}(\rho,T) + kT \int_0^\rho d\rho(P/\rho kT - 1)/\rho. \qquad (3)$$

For a one-component system, μ follows directly from F because $\mu = F + PV$. In principle, the applicability of the thermodynamic integration method sketched above is not limited to the computation of the chemical potential of the isotropic fluid phase. All that is needed is the existence of a reversible path from the ideal gas phase to the phase of interest. In the real world it is perfectly admissible for such a path to cross first order phase transitions, as long as the phase transformation can be carried out reversibly. The problem is that this latter condition is rarely satisfied in computer simulations of first order phase transitions. For the system sizes typically studied by computer simulation, first order phase transitions exhibit strong hysteresis. As a consequence, the transition from one phase to the other usually requires appreciable over-compression/overexpansion and then proceeds irreversibly if at all. Fortunately, in computer simulation one is not restricted to thermodynamic integration along 'natural' paths. Apart from changing the density or temperature one may modify the Hamiltonian H of the system through application of artificial external fields or through gradual modification of the intermolecular interactions between the particles. The idea is to modify the Hamiltonian in such a way that the resulting model system is sufficiently simple that its free energy can be computed analytically. Let us call this reference free energy F_{ref}. By gradually changing the Hamiltonian back to its original form, we have constructed a path that joins the phase of interest to a state of known free energy. Of course, such a path is not necessarily reversible. However, by a judicious choice of the parameters that control the variation of the artificial Hamiltonian one can often create reversible paths through 'parameter space'. To give an example, the free energy of a crystal of hard spheres can be computed by constructing a reversible path from the hard-sphere crystal to an Einstein crystal of the same structure [20]. In this case, the modification of the Hamiltonian is simply that in addition to the intermolecular interactions all atoms are bound to their lattice sites $(\vec{r}_i^0)$ by harmonic springs of strength λK

$$H(\lambda) = H_0 + \lambda K \sum_i (\vec{r}_i - \vec{r}_i^0)^2 . \tag{4}$$

For $\lambda = 0$, the Hamiltonian corresponds to that of the normal hard-sphere system, while for $\lambda = 1$ the system behaves as an (almost) ideal Einstein crystal. The derivative of the free energy with respect to λ equals:

$$dF/d\lambda = \langle dH(\lambda)/d\lambda \rangle = K \langle \sum_{i=1}^{N} (\vec{r}_i - \vec{r}_i^0)^2 \rangle . \tag{5}$$

As the last term on the right hand side of the above equation can be evaluated by Monte Carlo simulation, the free energy difference

$$F(\lambda = 1) - F(\lambda = 0) = \int_0^1 dF/d\lambda \ d\lambda \tag{6}$$

can be computed to any desired degree of accuracy by computing $\langle \sum_i (\vec{r}_i - \vec{r}_i^0)^2 \rangle$ at a sufficiently large number of intermediate values of the coupling parameter λ.

In order to find the phase transitions for a system of hard ellipsoids, free energy calculations of the type described above must be carried out both for the solid and the nematic phase. In the case of the solid, the reference system is an Einstein crystal with the same structure as the original solid. In this Einstein crystal, harmonic restoring forces reduce the fluctuations of both the translational and the orientational degrees of freedom around their equilibrium values. The situation is slightly more complex for the nematic phase, because in this case the reference state clearly cannot be a crystal. One possible approach is described in Ref. [18]. The reference state is an ideal gas in a strong aligning field. The nematic phase is reached in two stages. First the aligned gas is compressed to a density at which the system without field is in a nematic phase, and then the aligning field is slowly switched off. This approach makes use of the fact that the first-order isotropic-nematic transition is suppressed in a sufficiently strong field, in much the same way that the liquid-vapour transition is suppressed along a supercritical isotherm. For this reason the path from the reference state (the dilute gas in a field) to the nematic phase, is free of phase transitions.

III.5 Phase Diagram

Using the methods described above, the free energy of all phases of the system of hard ellipsoids was computed and the phase transitions located. The resulting 'phase diagram' is shown in Figure 6. In this figure the molecular shape is varied from left to right from extremely oblate ellipsoids ('platelets'), through spheres to extremely prolate

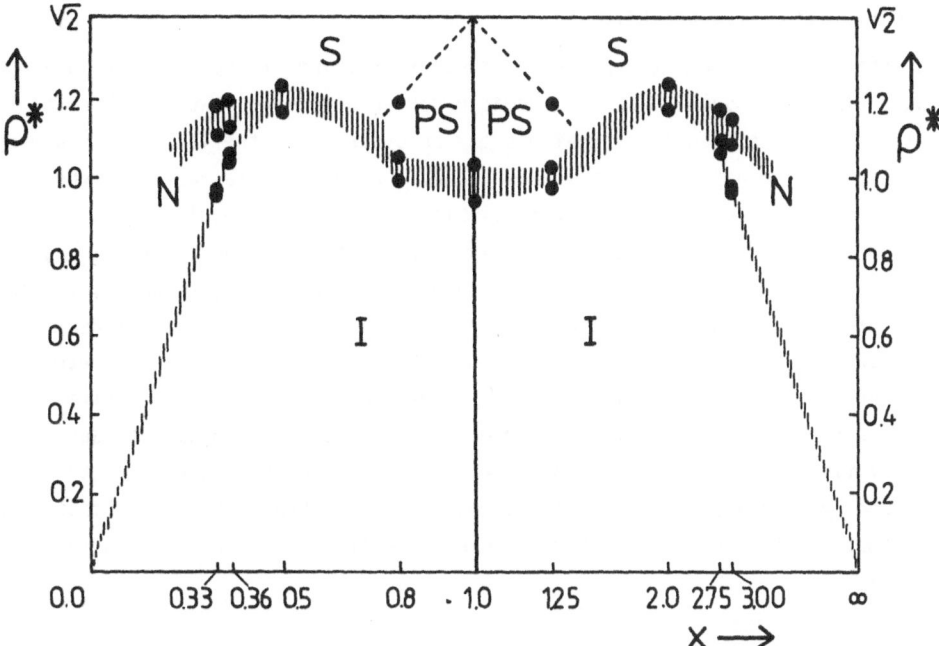

Fig. 6: 'Phase diagram' of hard ellipsoids of revolution. Vertical axis: density in units $(8ab^2)^{-1}$. Horizontal axis: Length-to-breadth ratio, x. The shaded areas correspond to two-phase regions. The dots are the computed coexistence points. The points for x = 1 were taken from Ref. [16]. The following phases can be distinguised: isotropic fluid (I), nematic liquid crystal (N), orientationally ordered solid (S) and plastic solid (PS).

ellipsoids ('needles'). The ordinate measures the density which, in the present units, varies between 0 and $\sqrt{2}$ (regular close packing). Four distinct phases can be identified, the isotropic fluid (I), nematic fluid (N), orientationally ordered solid (S) and orientationally disordered (plastic) solid (PS). The shaded areas represent two-phase coexistence regions. Several aspects of Figure 6 are worth noting. First of all the overall phase diagram has a high degree of symmetry under the interchange of oblate and prolate ellipsoids. As in Figure 2 above, this symmetry is not perfect. Still, the physical reason for this near symmetry is not understood at present. Figure 6 also provides us with an answer to the old question: what is the minimum non-sphericity needed to form a stable nematic phase? Clearly, for ellipsoids with axial ratios approximately between 1/2.5 and 2.5, no stable nematic phase is possible. As a bonus, we obtain an estimate of the range of stability of the orientationally disordered solid phase (approximately between a/b = 1/1.5 and a/b = 1.5) although in this case the boundaries are less well-determined.

Now that we know the range of stability of the nematic phase in this model system, let us look in more detail at the physical consequences of the onset of orientational ordering. In the isotropic phase, the presence of a nematic is betrayed by an increase in the amplitude of collective orientational fluctuations. Below we shall illustrate what computer simulation can tell us about such pretransitional phenomena.

III.6 Orientational Precursor Effects

At the transition to the nematic phase spontaneous ordering of the molecular orientations takes place. Although this macroscopic ordering occurs at a well-defined state point, strong pretransitional fluctuations may be observed in the isotropic phase as the transition to the nematic phase is approached. Actually, the fact that such pretransitional fluctuations can be observed at all is due to the peculiar nature of the isotropic-nematic transition. After all, this is a first-order phase transition, and usually there are no precursor effects near such phase transitions (such as, for instance, the freezing transition). The isotropic-nematic transition is different because it is a 'weak' first-order transition; it exhibits most of the 'critical' pretransitional behaviour typical for a continuous phase transition, but before the 'critical' point is reached, a first-order transition intervenes. The classical language to describe pretransitional fluctuations in the vicinity of an (almost)-second-order phase transition was developed by Landau [21]. The basic assumption in the Landau theory of phase transitions is that the free energy density F can be written as a power series in the order parameter S:

$$F(S) = F_0 + AS^2 + BS^3 + CS^4 + \dots , \qquad (7)$$

where A, B, C, etc. are coefficients that depend on temperature and/or density. For a second-order phase transition ($B = 0$), the coefficient A changes sign at the transition density ρ^*. In the vicinity of the transition A is assumed to be a linear function of $(\rho^* - \rho)$:

$$A(\rho) = a \, (\rho^* - \rho). \qquad (8)$$

For a weakly first-order phase transition ($B \neq 0$), A still goes to zero, at a density ρ^* which is, however, slightly higher than the actual transition point ρ_{IN}. The most likely value of the order parameter is found by minimizing $F(S)$ with respect to S. Depending on the values of A, B and C this minimum may be situated at $S = 0$ (isotropic phase) or at $S > 0$ (nematic phase). For $B^2 = 4AC$, $F(S)$ has two equivalent minima, one at $S = 0$, the other at $S = -B/2A$. This is the point where isotropic and nematic phases are in equilibrium. In a Monte Carlo simulation it is possible to 'measure' $F(S)$ directly, because $P(S)$, the probability density for observing a particular value of the order parameter S is proportional to $\exp(-\beta F(S))$. $P(S)$ is obtained by collecting a histogram of the fluctuations in S during a Monte Carlo run. $F(S)$ then follows from $\beta F(S) = \text{constant} - \ln(P(S))$. Figure 7 shows an example of the behaviour of $F(S)$ as the isotropic-nematic transition is crossed. Note that information about large values of $F(S)$ resides in the wings of $P(S)$ where the statistics tend to be poor. As a consequence, only the lowest few coefficients of the Landau-de-Gennes expansion can be determined by Monte Carlo simulation. In order for the Landau expansion to be useful as a phenomenological description, the coefficients B and C should be

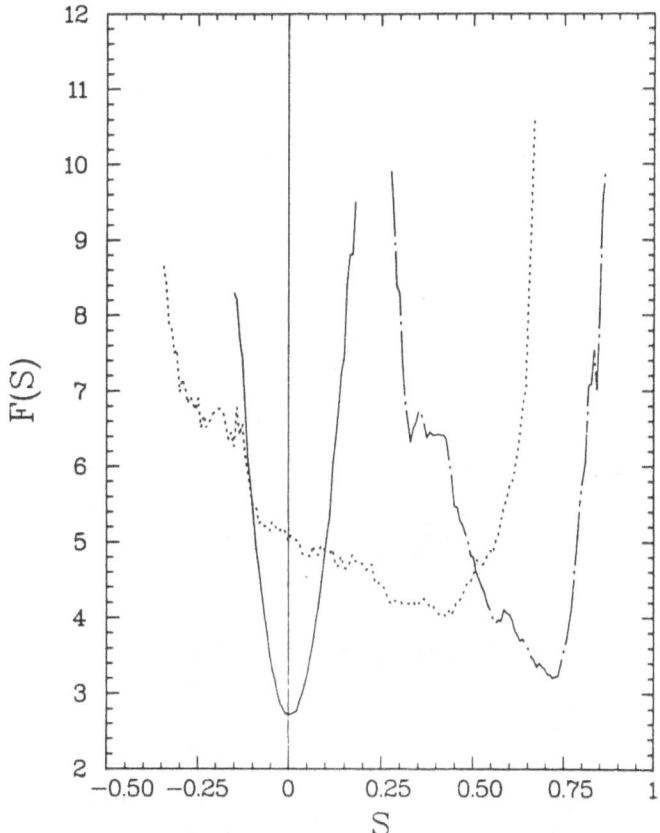

Fig. 7: The Landau free energy F(S) (in units kT) associated with fluctuations of the nematic order parameter S in a system of 100 hard platelets of diameter σ (see Ref. [17]). Drawn curve: low density isotropic phase. Dotted curve: higher density, just beyond the isotropic-nematic transition. Note that the minimum of F(S) is shifting to a non-zero value of S. Dash-dot curve: high density nematic. Now only small fluctuations around a non-zero average value of S are possible.

slowly varying functions of ρ, while A should change sign close to the thermody-namic transition point. Figure 8 shows that A does indeed behave more or less as pre-dicted by Eq. (8). The statistical errors in B (and even more in C), are large. Yet the data shown in Figure 8 seem to suggest that B (and possibly C) is in fact changing appreciably in the vicinity of the isotropic-nematic transition. This unex-pected behaviour of the Landau coefficients was also observed in a simulation of a very different model system [22]. The numerical results suggest that the Landau-de-Gennes theory may not correctly describe the non-Gaussian order-parameter fluctuations in the vicinity of the isotropic-nematic transition. However, more detailed calculations are needed to clarify this point. For a discussion of the Landau-de-Gennes theory in the context of recent experiments on real nematogens, the reader is referred to a review by Gramsbergen, Longa and de Jeu [23]. This review also discusses experimental evidence for non-Gaussian order-paramter fluctuations in the vicinity of the isotropic-nematic transition.

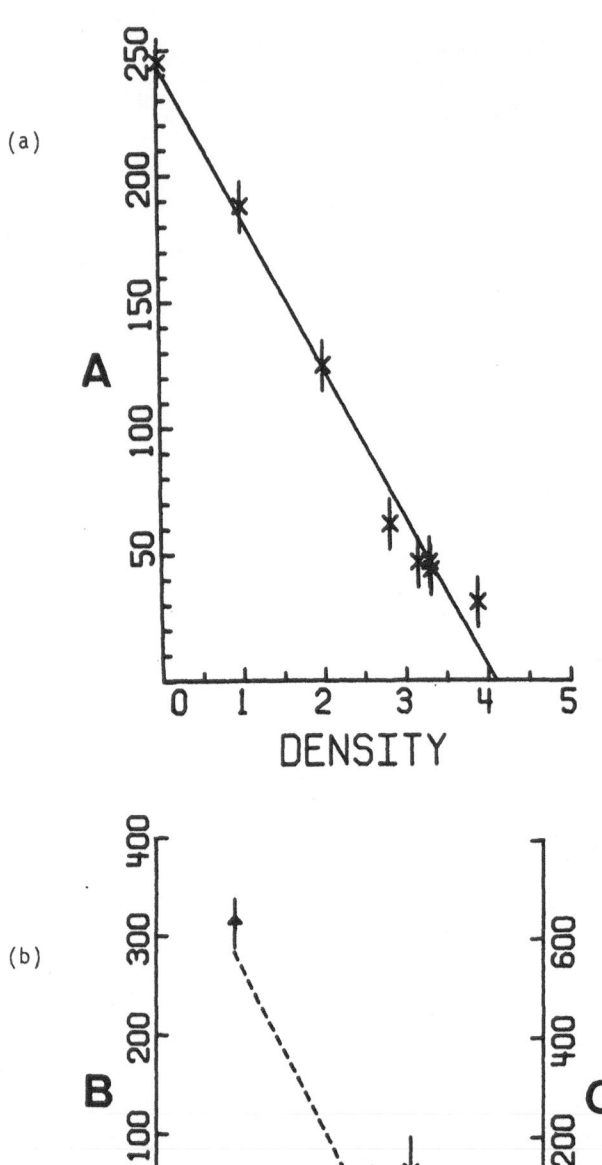

(a)

(b)

Fig. 8: Density dependence of the coefficients of the lowest few powers of S in the Landau expansion: $F(S) = constant + AS^2 + BS^3 + CS^4 +$ higher order terms. The model is a system of 100 infinitely thin hard platelets (see Ref. [17]). Note that A ((a)) extrapolates to zero at the I-N transition. B(x) appears to change sign in the vicinity of the I-N transition. The statistical error in C(▲) is very large.

Although interesting from a fundamental point of view, non-Gaussian effects play a relatively minor role in the pretransitional behaviour of nematogens. The most striking effect that is observed as the isotropic-nematic transition is approached from the iso- tropic side is the strong increase in the amplitude of order-parameter fluctuations, which results, among other things, in enhanced depolarized light scattering. To a first approximation, the intensity of depolarized light scattering, I_d, is proportional to the average trace of the square of the collective orientational tensor $\vec{Q}$ defined in Eq. (1) above:

$$I_d \sim \langle Q_{\alpha\beta} Q_{\beta\alpha} \rangle = 3/2 \langle S^2 \rangle . \tag{9}$$

From the Landau-de-Gennes expansion (7) and the relation between $P(S)$ and $F(S)$ it follows that, in the Gaussian approximation, $\langle S^2 \rangle \sim A^{-1}$ and hence:

$$I_d \sim (\rho^* - \rho)^{-1} . \tag{10}$$

$\langle Q_{\alpha\beta} Q_{\beta\alpha} \rangle$ in Eq. (9) is a measure for the correlation of the orientation of different molecules. Using Eqs. (1) and (9), it is easy to show that $\langle S^2 \rangle = (1 + g_2)/N$, where g_2 is the static orientational correlation factor, defined as:

$$g_2 = \sum_{j \neq i} P_2(\vec{u}_i \cdot \vec{u}_j). \tag{11}$$

The proximity of the isotropic-nematic transition influences not just the static orien- tational properties of the fluid but also the rotational dynamics. In particular, the relaxation time τ_2^C of the collective orientational fluctuations diverges as the I-N transition is approached. The relation between the divergence of g_2 and τ_2^C is of particular interest. Using Mori theory, Keyes and Kivelson [24] derived the following expression relating g_2 to τ_2^C [25]:

$$\tau_2^S/\tau_2^C = (1 + j_2)/(1 + g_2). \tag{12}$$

Here τ_2^S is the single-particle correlation time and j_2 is the dynamic orientational correlation factor. The latter quantity can be expressed in terms of memory functions, but has no simple physical interpretation. A direct experimental determination of $(1 + g_2)$ and τ_2^S/τ_2^C is complicated by the fact that in real liquids part of the depolarized light scattering intensity is interaction induced. It is not trivial to separate out the purely orientational contribution to the scattering intensity [26]. As a consequence, there is a scarcity of reliable data from which j_2 can be determined. The available information (see e.g. [27]) suggests that j_2 is small compared to 1, and it is common practice among light scatterers to assume that $j_2 = 0$.

Numerical simulation offers a unique possibility to study the static and dynamic precursor effects to the I-N transition in simple model systems. As an example, we consider the pretransitional behaviour in a system of prolate and hard ellipsoids with $a/b = 3$ [28]. From the Monte Carlo simulations mentioned above, we know that this system has a transition from the isotropic to the nematic phase at 70% of the density of regular close packing [18]. For comparison, we also studied the orientational dynamics of hard ellipsoids with $a/b = 2$. The latter system has no stable nematic phase and hence we should expect it to behave rather differently from the more anisometric $a/b = 3$ system. Figure 9 shows typical examples of $C_1^S(t) \equiv \langle P_1(\vec{u}(0)\cdot\vec{u}(t))\rangle$ and $C_2^S(t) \equiv \langle P_2(\vec{u}(0)\cdot\vec{u}(t))\rangle$, the correlation functions of the first and second Legendre polynomials of the molecular orientation vectors. Both $C_1^S(t)$ and $C_2^S(t)$ are single-particle correlation functions [19]. In the rotational diffusion limit, the ratio of the relaxation times of $C_1^S(t)$ and $C_2^S(t)$, τ_1^S/τ_2^S equals 3. We find that this behavior well obeyed at not too low densities. Both single particle relaxation times grow as the density is

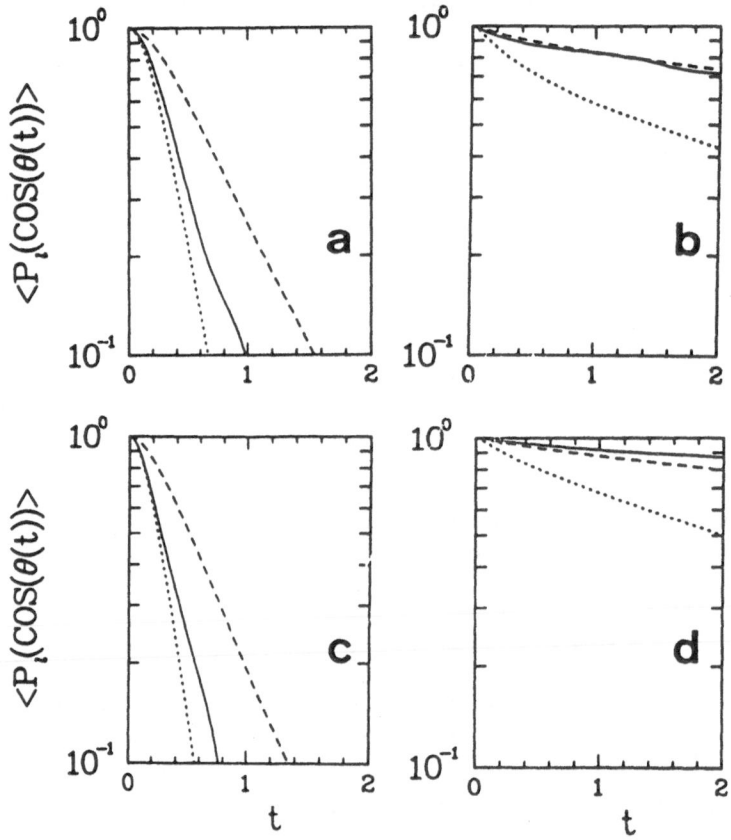

Fig. 9: Density and shape dependence of single-particle and collective orientational correlation function for hard ellipsoids of revolution. Figure A: $a/b = 2$, $\rho/\rho_0 = 0.5$, figure B: $a/b = 2$, $\rho/\rho_0 = 0.8$, figure C: $a/b = 3$, $\rho/\rho_0 = 0.3$ and figure D: $a/b = 3$, $\rho/\rho_0 = 0.7$. (---) Single-particle first-rank orientational correlation function, $C_1^S(t) = \langle P_1(\vec{u}(0)\cdot\vec{u}(t))\rangle$. (•••) Single-particle second-rank orientational correlation function, $C_2^S(t) = \langle P_2(\vec{u}(0)\cdot\vec{u}(t))\rangle$. (——) Collective second-rank orientational correlation function, $C_2^C(t) = \Sigma_j\langle P_2(u_1(0)\cdot u_j(t))\rangle$. The single-particle orientational variable has been made orthogonal to the collective one by projection. Note that at high densities the decay of $C_2^C(t)$ becomes much slower than that of $C_2^S(t)$.

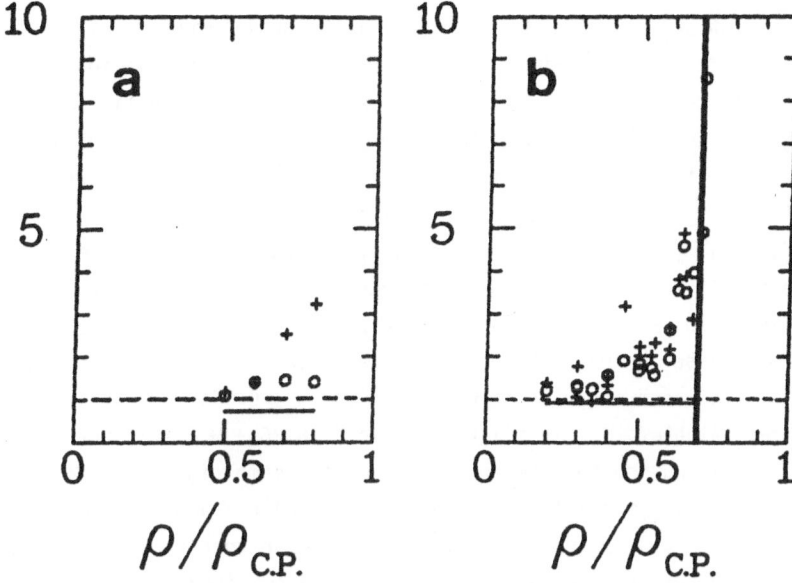

Fig. 10: Density dependence of the ratio $\tau_2{}^C/\tau_2^S$ (pluses), for hard ellipsoids of revolution with $a/b = 2$ (figure A) and $a/b = 3$ (figure B). Also shown is the behavior of the static orientational correlation factor $(1 + g_2)$ (open circles). The best estimate for the dynamic orientational correlation factor $(1 + j_2)$ is shown as a horizontal line. In figure B the isotropic-nematic coexistence region is bordered by parallel vertical lines.

increased. This is simply a consequence of the rapid increase of the viscosity as the the density of the fluid is raised, and the effect is observed for both shapes of ellipsoids. The density dependence of the collective orientational correlation function, $C_2^C (t) = \Sigma_j \langle P_2(\vec{u}_1(0) \cdot \vec{u}_j(t)) \rangle$ is more interesting. At low densities, the decay of $C_2^C (t)$ closely follows that of $C_2^S (t)$, as can be seen from Figure 9. However, as the density is increased, the decay of the collective orientational correlation function becomes much slower than the corresponding single particle one. Hence, cooperative effects in the rotational dynamics become important at high densities, both for the $a/b = 2$ and for $a/b = 3$ case. Figure 10 shows the density dependence of the ratio τ_2^C /τ_2^S for both shapes of ellipsoids. In the same figure we have also plotted the corresponding values of $(1 + g_2)$. The difference in the behaviour of the $a/b = 3$ and $a/b = 2$ ellipsoids is immediately apparent. For $a/b = 2$ there is evidence for some cooperative behaviour, but even at the freezing density (at 84% of close packing), there is no sign of diverging fluctuations or critical slowing down. In contrast, the results for $a/b = 3$ clearly show pretransitional effects. Both τ_2^C/τ_2^S and $(1 + g_2)$ appear to diverge as the transition to the nematic phase is approached. If the dynamic orientational correlation factor j_2 vanishes, then the data points for τ_2^C /τ_2^S and $(1 + g_2)$ should superimpose. On this point the simulations appear inconclusive. The statistics on τ_2^C /τ_2^S and $(1 + g_2)$ are poor, even though these are fairly long runs ($O(10^6)$ collision for a 144 particle system). The reason for the large scatter in the data points is that the quantities shown are collective properties. The signal-to-noise ratio for such quantities is of order $\sqrt{\tau/T}$, where τ is a characteristic relaxation time, and T is the duration of the simulation [30]. Hence, as the correlation times of the collective orientational fluctuations grow larger, the statistics (for a fixed value of T) become worse. Clearly, it is not meaningful to extract estimates for j_2 from the individual data points. We therefore attempted to estimate j_2 in the following way. We assumed $(\tau_2^s/\tau_2^c)(1+g_2)$ $(=(1+j_2))$ was independent of density, and carried out a linear least squares fit to the data points. The results of these fits are indicated as drawn lines in Figure 10. We see that for both shapes the estimate for $(1+j_2)$ is less than 1. For the $a/b = 3$ case, where we have the largest number of data points, the fit suggests that j_2 is small and negative (typically -0.08 ± -0.05). For the $a/b = 2$ ellipsoids the estimated value of j_2 is slightly larger (-0.26 ± -0.16), but the estimated error is larger. It is amusing to note that a j_2 of the same sign and comparable magnitude followed from the light-scattering experiments of Gierke and Flygare on a real nematogen ($j_2 = -0.3 \pm -0.3$ for MBBA) [27]. We do not wish to attach too much significance to the actual numbers that we have obtained for j_2: clearly, much longer simulations are needed to carry out a direct comparison with the relevant theoretical predictions. In fact, such simulations are now in progress. The importance of the preliminary results presented here is that they provide the first numerical data on pretransitional effects in the orientational dynamics of any model nematogen. In addition, our results indicate that the numerical determination of j_2, although time-consuming, is certainly feasible.

IV. Beyond Nematics

IV.1 Introduction

In the previous sections we have shown that sufficiently anisometric hard ellipsoids of revolution can form thermodynamically stable nematic phases. And that, moreover, these simple model systems exhibit many of the static and dynamic properties of real nematogens. Naturally, the question arises whether the nematic phase is the only type of liquid crystal that can be simulated with a hard-core model. For instance, one might wonder whether hard ellipsoids of revolution can also form a smectic phase, i.e. a fluid that has (at least) one-dimensional translational order, in addition to the orientational order already present in nematics (see e.g. Ref. [8]). The answer to this question is almost certainly: no. The reason is the following: smectic phases tend to have a large degree of orientational order (i.e. the order parameter S, defined in Section III.3 above, is close to 1). Hence, to a first approximation, we can assume that a smectic consists of perfectly aligned, non-spherical molecules. However, a system consisting of hard ellipsoids of revolution with axial ratio a/b, all parallel to the z-axis (say), can be mapped onto the hard-sphere fluid by a simple scaling of all z-coordinates with a factor b/a. As hard spheres apparently do not form smectics, parallel hard ellipsoids cannot do so either. So, unless the orientational degrees of

freedom stabilize the smectic phase (and this seems unlikely), hard ellipsoids of revolution cannot form smectics.

The question then arises whether smectic phases can be formed by other rigid
hard-core models. This is not obvious a priori. In fact, to my knowledge, most text-
books on liquid crystals do not even seriously consider the possibility (see, however,
Ref. [31]). Even more than in the case of nematics, dispersion forces [32] and the effects of flexible tails [12] are held responsible for the formation of smectic phases.
Although we do not contest that these factors must have a pronounced effect on the
stability of smectics, the question that must be answered is whether they are essential
in the same way that attractive forces are essential to explain the liquid-vapour transition [33].

IV.2 Computer Simulations

In order to explore the possibility of smectic order in rigid hard-core systems, we
carried out Monte Carlo and Molecular Dynamics simulations on model systems consisting
of parallel spherocylinders with diameter D and length L (i.e. the hemispherical
caps were separated by a straight cylindrical segment of length L) [34]. As the particles in this system are always perfectly aligned, the low-density phase is a "nematic"
fluid. The parallel spherocylinder fluid can be thought of as a model for a fluid of
rodlike particles in a strong magnetic field. As mentioned above, we know that the corresponding hard-ellipsoid model will not exhibit smectic order. Simulations were carried
out for systems of parallel spherocylinders with L/D ratios of 0.25, 0.5, 1, 2, 3
and 5. The well-known case $L/D = 0$ (i.e. hard spheres) was also studied, as a check.
System sizes varying from 90 to 1080 particles were studied. Initially, the system was
prepared in a regular close-packed lattice. We prepared this lattice by expanding a
close-packed face-centered cubic hard-sphere crystal by a factor $(L/D + 1)$ along the
[111]-axis. In order to avoid spurious translational order due to incomplete melting,
the crystals were expanded to low densities (typically, 30% of close-packing), where
the solid rapidly melted to form a translationally disordered fluid. Fluid configurations at higher densities were subsequently generated by slow compression. In contrast,
solid configurations were generated by gradual expansion from the close-packed lattice.
During the simulations of the solid phase we allowed for changes in the shape of the
crystal unit cell, using an isotropic-stress Monte Carlo method, as described in Ref.
[18]. In addition, we computed the absolute free energies of both the fluid and the
solid phases, employing the techniques described in Section III.4 and Ref. [34]. Combining the free energy and equation of state data, we can determine the dependence of
the melting point of the spherocylinder crystal on L/D (see Table I). Having determined the limits of thermodynamic stability of the solid phase, the next step is to
look for possible fluid-fluid phase transitions. No discontinuities or van der Waals
loops are observed in the fluid branch of the equation of state [34], but for the larger
L/D ratios we observe a rather sudden change in the compressibility at densities

Table I: Densities and pressures of the solid-liquid coexistence points of parallel spherocylinders with length-to-width ratios L/D between 0.25 and 5.0.

L/D	ρ/ρ_0 (fluid)	ρ/ρ_0 (solid)	P_{FS}
0.25	0.610	0.685	4.75
0.50	0.612	0.665	4.58
1.00	0.624	0.650	4.50
2.00	0.751	0.802	9.99
3.00	0.771	0.818	11.22
5.00	0.693	0.759	7.83

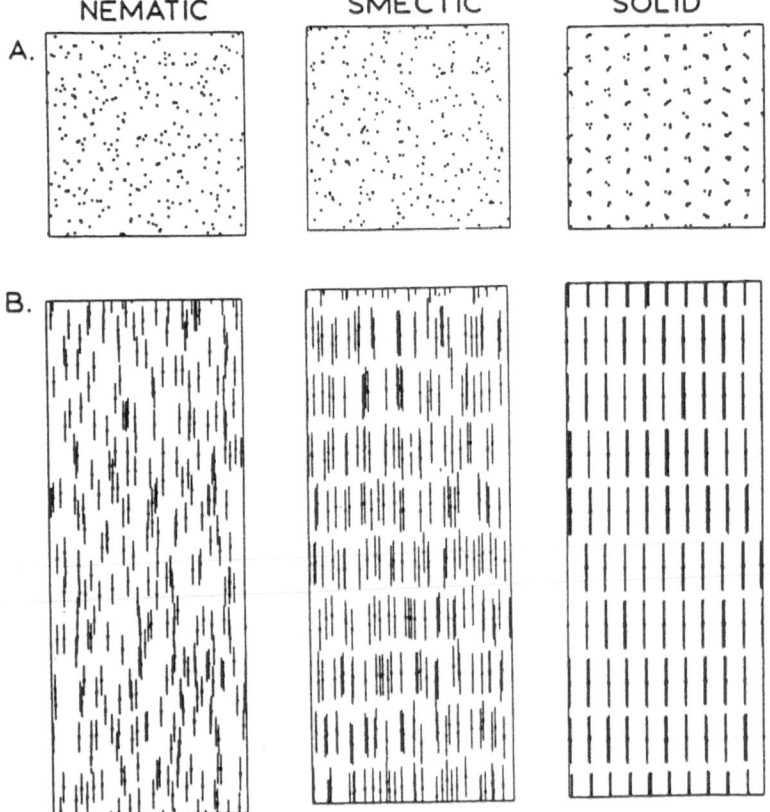

Fig. 11: Snapshots of typical configurations of a system of 270 parallel hard spherocylinders with L/D = 5. Left: nematic phase (ρ/ρ_0 = 0.24). Middle: smectic phase (ρ/ρ_0 = 0.62). Right: solid phase (ρ/ρ_0 = 0.89). The upper figures (A) show a projection in the plane perpendicular to the molecular axes, the lower figures (B) show a projection in a projection in a plane through the molecular axes. For the sake of clarity, the spherocylinders are indicated by a line segment of length L .

between 40% and 60% of close packing. Snapshots of typical molecular configurations both below and above this 'cusp' in the equation of state (Figure 11) show a dramatic change in the structure of the fluid. At low densities, the fluid is translationally disordered. Then, at a higher density (but well before the freezing point), the fluid orders into parallel layers, but there is no order within the layers. In other words, the parallel spherocylinders form a smectic A-phase [35]. For comparison, the crystalline phase, at still a higher density, is also shown. In the latter case there is clearly translational order within the layers.

A more quantitative method to locate the transition to the smectic phase is to study the static and dynamic behaviour of the longitudinal component of the intermediate scattering function $F(k_z,t)$, defined as:

$$F(k_z,t) = <\rho(k_z,0)\rho(-k_z,t)> \, , \qquad (15)$$

where $\rho(k_z,t)$ is the instantaneous amplitude of a longitudinal density fluctuation with wavevector k_z. $F(k_z,t=0)$ is the longitudinal part of the static structure factor, $S(k_z)$, which determines, for instance, the x-ray scattering intensity. As the smectic phase is approached from lower densities, there will be smectic precursor fluctuations. These will show up as peaks in $S(k_z)$, for values k_z equal to (multiples of) $2\pi/d$, where d is the spacing of the incipient smectic layers. If the transition to the smectic phase is continuous, the peaks in $S(k_z)$ will diverge at the transition. An example of this behaviour is shown in Figure 12. If we define the smectic layer

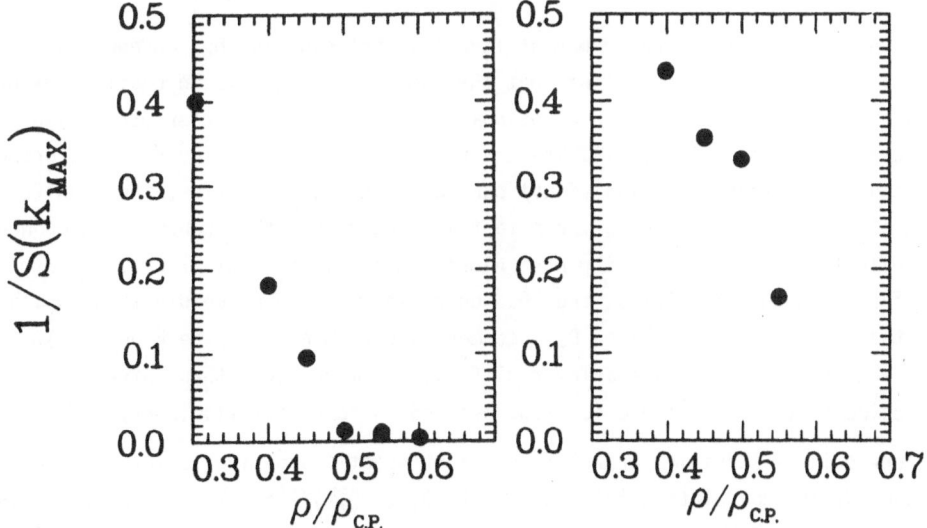

Fig. 12: Density dependence of the first maximum in the longitudinal component of the structure factor of parallel hard spherocylinders with L/D = 5 (left) and L/D = 0.5 (right). The ordinate shows $1/S(k_z^{max})$. In the dilute gas, this quantity approaches one. In a smectic phase it should vanish in the thermodynamic limit. The present results apply to a 270-particle system, hence $1/S(k_z^{max}) \geq 1/270$. From the behaviour of $S(k_z^{max})$ one may deduce the density dependence of the longitudinal correlation length of $\xi_{||}$ of smectic precursor fluctuations.

spacing d , by $d = 2\pi/k_z(max)$, where $k_z(max)$ is the wavevector corresponding to the first (and largest) maximum in $S(k_z)$, then we can measure how many smectic layers fit into the simulation box. Figure 13 shows how the number of layers depends on density

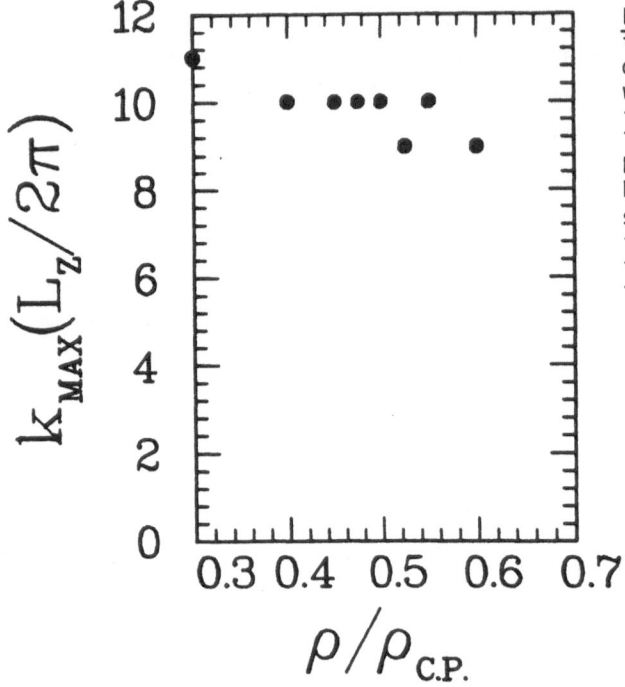

Fig. 13: Density dependence of the 'number of layers' in a system of 270 parallel spherocylinders with $L/D = 5$. The number of layers is defined as the number of oscillations of the largest longitudinal Fourier component in the periodic box: $k_{max}L_z/2\pi$. Note that the observed layering is not a metastability effect associated with the insufficient equilibration of the initial configuration.

for $L/D = 5$. The fact that the number of layers is not constant but changes on compression, is gratifying. It indicates that the number of layers is an equilibrium property, and not the consequence of some residual, metastable solid order (the solid phase, in this case, consists of 9 layers). The ordering of the fluid in layers has a pronounced effect on the diffusion of individual particles. At low densities, the longitudinal component of the self-diffusion, D_ℓ, constant is larger than the transverse component, D_t. This effect is strongest for the most elongated particles. But as smectic layers start to form at higher densities, the D_ℓ decreases much more rapidly than D_t. So much so, that in the smectic phase D_t becomes larger than D_ℓ (see Figure 14). Hence, in the smectic phase, intra-layer diffusion is more rapid than inter-layer diffusion. Incidentally, this effect is also observed in some real liquid crystals [36].

A particularly sensitive probe of incipient smectic ordering is the critical slowing down of the correlation function $<\rho(k_z,0)\rho(-k_z,t)>$ at $k_z = k_z(max)$. As can be seen from Figure 15, the decay rate of $F(k_z(max),t)$ first increases as the density is raised, because the fluid is less compressible at higher densities. But beyond a certain density the decay rate drops precipitously to zero. This provides us with an estimate for the nematic-smectic transition density. For comparison, Figure 15 also shows an example of a fluid which does not become smectic ($L/D = 0.25$). In this case the decay rate of $F(k_z(max),t)$ depends only weakly on density.

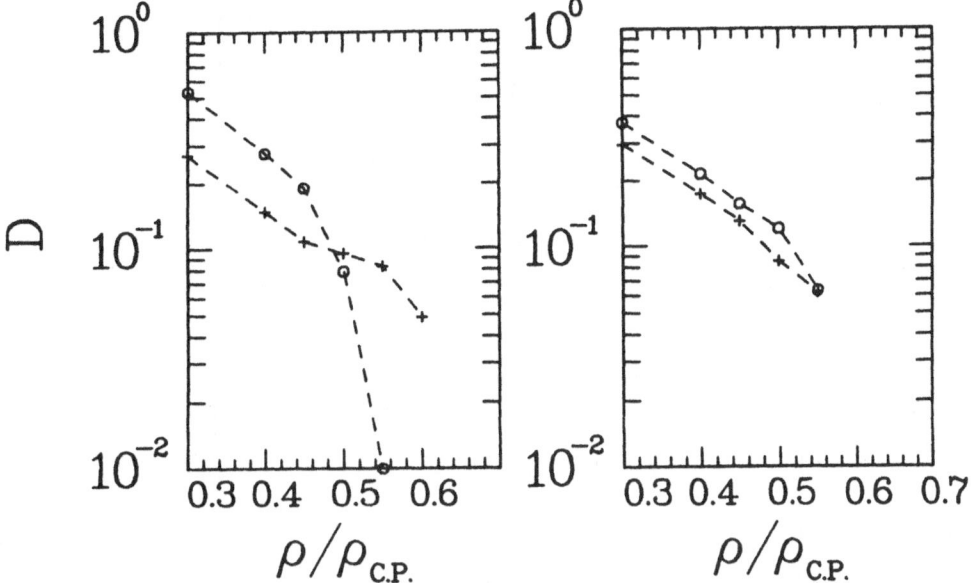

Fig. 14: Typical example of the density dependence of the longitudinal (O) and trans-verse (+) components of the self-diffusion tensor in a system of parallel spherocylinders with L/D=1(left) and L/D = 0.25 (right). Note that the relative magnitudes of the parallel and perpendicular components of D change as the fluid goes from the nematic to the smectic phase.

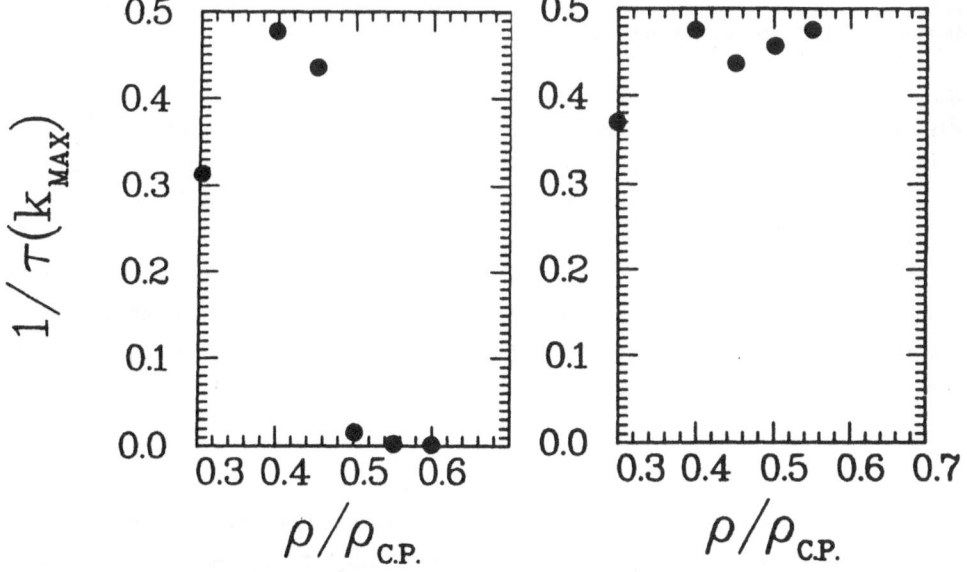

Fig. 15: Relaxation rate of the longitudinal density fluctuations with a wavevector corresponding to the first maximum in $S(k_z)$. Right figure: parallel spherocylinders with L/D = 0.25, this system does not form a stable smectic. The figure shows no evidence for critical slowing down. Left hand figure: parallel spherocylinders with L/D = 1. This system does form a smectic. As the transition is approached, the decay rate of the smectic precursor fluctuations drops dramatically.

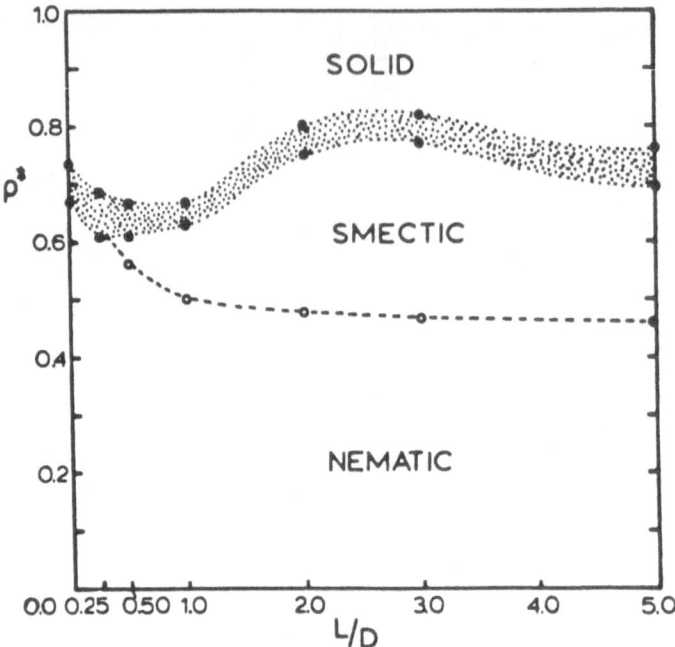

Fig. 16: Schematic 'phase diagram' of hard parallel spherocylinders, as a function of length-to-width ratio L/D. The shaded area corresponds to the fluid-solid two-phase region. Black circles: densities of the coexisting fluid and solid phase. Open circles: densities at which smectic ordering sets in. The dashed line indicates the estimated nematic-smectic boundary.

Combining the information about the smectic order parameter fluctuations with the data in Table I on the melting transition, we can construct a 'phase diagram' of hard parallel spherocylinders (Figure 16). This phase diagram shows the regions of stability of the nematic, smectic and solid phases as a function of L/D. For L/D ≥ 0.5, a smectic phase is found between the crystalline solid and the 'nematic' fluid. As the non-sphericity of the particles is increased, the smectic range initially grows and then saturates [37]. It should be stressed that this phase diagram pertains to a system of parallel spherocylinders. The effect of the orientational degrees of freedom on the phase diagram is currently under investigation [38].

V. Conclusions

In this paper I have tried to demonstrate that model systems consisting of non-spherical hard-core particles have a surprisingly rich phase diagram. In the case of hard ellipsoids, we found that the stability of the nematic phase is simply related to the degree of non-sphericity of the molecules. Moreover, even though parallel ellipsoidal particles cannot form stable smectics, parallel spherocylinders can. This result is quite unexpected because of the apparent similarity of long spherocylinders and needle-like ellipsoids.

In addition I have shown how many of the physically interesting static and dynamic properties of liquid-crystal-forming fluids can be studied by computer simulation. The interest of such calculations lies not in the possibility to reproduce the behaviour of real liquid crystals (although it is encouraging to find qualitative agreement). The use of these, and future, calculations is that they can be used to test approximate theoretical expressions for a wide variety of equilibrium and transport properties of molecular liquids and liquid crystals. Moreover, now that we have demonstrated the existence of stable hard core nematics and (aligned) smectics for 'realistic' molecular shapes, we can prepare for the next step: thermodynamic perturbation theory.

There is more to liquid crystals than hard repulsive interactions. In the next few years we should be able to decide whether hard-core models are useful reference systems for real liquid crystals.

Acknowledgements

The material presented in this article is the result of collaboration with: Mike Allen, Rob Eppenga, Henk Lekkerkerker, Bela Mulder and Alain Stroobants. The work of the FOM Institute for Atomic and Molecular Physics is part of the research program of FOM and is supported by the "Nederlandse Organisatie voor Zuiver Wetenschappelijk Onderzoek."

References

[1] J.D. van der Waals, Dissertation, Leiden 1873. For a discussion of the van der Waals model in the context of the modern theory of liquids, see: B.J. Alder and W.G. Hoover, in: The Physics of Simple Liquids, H.N.V. Temperley, J.S. Rowlinson and G.S. Rushbrooke, eds., North Holland, Amsterdam 1968, p. 79

[2] J.A. Barker and D. Henderson, J. Chem. Phys. 48, 4714 (1967)

[3] J.D. Weeks, D. Chandler and H.C. Andersen, J. Chem. Phys. 54, 5237 (1971)

[4] H.S. Kang, T. Ree and F.H. Ree, J. Chem. Phys. 84, 4547 (1986)

[5] B.J. Alder and T.E. Wainwright, J. Chem. Phys. 27, 1208 (1957)

[6] See, for instance, C. Zannoni in: The molecular physics of liquid crystals. G.R. Luckhurst and G.W. Gray, editors, Academic Press, London, 1979, p. 51

[7] S.J. Picken, Internal report, University of Groningen (1984)

[8] P.G. de Gennes, Physics of Liquid Crystals, Oxford University Press, 1974

[9] W. Maier and A. Saupe, Z. Naturf. A13, 564 (1958)

[10] W.M. Gelbart and A. Gelbart, Mol. Phys. 33, 1387 (1977)

[11] This holds for thermotropic liquid crystals. For lyotropic liquid crystals flexible tails play no role, although here the flexibility of the particle as a whole may be important, see: T. Odijk, Polymer Comm. 26, 197 (1985)

[12] F. Dowell and D.E. Martire, J. Chem. Phys. 68, 1088 (1978), ibid.: 68, 1094 (1978), ibid.: 69, 2322 (1978)
       F. Dowell, Phys. Rev. A28, 3526 (1983)

[13] L. Onsager, Ann. N.Y. Acad. Sci. 51, 627 (1949)

[14] See e.g. G.W. Gray in: The molecular physics of liquid crystals. G.R. Luckhurst and G.W. Gray, editors, Academic Press, London, 1979, pp. 1 and 263

[15] C. Destrade, P. Fouchet, H. Gasparoux, N.H. Tinh, A.M. Levelut and J. Malthete, Mol. Cryst. Liq. Cryst. 106, 121 (1984)

[16] W.G. Hoover and F.H. Ree, J. Chem. Phys. 49, 3609 (1968)

[17] R. Eppenga and D. Frenkel, Mol. Phys. 52, 1303 (1984)

[19] A. Isihara, J. Chem. Phys. 19, 1142 (1956)

[20] D. Frenkel and A.J.C. Ladd, J. Chem. Phys. 81, 3188 (1984)

[21] L.D. Landau, Phys. Z. Sowjetunion 11, 26 (1937)
For a discussion of Landau theory in the context of the isotropic-nematic transition see: P.G. de Gennes, Ref. [8]

[22] G.R. Luckhurst, personal communication

[23] E.F. Gramsbergen, L. Longa and W.H. de Jeu, Physics Reports 135, 197 (1986)

[24] T. Keyes and D. Kivelson, J. Chem. Phys. 56, 1057 (1974)

[25] Unfortunately, there is some confusion in the literature about the notation used for the static and dynamic correlation factors $(1+g_2)$ and $(1+j_2)$. Here we have chosen a definition of g_2 and j_2 such that both quantities vanish in the absence of correlations.

[26] For a discussion, see: D. Kivelson and P.A. Madden, Ann. Rev. Phys. Chem. 31, 523 (1980)

[27] G.R. Alms. T.D. Gierke and W.H. Flygare, J. Chem. Phys. 61, 4083 (1974)
T.D. Gierke and W.H. Flygare, J. Chem. Phys. 61, 2231 (1974)

[28] M.P. Allen and D. Frenkel, Phys. Rev. Lett. 58, 1748 (1987)

[29] Actually, the correlation functions shown are the orthogonalized single-particle and collective orientational correlation functions.

[30] R. Zwanzig and N.K. Ailawadi, Phys. Rev. 182, 280 (1969)

[31] M. Hosino, H. Nakano and H. Kimura, J. Phys. Soc. Japan 46, 1709 (1979)

[32] W.L. McMillan, Phys. Rev. A4, 1238 (1971)

[33] Actually, transitions from a dilute to a condensed fluid phase need not always be caused by attractive interactions. The crucial point is that hard-core repulsion alone cannot explain such phase transitions.

[34] A. Stroobants, H.N.W. Lekkerkerker and D. Frenkel, Phys. Rev. Lett. 57, 1452 (1986)

[35] We also checked for long-range bond-orientational order within the layers, but did not find any. Hence the fluid is not a hexatic-B phase.

[36] M. Hara, H. Tenmei, S. Ichikawa, H. Takezoe and A. Fukuda, Jap. J. Appl. Phys. 24, L777 (1985)

[37] More recent evidence indicates that the phase diagram is even richer than shown in the figure. A. Stroobants, H.N.W. Lekkerkerker and D. Frenkel, submitted for publication

[38] D. Frenkel, J. Phys. Chem., in press

MOLECULAR ORDER AND DYNAMICS OF LIQUID CRYSTAL POLYMERS
STUDIED BY MULTIPULSE DYNAMIC NMR TECHNIQUES

K. Müller and G. Kothe

Institut für Physikalische Chemie, Universität Stuttgart,
Pfaffenwaldring 55, D-7000 Stuttgart 80, West-Germany

Introduction

Pulsed nuclear magnetic resonance (NMR) has been established as a valuable tool to study molecular dynamics in complex chemical and biological systems [1-8]. Until recently, however, most investigations have been concerned with conventional relaxation rates and single quantum spectra ignoring phenomena, which arise in multipulse sequences. In this article we present a more comprehensive study, employing multipulse dynamic NMR techniques [9,10]. Variation of pulse sequence and pulse separation provides the large number of independent experiments, necessary for a proper molecular characterization of the systems.

Analysis of these experiments in terms of molecular order and dynamics is often complicated by the various couplings in a multispin system. Thus, in order to simplify the analysis, nuclear spinlabels with isolated magnetic interactions are introduced. For the studies presented in this paper, only deuteron spinlabels have been employed. The technique, however, is easily extended to other nuclei, appropriate for dynamic NMR investigations.

In the first Section, the theoretical method is developed. Then typical examples are given to illustrate the applicability of the model and to determine the limiting conditions under which the simpler approaches used previously are valid. In the main Section, multipulse dynamic NMR experiments of specifically deuterated liquid crystal polymers are presented. Computer simulations provide the orientational distributions and conformations of the polymer chains and the correlation times of the various motions. The Discussion clearly shows the advantage of multipulse dynamic NMR in characterizing complex chemical and biological systems.

Techniques

Multipulse dynamic NMR is a time domain technique. The spin system is subject to a sequence of non-selective rf pulses and the response after the last pulse is used to characterize the molecular order and dynamics of the system. Different NMR responses are obtained according to the sequences which are used. Moreover, significant signal changes occur when the pulse separations are varied.

Theoretical Background

Analysis of these experiments in terms of molecular order and dynamics requires a comprehensive model. We have developed such a model, based on the density matrix formalism [9,10]. The action of the different pulses on the time-dependent spin density matrix $\rho(t)$ is considered by unitary transformations, employing Wigner rotation matrices. Between the pulses the density matrix is assumed to obey the stochastic Liouville equation [11,12]

$$\frac{\delta}{\delta t} \, \rho(\Omega,t) = -(i/\hbar)H^X(\Omega)*\rho(\Omega,t) - \Gamma_\Omega*[\rho(\Omega,t)-\rho_{eq}(\Omega)] \tag{1}$$

which we solve using a finite grid point method [13,14]. Here $H^X(\Omega)$ is the Hamiltonian superoperator of the spin system, depending on the orientation and conformation of the molecule specified by its Euler angles Ω. Γ_Ω is the time-independent Markov operator for the various rotational processes with the equilibrium distribution $P_{eq}(\Omega)$ obeying

$$\Gamma_\Omega * P_{eq}(\Omega) = 0 \tag{2}$$

and $\rho_{eq}(\Omega)$ is the equilibrium density matrix.

In the finite grid point method the Markov operator is represented by a transition probability matrix $W(\Omega_m,\Omega_n)$ whose elements give the transition probability between the m-th and n-th discrete site of Ω. $W(\Omega_m,\Omega_n)$ allows for various intermolecular motions, depicted in Figure 1. The intermolecular motion is the motion of the molecule as a whole. It is assumed that the chain molecule undergoes anisotropic rotational diffusion in an orienting potential. The intramolecular motion consists of local isomerization, which is represented by a jump process. The dynamics of the systems are thus characterized by three motional correlation times. These are $T_{R\parallel}$ and $T_{R\perp}$ for rotation about the diffusion tensor axis and rotation of this axis, respectively, and T_J the correlation time for trans-gauche isomerization.

The equilibrium distribution $P_{eq}(\Omega)$ is described in terms of internal and external coordinates. The internal part accounts for different conformations and the external part for different orientations. Generally, there are only four conformational states of a particular aliphatic chain segment [15]. The corresponding populations n_1, n_2, n_3 and n_4 may be used to set up a segmental order matrix, which on diagonalization yields the segmental order parameters $S_{Z'Z'}$ and $S_{X'X'} - S_{Y'Y'}$ [10]. They express the degree of order of the most-ordered segmental axis Z' and the anisotropy of that order, respectively. Within the limits of a completely disordered segment all n_K are equal to 1/4, resulting in an order parameter of $S_{Z'Z'} = 0$. At the other extreme, a fully extended chain is fixed to its all-trans conformation, where n_1 equals

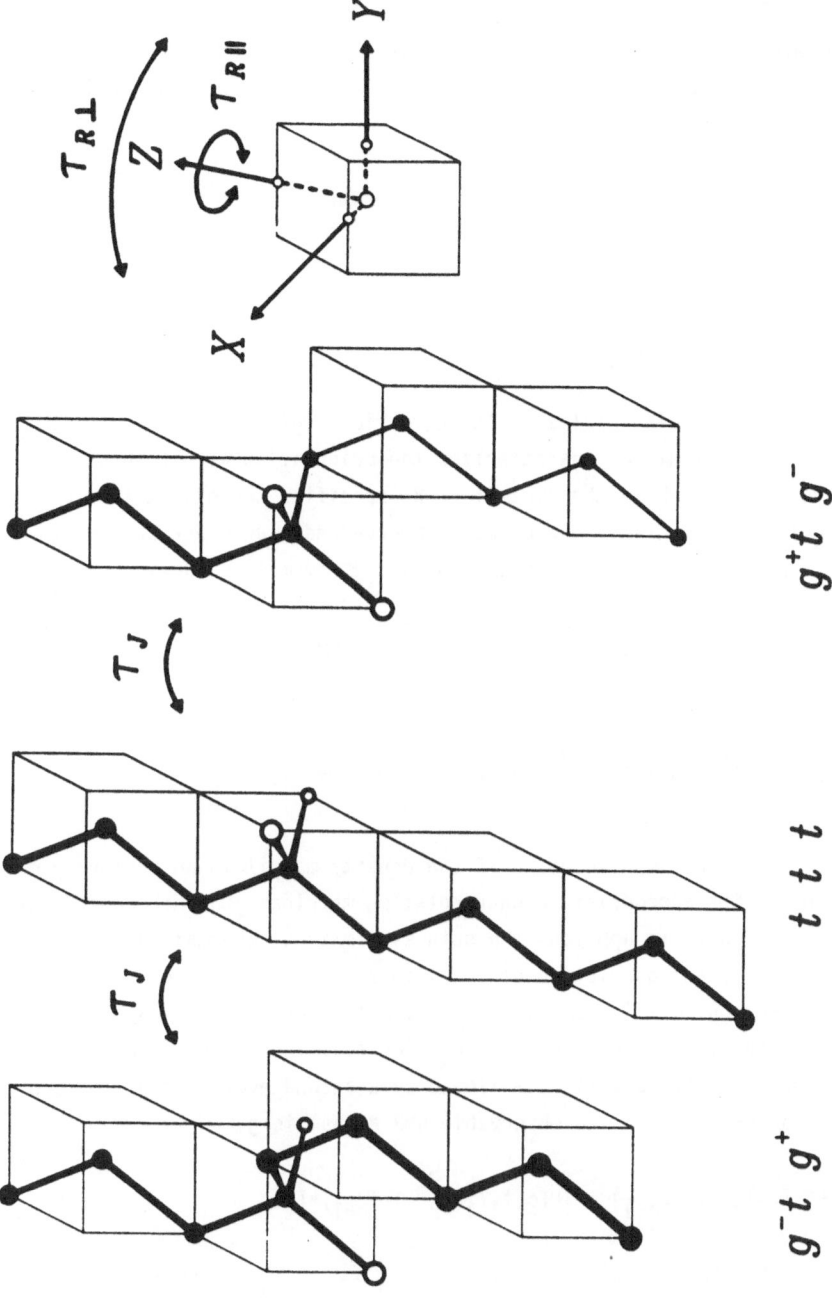

Figure 1: Conformation of an aliphatic chain in its all-trans state (ttt) and with a single kink (g^+tg^- or g^-tg^+) at the point of attachment of two deuteron spinlabels. Significant motional modes are represented by arrows. They refer to chain rotation ($T_{R\parallel}$), chain fluctuation ($T_{R\perp}$) and trans-gauche isomerization (T_J), respectively.

unity for each segment, and the order parameter becomes $S_{z'z'} = 1$.

The orientational distribution of the chain molecules is described in the mean-field approximation, using an orienting potential such as is common in molecular theories of liquid crystals [16]:

$$f(\phi,\theta,\Psi) = N_1 \exp A(\cos\theta\cos\xi - \sin\theta\cos\Psi\sin\xi)^2$$

$$\cos\xi = \cos\delta\cos\rho - \sin\delta\cos\epsilon\sin\rho$$

$$f(\delta,\epsilon) = N_2 \exp(B\cos^2\delta) . \tag{3}$$

Here ϕ, θ, Ψ, δ, ϵ, ρ are Euler angles, relating various molecular and laboratory systems [9,10]. The coefficient A characterizes the orientation with respect to a local director (microorder), while the parameter B specifies the orientation of the director axes in a laboratory frame (macroorder). Micro- and macroorder parameters S_{ZZ} and $S_{z''z''}$ are related to the coefficients A and B by mean value integrals [17]:

$$S_{ZZ} = (1/2) N_1 \int_0^\pi (3\cos^2\beta-1)\exp(A\cos^2\beta)\sin\beta d\beta$$

$$\tag{4}$$

$$S_{z''z''} = (1/2) N_2 \int_0^\pi (3\cos^2\delta-1)\exp(B\cos^2\delta)\sin\delta d\delta .$$

Let us now evaluate the time evolution of the density matrix in an arbitrary multi-pulse sequence by using the appropriate Wigner rotation matrices and the equation of motion. Before any rf pulse is applied, the spin system is at thermal equilibrium, ρ_{eq}, according to a Boltzmann population of the spin states. Application of the first pulse creates a defined non-equilibrium state $\rho(0)$. After the pulse the density matrix evolves according to the stochastic Liouville equation (1). Then a second pulse is applied, preparing a new initial condition, followed by a second evolution period and so on. Finally, after the n-th pulse, the observable NMR signal is given by

$$L(t,T_1,T_2 \cdots T_{n-1}) = Tr[\rho(t,T_1,T_2 \cdots T_{n-1})*I_+] \tag{5}$$

where I_+ is the nuclear spin raising operator and $T_1,T_2 \cdots T_{n-1}$ are the various pulse separation times.

Fourier transformation of $L(t,T_1,T_2 \cdots T_{n-1})$ starting from $t = T_1 + T_2 \cdots + T_{n-1} + T_i$ (spin echo) yields single quantum frequency spectra, which sensitively depend on the actual pulse sequence. From the decay of the echo amplitude as function of a chosen T_i various relaxation times can be evaluated. Finally, by Fourier transforming $L(t,T_1,T_2 \cdots T_{n-1})$ for a fixed $t > T_1 + T_2 \cdots + T_{n-1}$ with respect to a particular T_i multiple quantum spectra are obtained [18].

Characterization of Motions

There are four pulse sequences mainly employed in dynamic NMR of $I = 1$ spin systems. The quadrupole echo sequence [19-21] consists of two $\pi/2$ pulses, which are separated by a time T_1 and have a $\pi/2$ relative phase shift

$$(\pi/2)_x - T_1 - (\pi/2)_y \ .$$

At another time T_1 later a refocussing of the transverse magnetization occurs. Measurement of the echo amplitude as a function of T_1 provides the spin-spin relaxation time T_{2E}. Since T_{2E} is most sensitive to motions with correlation times $T_R \cong (e^2qQ/\hbar)^{-1}$, where e^2qQ/h is the quadrupolar coupling constant, this pulse sequence offers a means to study molecular dynamics in the range of $10^{-8}\,s < T_R < 10^{-4}\,s$.

Faster motions are accessible by employing the inversion recovery [22]

$$\pi - T_1 - (\pi/2)_x - T_2 - (\pi/2)_y$$

or saturation recovery sequence

$$(\pi/2) - T_1 - (\pi/2)_x - T_2 - (\pi/2)_y$$

in a high magnetic field. From measurements of the echo amplitude at time $t = T_1 + 2T_2$ as a function of T_1, the spin lattice relaxation times T_{1Z} can be determined. They are particularly sensitive to motions with correlation times $T_R \cong \omega_0^{-1}$, where ω_0 is the Larmor frequency. Using a magnetic field of $B = 7.0\,T$ fast molecular dynamics in the range $10^{-11}\,s < T_R < 10^{-7}\,s$ can be detected.

In contrast, spin alignment permits the study of extremely slow molecular motions. It is created through the Jeener-Broekaert sequence [23]

$$(\pi/2)_x - T_1 - (\pi/4)_y - T_2 - (\pi/4)_y \ .$$

The decay of the alignment echo at $t = 2\tau_1 + \tau_2$ as a function of T_2 generally deviates from a single exponential. Analysis of this relaxation curve yields information about type and time scale of extremely slow motions [24] with correlation times as long as $T_R = 10\,s$, limited only by the condition $T_R < T_{1Q}$, where T_{1Q} is a time constant being comparable but not identical with the spin lattice relaxation time T_{1Z}. Thus by combining analysis of quadrupole echo, inversion recovery and spin alignment studies, it is possible to follow dynamic processes over 12 orders of magnitude of correlation times.

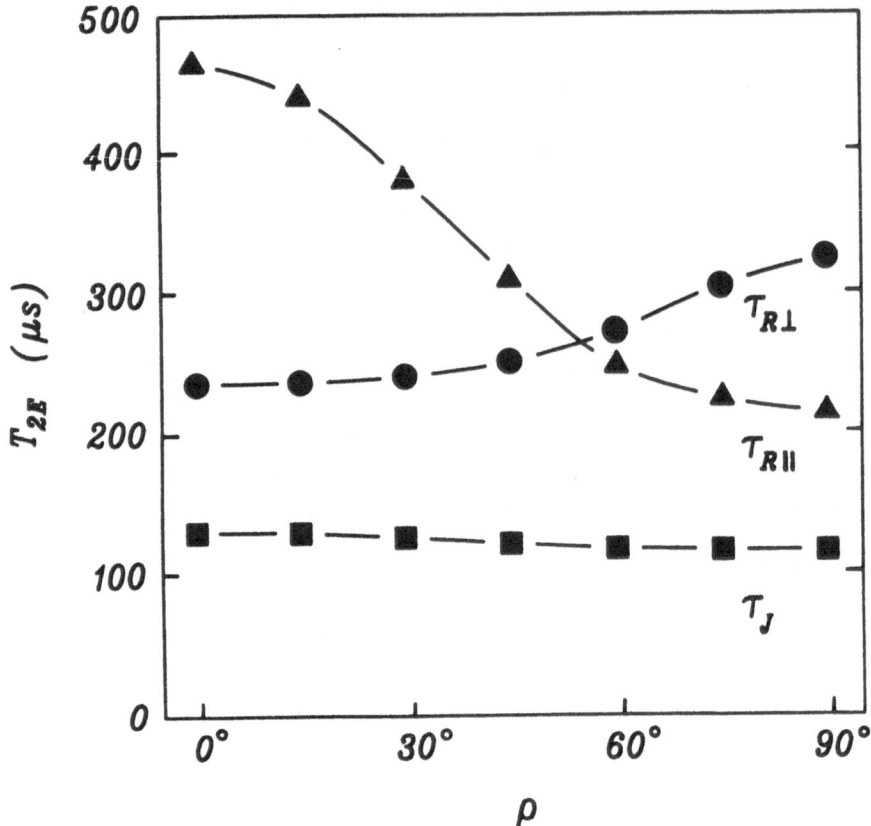

Figure 2: Angular dependence of ^{2}H spin-spin relaxation times T_{2E} for various motional modes of partially ordered aliphatic chains. All relaxation times were obtained with e^2qQ/h = 165 kHz, A = 3.0 (orientational order), n_1 = 0.7 and n_4 = 0 (conformational order). Full triangles denote chain rotation ($\tau_{R\parallel}$ = 3 × 10^{-7} s, $\tau_{R\perp}$ = τ_J > 10^{-2} s) full circles denote chain fluctuation ($\tau_{R\perp}$ = 3 × 10^{-7} s, $\tau_{R\parallel}$ = τ_J > 10^{-2} s), and full squares indicate trans-gauche isomerization (τ_J = 3 × 10^{-7} s, $\tau_{R\parallel}$ = $\tau_{R\perp}$ > 10^{-2} s). ρ is the angle between alignment axis and magnetic field.

Complex chemical systems are often characterized by a variety of different motions, which can be sorted out by multipulse dynamic NMR techniques [9,10]. A representative example is given in Figure 2, which shows the angular dependence of the spin-spin relaxation times T_{2E} in consideration of different types of motion. All T_{2E} values were obtained with the same correlation time of $\tau_{R\parallel}$ = $\tau_{R\perp}$ = τ_J = 3 × 10^{-7} s. The observed changes of T_{2E} with orientation are characteristic for the motional modes involved. Thus, angular dependent spin-spin relaxation times are highly indicative of the type of motion. Similar is true for the spin lattice relaxation times T_{1Z} and T_{1Q}, which sensitively depend on the motional anisotropy, likewise [25].

Powder lineshape obtained via multipulse sequences may be distorted as compared with those obtained from the FID due to angular dependent relaxation times [22,26]. Generally, one observes with increasing T_i significant spectral changes, which depend only on the type of motion [9,10,21]. Therefore, the distortion of powder lineshapes can also be used to determine and to separate different motional processes.

A further method for discriminating motional modes is based on the decay of the quadrupole and spin alignment echo in the slow motional region [9,10,27]. Again, their non-exponential decay depends markedly on the type of motion involved. Thus, by using different multipulse techniques, the various motions can be differentiated over an extremely wide dynamic range, extending from 10^{-11} s in the fast motional to 10 s in the ultraslow motional regime.

Results and Discussion

The liquid crystal polymers and monomers which are considered in this paper have the general structures shown in Figure 3 [28-31]. The Roman numerals, I-IV, refer to different samples, deuterated at the specific sites, as indicated in the formula [32]. The polyesters exhibit a glass temperature (T_g) at 303 K, a melting point (T_m) at 433 K and a clearing temperature (T_{ni}) at 553 K, forming a stable nematic melt over the latter temperature range. All clearing temperatures slightly depend on the average molecular weight $\overline{M}_n$ of the samples, varying between $5000 < \overline{M}_n < 30000$. The corresponding low molecular mass liquid crystal IV exhibits a nematic mesophase between $T_m = 368$ K and $T_{ni} = 453$ K [30,31]. Macroscopic alignment of the samples was achieved with electric or magnetic fields. In addition, the liquid crystal polyesters I-III were oriented by melt-spinning methods [33].

Lineshapes and Relaxation

The specifically deuterated liquid crystal polymers I-III (Figure 3) have been studied over a wide temperature range, employing multipulse dynamic NMR techniques [9,34,35]. The ^{2}H NMR measurements were performed at $\omega_0/2\pi = 46.1$ MHz (B = 7.0 T), using quadrupole echo, inversion recovery, saturation recovery and Jeener-Broekaert sequences. Typical results for polymer II are shown in Figures 4-8. The observed ^{2}H NMR lineshapes and relaxation curves, varying drastically with magnetic field orientation, demonstrate the power of the method.

Figure 4 depicts the angular variation of the quadrupole echo lineshapes at T = 124 K. The spectra refer to five different orientations of alignment axis and magnetic field. Drastic spectral changes are observed when the sample is rotated showing that liquid crystal order is maintained in the solid and glassy state of the polymer.

Experimental saturation recovery and quadrupole echo lineshapes at different pulse separations T_1 are shown in Figure 5. The spectra refer to the same temperature

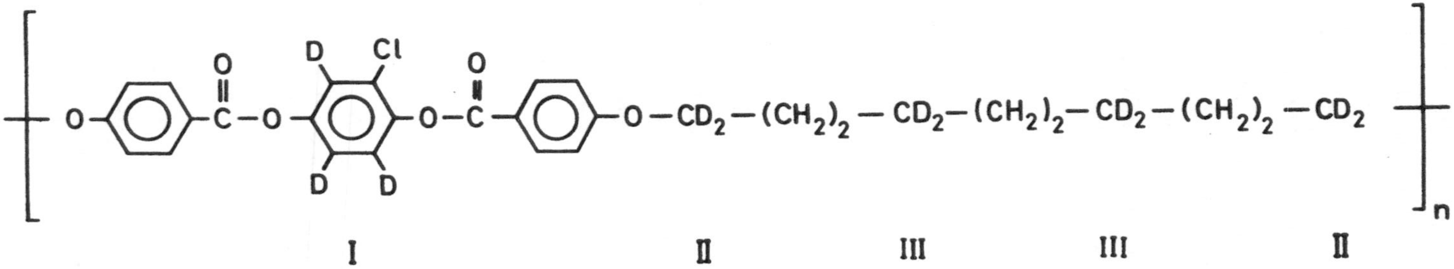

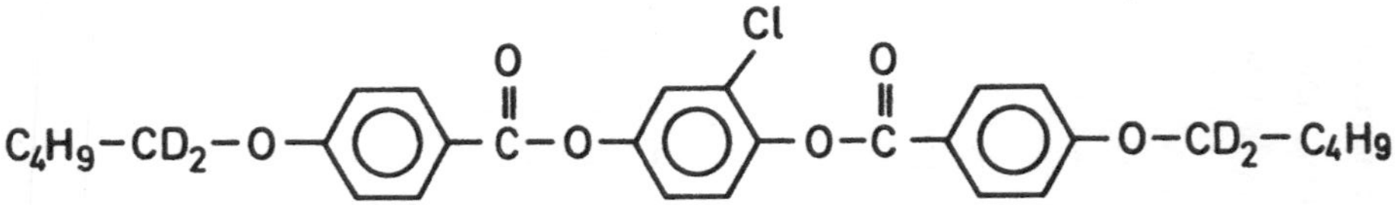

Figure 3: Molecular structures of selected liquid crystal polymers and monomeric analogues. The Roman numbers refer to four different liquid crystals, specifically deuterated at the sites indicated in the formula.

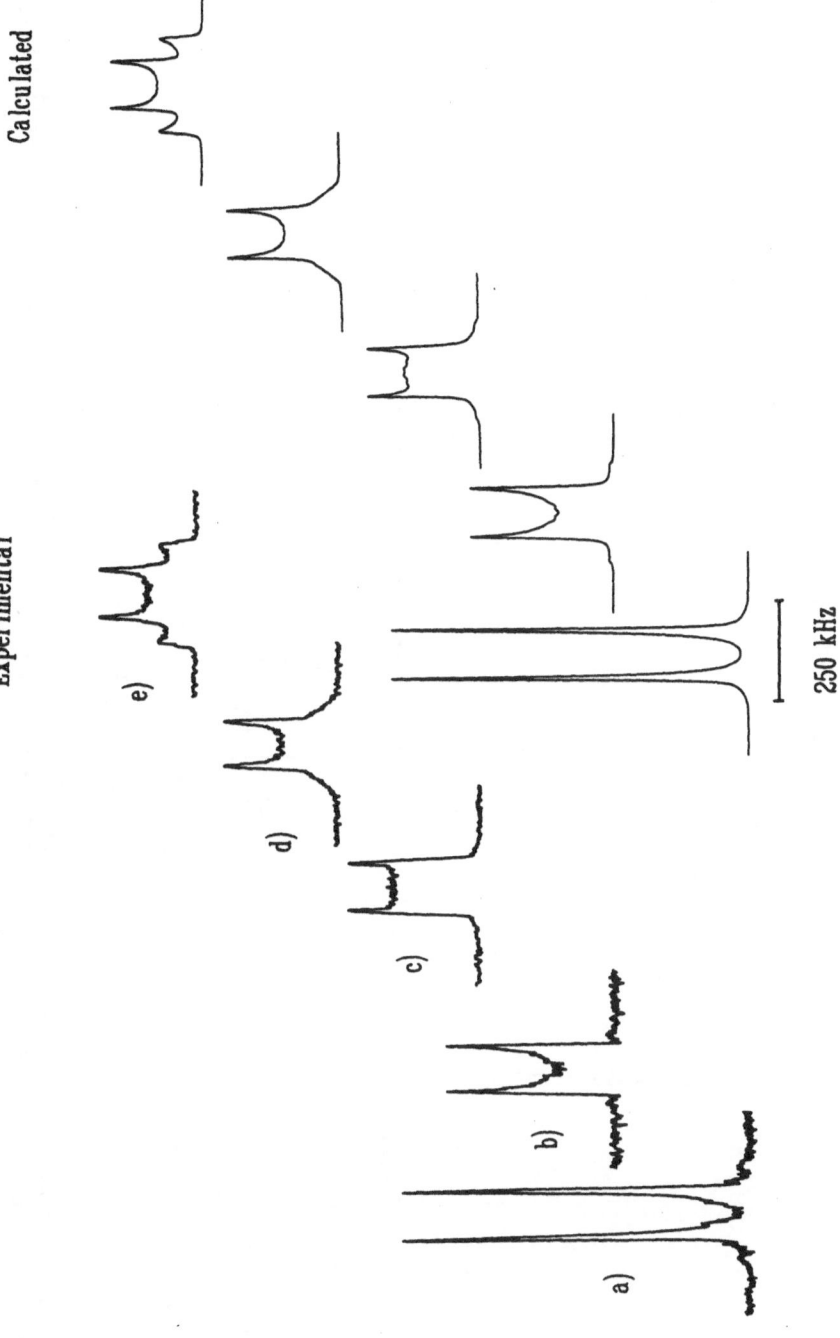

Figure 4: Experimental and calculated ^2H NMR spectra of polymer II at $T = 124$ K and five different angles ρ between alignment axis and magnetic field. The spectra refer to quadrupole echo sequences and a fixed pulse separation time of $T_1 = 20$ µs. The calculations were obtained with $e^2qQ/h = 165$ kHz, $T_{R\perp} = T_{R\parallel} = T_J > 10^{-4}$ s, $A = 15.5$, $n_1 = 0.8$, $n_4 = 0$ and refer to (a) $\rho = 0°$, (b) $\rho = 30°$, (c) $\rho = 45°$, (d) $\rho = 60°$, (f) $\rho = 90°$.

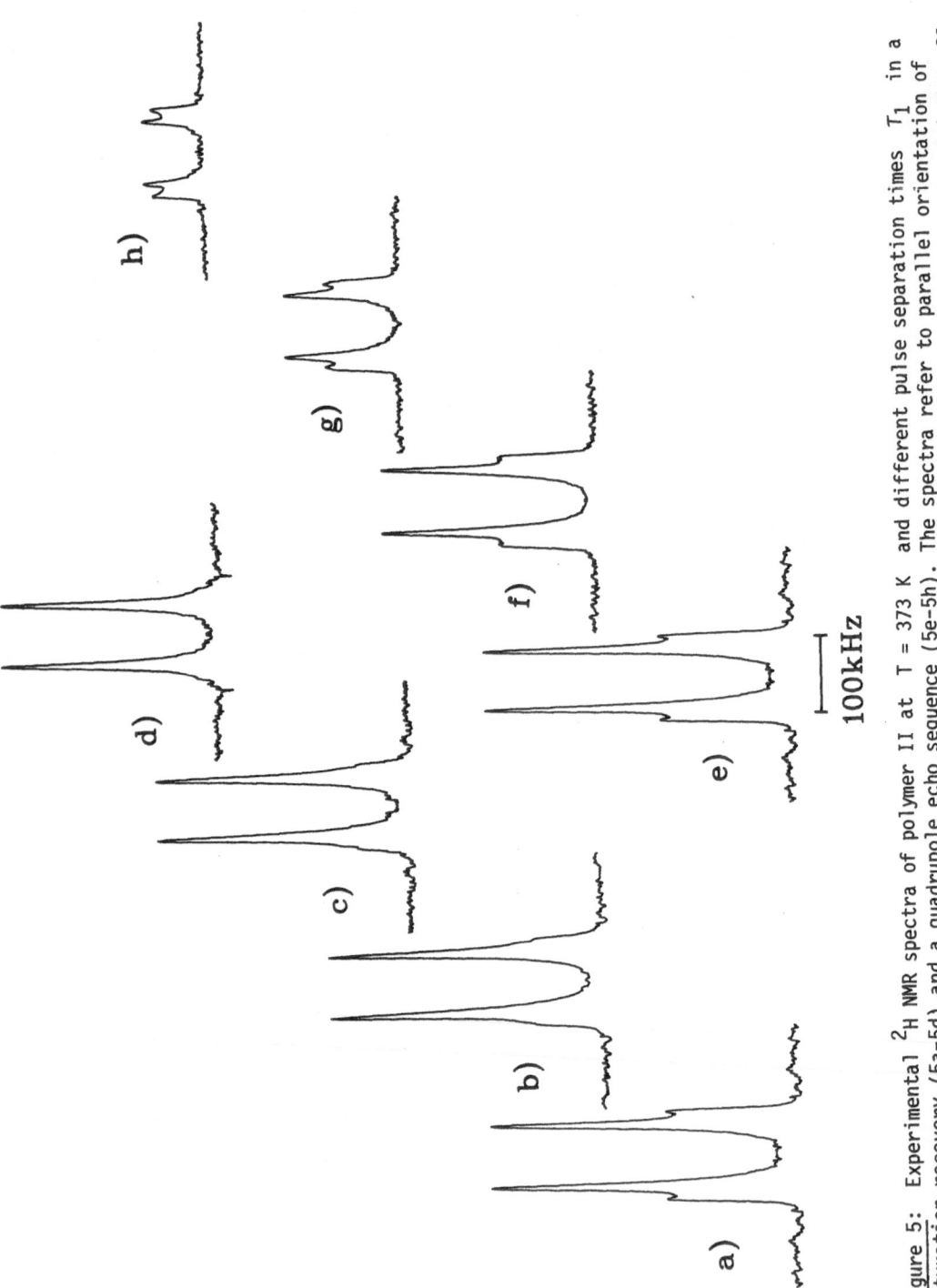

Figure 5: Experimental ^{2}H NMR spectra of polymer II at T = 373 K and different pulse separation times T_1 in a saturation recovery (5a–5d) and a quadrupole echo sequence (5e–5h). The spectra refer to parallel orientation of alignment axis and magnetic field and (a) T_1 = 1000 ms, (b) T_1 = 150 ms, (c) T_1 = 80 ms, (d) T_1 = 5 ms, (e) T_1 = 20 μs, (f) T_1 = 40 μs, (g) T_1 = 80 μs and (h) T_1 = 120 μs.

100kHz

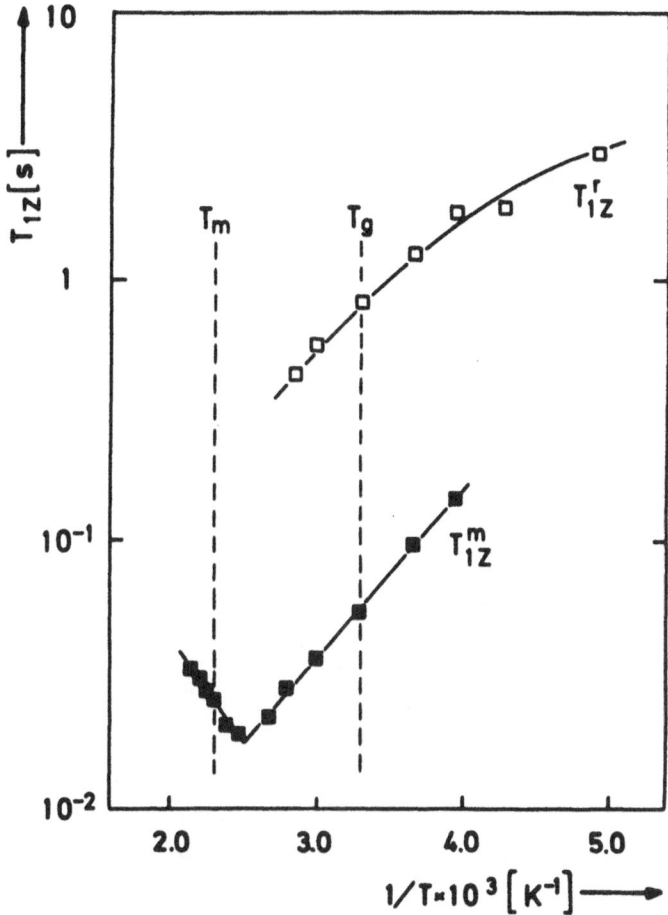

Figure 6: Temperature dependence of ^{2}H spin lattice relaxation times T^m_{1Z} (full squares) and T^r_{1Z} (open squares) of polymer II. All relaxation times refer to parallel orientation of alignment axis and magnetic field. The T^m_{1Z} values are the short decay times of the mobile component, while the T^r_{1Z} values represent relaxation times of the rigid component. Dashed lines indicate different phase transitions. The solid lines are best fit simulations of the relaxation times, employing the NMR model (Redfield limit) and the parameters of Figures 9 and 10.

(T = 373 K) and parallel orientation of alignment axis and magnetic field. Two spectral components are observed. The central peaks refer to the mobile fraction of the polymer while the outer peaks correspond to the rigid part. The spectral changes with increasing T_1 are due to different T_{1Z} and T_{2E} relaxation times of the two components. As can be deduced from Figure 5 the mobile fraction relaxes faster than the rigid one.

In Figure 6 the spin lattice relaxation times of the mobile and rigid components, T^m_{1Z} and T^r_{1Z}, are plotted as a function of 1/T. All relaxation times refer to parallel orientation of alignment axis and magnetic field. The values for T^m_{1Z} and T^r_{1Z} were obtained by recording the echo amplitude of an inversion or saturation recovery sequence as a function of T_1, and by decomposing this curve into two exponentials. One sees that T^m_{1Z} passes through a sharp minimum at T = 390 K.

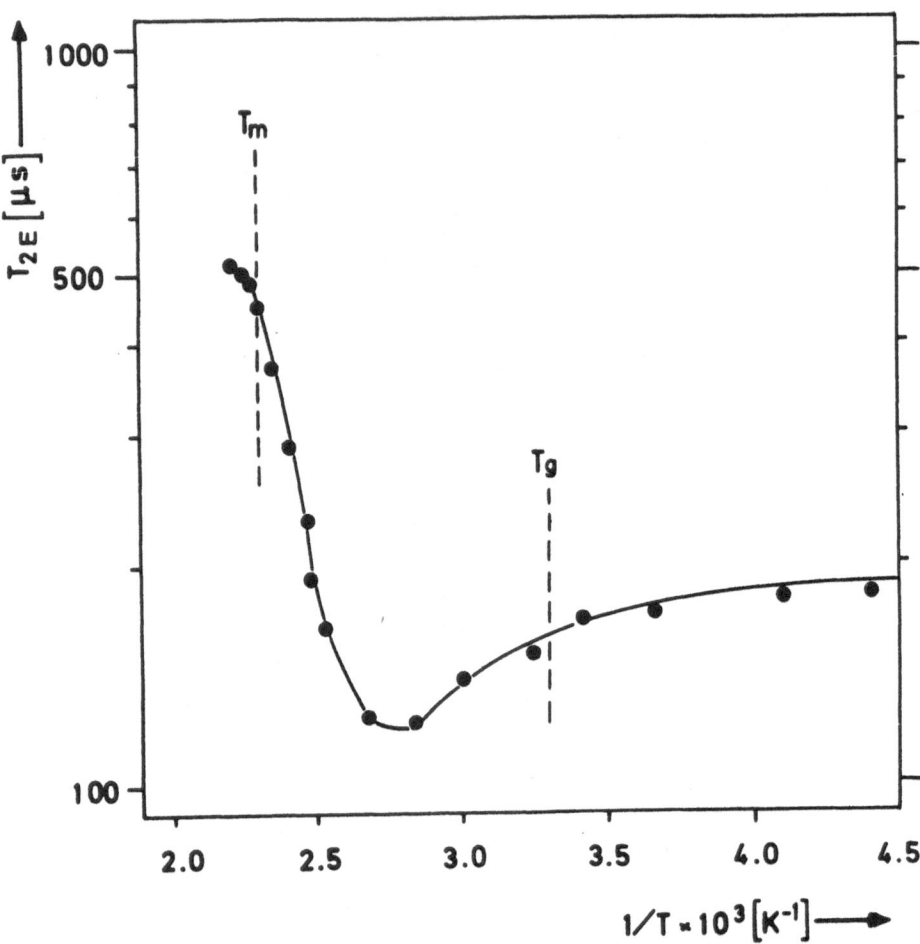

Figure 7: Temperature dependence of ^{2}H spin-spin relaxation times T_{2E} of polymer II. All relaxation times refer to parallel orientation of alignment axis and magnetic field and represent bulk properties, which are an average of relaxation times for mobile and rigid deuterons. Dashed lines indicate different phase transitions. The solid lines are best simulations of the relaxation times, employing the NMR model and the parameters of Figures 9 and 10.

The spin-spin relaxation times T_{2E} in Figure 7 were obtained from the decay of the echo amplitude in a quadrupole echo sequence. Since a unique decomposition into two exponentials was impossible the given T_{2E} values represent bulk relaxation times and refer to both mobile and rigid deuterons. Note, that T_{2E} decreases abruptly at the melting point and passes through a minimum at $T = 343$ K. This striking behavior, expected on theoretical grounds, has also been observed in a lyotropic liquid crystal [10].

Figure 8 depicts the decay of the spin alignment echo as a function of the waiting time T_2. The observed decay curves refer to the same temperature (T = 233 K), parallel orientation of alignment axis and magnetic field and three different evolution

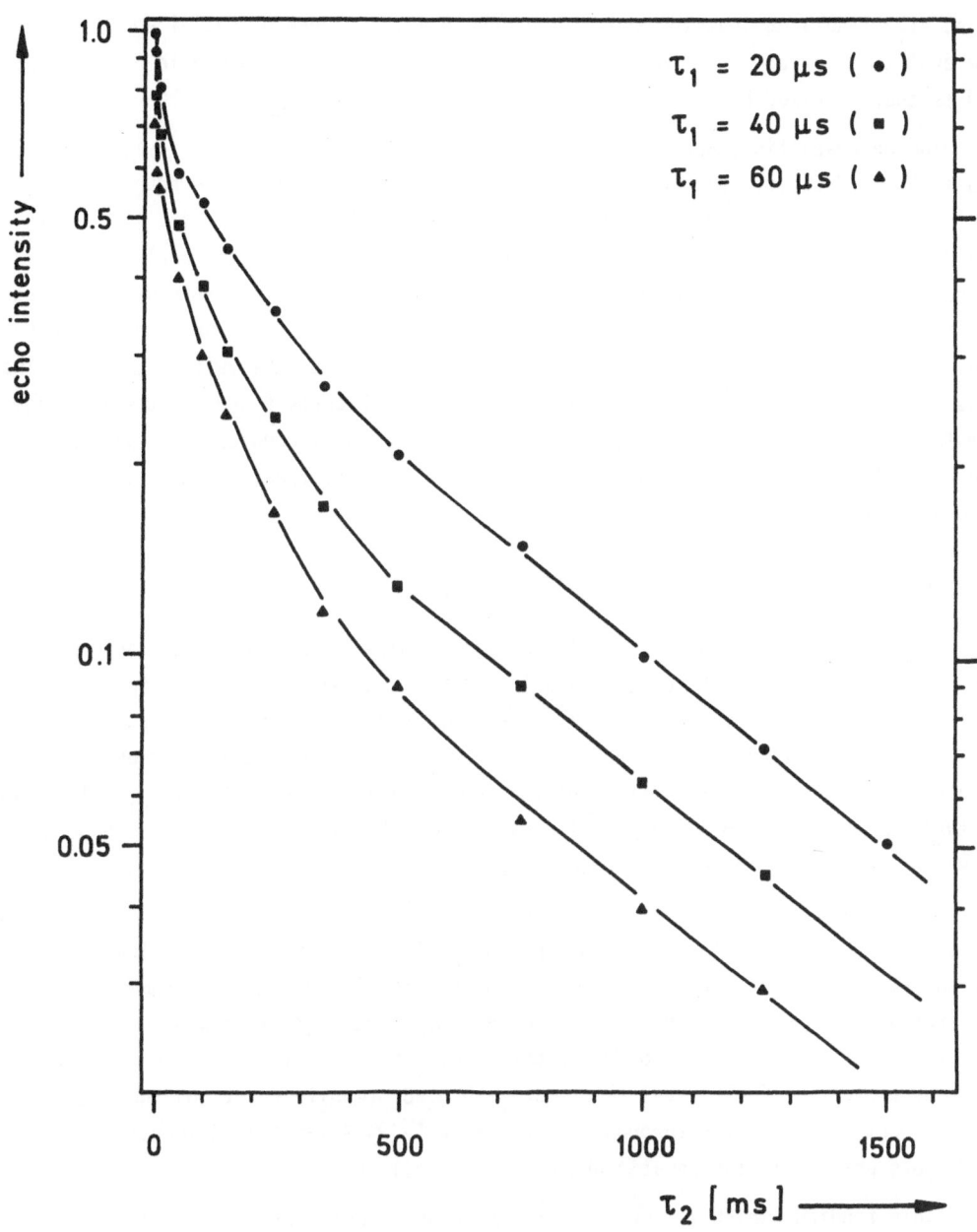

<u>Figure 8:</u> Decay of the ^{2}H spin alignment echo as a function of the waiting time τ_2. The decay curves of polymer II refer to T = 233 K, parallel orientation of alignment axis and magnetic field, and three different evolution times T_1.

times T_1 . One sees that the short time echo decay is non-exponential and depends on the evolution period T_1 . In contrast, the long time decay is a single exponential and is characterized by the spin lattice relaxation time T_{1Q} .

The observed lineshapes and relaxation curves were simulated employing the NMR model outlined above. Numerical solution of the stochastic Liouville equation was achieved employing either the Rutishauser [36] or the Lanczos algorithm [37]. Within the Redfield limit [38], analytical expressions are employed in the analysis. An iterative fit of several angular and pulse dependent experiments for any given temperature provided reliable values for the simulation parameters, i.e. the parameters of conformational and orientational order and the correlation times of the various motions. The lower row spectra in Figure 4 and the solid lines in Figures 6 and 7 represent best fit simulations. They agree favourably with their experimental counterparts. In the following we would like to describe the results in more detail, treating the various simulation parameters separately.

Molecular Order

The molecular order of semiflexible thermotropic polymers comprises the conformational order of the spacer and the orientational order of the repeating untis. Conformational order in these systems is conveniently discussed in terms of segmental order parameters $S_{Z'Z'}$ (see Theoretical Background) [10]. In Figure 9 these segmental order parameters (squares) are plotted as a function of the reduced temperature $T^* = T/T_{ni}$. They refer to the α-methylene group of polymer II (Figure 3). The rigid and mobile components, observed below the melting point, are distinguished by open and full symbols.

At the isotropic-nematic transition the segmental order parameter of the α-segment is $S_{Z'Z'} = 0.46$, while the δ-segment (polymer III) exhibits a somewhat lower value [9]. Decreasing the temperature increases the conformational order of all segments. However, the central units order faster than the outer ones. Thus, at $T^* = 0.8$ a uniform conformational order of $S_{Z'Z'} = 0.60$ is obtained. Further cooling increases the segmental order parameter to limiting values of $S_{Z'Z'} = 0.68$ (mobile fraction) and $S_{Z'Z'} = 0.73$ (rigid fraction) exhibited throughout the spacer [9,34,35,39,40]. Evidently, extended conformers prevail in the nematic phase of these polymers.

This finding, which is corroborated by ^{2}H NMR studies of other thermotropic polyesters [41,42], is the most prominent feature distinguishing liquid crystal main chain polymers from side chain polymers [43] and their monomeric analogues [31,44-46]. Predictions of statistical mechanical theories [47,48], developed for these polymers, are in qualitative agreement with the NMR results. It appears that a number of unique properties, exhibited by main chain polymers, can be attributed to the conformational order of the spacer.

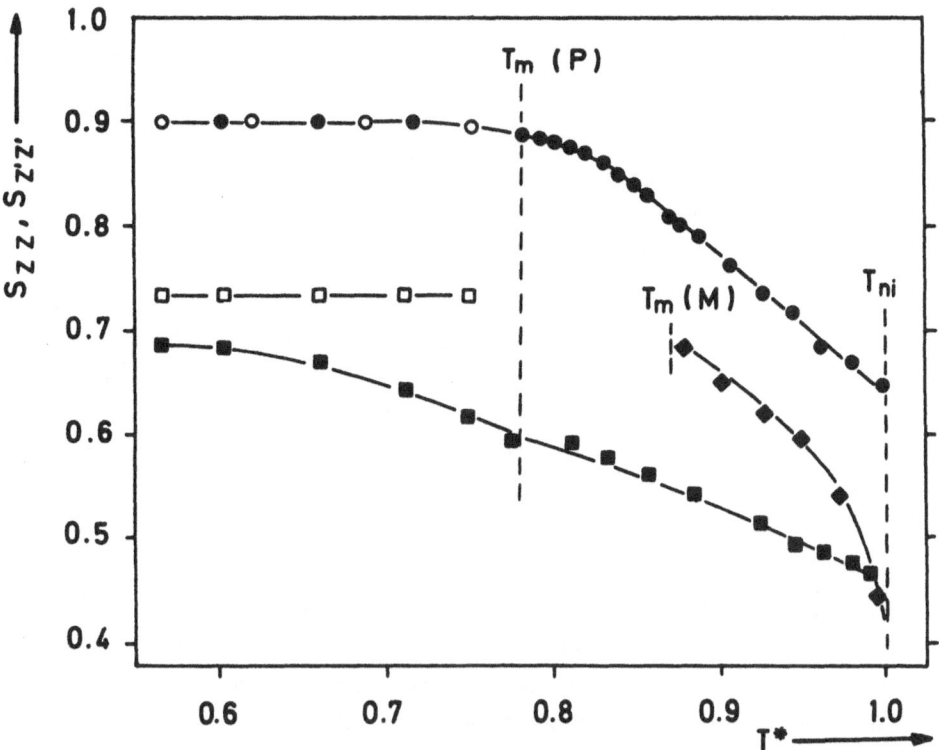

Figure 9: Temperature dependence of the conformational and orientational order of the polymers I-III and the low molecular weight analogue IV. Squares refer to the segmental order parameter $S_{Z'Z'}$ of polymer II, circles denote the orientational order parameter of S_{ZZ} of polymers I-III and diamonds denote S_{ZZ} of the monomeric liquid crystal IV. Dashed lines indicate different phase transitions. The mobile and rigid component, observed below the melting point, are distinguished by full and open symbols. Reduced temperature $T^* = T/T_{ni}$.

We now discuss the orientational order of the polymers in terms of the familiar order parameter S_{ZZ}, characterizing the average orientation of the repeating units with respect to the director. Figure 9 shows the temperature dependence of these order parameters of the polymeric (circles) and monomeric (diamonds) liquid crystals.

In the isotropic phase $S_{ZZ} = 0$, indicating a random orientation of the molecules. At the isotropic-nematic transition all order parameters jump to finite values and then continuously increase with decreasing temperature. Note, however, that the order parameter of the main chain polymers are considerably larger than those exhibited by their corresponding low molecular weight analogue [31,44-46]. As predicted [47-49], the polymer chains are highly ordered on a molecular level in agreement with ESR [50] and ^{1}H NMR studies [51,52].

Interestingly, the order parameter of the polymers is retained quantitatively when the sample is cooled below the melting point and glass transition. No change in S_{ZZ} is observed even after keeping the polymer at room temperature for a longer time period. Thus, in contrast to conventional mesogens the long range orientational order is preserved even upon crystallization. In all systems studied $S_{\overline{ZZ}}$ is independent of the molecular weight of the polymers within the range $5000 < \overline{M}_n < 30000$, according to a plateau effect, observed for liquid crystal main [52,53] and side chain polymers [54,55].

Orientational order of liquid crystal polymers in bulk samples is conveniently described in terms of micro- and macroorder parameters, characterizing the average orientation of the repeating units relative to the director and the macroscopic alignment of the directors. Determination of the macroorder is always possible, provided the molecular motion is fast.

The degree of macroscopic alignment $S_{z''z''}$ depends on the orientation method. Because the anisotropic permittivity of the polyesters I-III is negative, only a two-dimensional distribution of director axes is achieved, using high electric fields. However, a uniform alignment of all director axes is obtained by a magnetic field of B = 7.0 T. Likewise, melt-spinning of the system produces fibres with practically all of the domains aligned, $S_{z''z''} = 1.0$. High modulus and strength results from this highly oriented chain configuration [33].

Generally, the macroorder of the liquid crystal polymers is frozen in at the melting point and glass transition, respectively. No external forces are required to maintain the director distribution in the sample. Therefore, liquid crystal polymers can be used as storage materials. The information which is inserted in the nematic state by an electric field can be stored permanently in the solid state of the material. This exceptional property of liquid crystal polymers has recently been applied in a new laser-addressed thermo-optic storage device [56]. Further applications in the display technology are currently being developed [57].

Molecular Dynamics

In Figure 10 the motional correlation times of liquid crystal polymer II are plotted as a function of $1/T$. They refer to chain fluctuation (full circles), chain rotation (full triangles) and trans-gauche isomerization (full and open squares) of the first spacer segment, respectively. For comparison, the correlation times (chain rotation) of the monomeric liquid crystal are depicted, likewise (full diamonds). There is an additional diffusive motion of the polymer, not shown in the Figure which occurs below the glass transition on a time scale of 10^{-2} s $< T_R < 1$ s. However, unambiguous assignment of this ultraslow motion is not yet possible.

The correlation times, varying by seven orders of magnitude, reflect the complex moelcular dynamics in the liquid crystal, solid and glassy state. Note, that this detailed information could only be obtained by employing multipulse dynamic NMR techniques. For any given temperature at least two different relaxation experiments were carried out. Moreover, variation of pulse separation and magnetic field orientation provided additional experiments for a proper dynamic characterization of the systems.

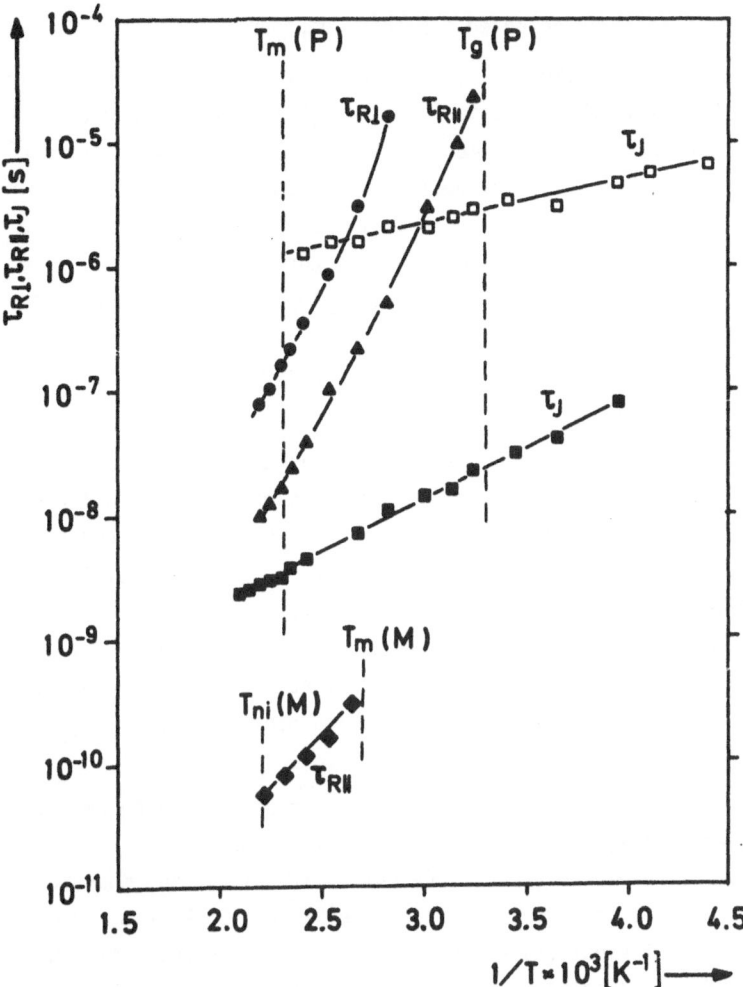

Figure 10: Arrhenius plot of various correlation times, characterizing the molecular dynamics of polymer II and monomer IV. Circles refer to chain fluctuation, triangles (polymer) and diamonds (monomer) denote chain rotation and squares denote trans-gauche isomerization. Dashed lines indicate different phase transitions. The mobile and rigid components, observed below the melting point, are distinguished by full and open symbols.

In the anisotropic polymer melt the correlation times for chain rotation and chain fluctuation are of the order of 10^{-8} s, while trans-gauche isomerization occurs even faster. Apparently, these rapid motions are responsible for the unusual rheological behavior of liquid crystal polymers [58]. Note, however, that chain rotation of the monomeric liquid crystal is still faster by two orders of magnitude.

At the melting point the polymer system becomes heterogeneous. Two components are now observed, which we assign to a liquid crystalline (full symbols) and a crystalline phase (open symbols) [59]. Decomposition of various relaxation curves into two components yields a crystallinity of (60±5)%, practically independent of temperature. In the crystalline component all motional processes are slowed down. Thus, only trans-gauche isomerization can be detected for this component.

In contrast, the dynamics of the liquid crystalline component continues into the solid region. Obviously, the Arrhenius plot for chain rotation and chain fluctuation is not linear, the apparent activation energies increasing with decreasing temperature. Thus, all intermolecular motions gradually freeze and at temperatures $T < T_g$ intramolecular motions are the dominant process. In fact, we have been able to detect trans-gauche isomerization even at $T = 130$ K with a correlation time of $T_J \cong 10^{-4}$ s. Similarly, ring flips in side chain polymers could be followed down to temperatures 100 K below the glass transition [43,60]. It should be noted that ESR spin probes, incorporated into the systems, are reliable indicators of the motional change at the glass transition [54,61].

Figure 10 clearly shows the coexistence of liquid crystalline and crystalline components differing drastically in molecular dynamics. In that respect thermotropic polymers resemble ordinary polymers, which exhibit amorphous and crystalline phases. However, a broad distribution of correlation times is generally evaluated in all but the lowest molecular weight systems [62]. In contrast, molecular reorientation in the liquid crystal polymers above T_g appears to occur essentially by single processes, where any distribution of correlation times must be restricted to less than one decade. This is particularly obvious from Figure 6, which depicts a sharp T_{1Z} minimum, characteristic of a single motional process.

Chain fluctuations in liquid crystals may occur as isolated or collective modes. For the latter mechanism, known as director fluctuations, a continuous distribution of correlation times is expected [63,64]. Recent [1]H T_{1Z}-dispersion measurements, carried out over a frequency range of six orders of magnitude (10^2 Hz $< \omega_0/2\pi < 10^8$ Hz) clearly show, that director fluctuations do not constitute a major relaxation mechanism in the MHz range [65]. Rather, isolated chain motions fully account for the observed T_{1Z} data.

In the kHz range, however, an additional dispersion step is observed, characteristic of order fluctuations. The frequency dependence of T_{1Z} of the monomeric liquid crystal ($T_{1Z} \propto \omega_0^{1/2}$) corresponds to the theoretical predictions. For the polymeric liquid crystal, however, a somewhat steeper slope is observed ($T_{1Z} \propto \omega_0^{3/4}$) [65]. Further relaxation studies in the low-field region should clarify this point.

Conclusions

Multipulse dynamic NMR has been established as a powerful tool for studying complex chemical and biological systems. Generally, variation of pulse sequence, pulse separation and magnetic field orientation provides the large number of independent experiments necessary for a proper molecular characterization of the systems. Analysis of these experiments in terms of molecular order and dynamics is conveniently achieved by employing a density matrix treatment, based on the stochastic Liouville equation. Arbitrary relaxation rates and lineshapes of single and multiple quantum transitions can be considered. In addition, our method of analysis is also applicable in the slow-motional and/or zero-field [66-68] region, where the conventional relaxation theories no longer apply.

The studies described here provide new information concerning the dynamical organization of liquid crystal polymers. There is evidence for high orientational and conformational order of the chain molecules in the anisotropic melt. From a dynamic point of view, the primary result is the detection of the dominant motions, such as chain rotation, chain fluctuation and chain isomerization in the various polymer phases. By employing multipulse dynamic NMR techniques, it is possible to follow these motions over an extremely wide dynamic range, extending from 10^{-10} s in the liquid crystalline melt to 10^{-2} s in the glassy state. T_{1Z} dispersion measurements, carried out over a frequency range of six orders of magnitude indicate that collective order fluctuations constitute a major relaxation mechanism in the kHz range.

Acknowledgements

It is a pleasure to thank Dr. B. Hisgen (University of Mainz), Dr. A. Schneller (University of Massachusetts) and Dr. C. Eisenbach (University of Freiburg) for advice and help in preparing the specifically deuterated polymers. The authors are also grateful to Dr. E. Ohmes (University of Stuttgart) for assistance in the numerical computations. Finally, we thank Prof. H. Ringsdorf (University of Mainz), Dr. P. Meier (University of Stuttgart) and Prof. F. Noack (University of Stuttgart) for many helpful discussions. Financial support of this work by the Deutsche Forschungsgemeinschaft and Fonds der Chemischen Industrie is gratefully acknowledged.

References

[1] J. Jeener. B.H. Meier, P. Bachmann and R.R. Ernst, J. Chem. Phys. **71**, 4546 (1979)

[2] H.W. Spiess, in NMR Basic Principles and Progress, Eds. P. Diehl, E. Fluck and R. Kosfeld, Vol. 15, p. 55, Springer, Berlin (1978)

[3] E. Meirovitch and J.H. Freed, Chem. Phys. Lett. **64**, 311 (1979)

[4] T. Mügele, V. Graf, W. Wölfel and F. Noack, Naturforsch. **35A**, 924 (1980)

[5] F. Winter and R. Kimmich, Mol. Phys. **45**, 33 (1982)

[6] R.R. Vold and R.L. Vold, Israel J. Chem. **23**, 315 (1983)

[7] G. Hertz, Progr. Nucl. Magn. Reson. Spectrosc. **16**, 115 (1983)

[8] J.W. Emsley, Ed., Nuclear Magnetic Resonance of Liquid Crystals, NATO Advanced Study Institute, D. Reidel Publ., Dordrecht (1985)

[9] K. Müller, P. Meier and G. Kothe, Progr. Nucl. Magn. Reson. Spectroscop. **17**, 211 (1985)

[10] P. Meier, E. Ohmes and G. Kothe, J. Chem. Phys. **85**, 3598 (1986)

[11] R. Kubo, in Stochastic Processes in Chemical Physics, Advances in Chemical Physics, Ed. K. Shuler, Vol. 16, p. 101, Wiley, New York (1969)

[12]   J.H. Freed, G.V. Bruno and C.F. Polnaszek, J. Phys. Chem. 75, 3385 (1971)

[13]   J.R. Norris and S.I. Weissman, J. Phys. Chem. 73, 3119 (1969)

[14]   G. Kothe, Mol. Phys. 33, 147 (1977)

[15]   P.J. Flory, Statistical Mechanics of Chain Molecules, Interscience Publ.,
       New York (1969)

[16]   M.A. Cotter, J. Chem. Phys. 66, 1098 (1977)

[17]   A. Saupe, Z. Naturforsch. 19A, 161 (1964)

[18]   G. Bodenhausen, Progr. Nucl. Magn. Reson. Spectrosc. 14, 137 (1980)

[19]   J.G. Powles and J.H. Strange, Proc. Phys. Soc. 82, 6 (1963)

[20]   J.H. Davis, K.R. Jeffrey, M. Bloom, M.F. Valic and T.P. Higgs, Chem. Phys.
       Lett. 42, 390 (1976)

[21]   P. Meier, E. Ohmes, G. Kothe, A. Blume, J. Weidner and H.-J. Eibl,
       J. Phys. Chem. 87, 4904 (1983)

[22]   K.R. Jeffrey, Bull. Magn. Reson. 3, 69 (1981)

[23]   J. Jeener and P. Broekaert, Phys. Rev. 157, 232 (1967)

[24]   H.W. Spiess, J. Chem. Phys. 72, 6755 (1980)

[25]   D.A. Torchia and A. Szabo, J. Magn. Reson. 49, 107 (1982)

[26]   H.W. Spiess and H. Sillescu, J. Magn. Reson. 42, 381 (1981)

[27]   M. Lausch and H.W. Spiess, J. Magn. Reson. 54, 466 (1983)

[28]   J.-I. Jin, S. Antoun, C. Ober and R.W. Lenz, Br. Polym. J. 132 (1980)

[29]   S. Antoun, R.W. Lenz and J.-I. Jin, J. Polym. Sci. Polym. Chem. Ed. 19
       1901 (1981)

[30]   J.P. Schroeder, Mol. Cryst. Liq. Cryst. 61, 229 (1980)

[31]   K. Kohlhammer, K. Müller and G. Kothe, to be published

[32]   K. Müller, C. Eisenbach, B. Hisgen, H. Ringsdorf, A. Schneller, R.W. Lenz and
       G. Kothe, to be published

[33]   K. Müller, A. Schleicher, E. Ohmes, A. Ferrarini and G. Kothe, to be published

[34]   K. Müller, B. Hisgen, H. Ringsdorf, R.W. Lenz and G. Kothe, Mol. Cryst. Liq.
       Cryst. 113, 167 (1984)

[35]   K. Müller and G. Kothe, Ber. Bunsenges. Phys. Chem. 89, 1214 (1985)

[36]   R.G. Gordon and T. Messenger, in Electron Spin Relaxation in Liquids, Eds.
       L.T. Muus and P.W. Atkins, p. 341, Plenum Press, New York 1972

[37]   G. Moro and J.H. Freed, J. Chem. Phys. 74, 3757 (1981)

[38]   A.G. Redfield, Adv. Magn. Reson., 1, 1 (1965)

[39]   K. Müller, C. Eisenbach, A. Schneller, H. Ringsdorf and G. Kothe, Progr.
       Colloid Polym. Sci. 69, 127 (1984)

[40]   K. Müller, B. Hisgen, H. Ringsdorf, R.W. Lenz and G. Kothe, in Recent Advances
       in Liquid Crystalline Polymers, Ed. L.L. Chapoy, p. 223, Elsevier Applied
       Science Publ., London 1984

[41]   E.T. Samulski, M.M. Gauthier, R.B. Blumstein and A. Blumstein, Macromolecules
       17, 479 (1984)

[42]   D.Y. Yoon, S. Bruckner, W. Volksen, J.C. Scott and A.C. Griffin, Faraday Discuss.
       Chem. Soc. 79, 41 (1985)

[43]   C. Boeffel, B. Hisgen, U. Pschorn, H. Ringsdorf and H.W. Spiess, Israel J.
       Chem. 23, 388 (1983)

[44]   S. Hsi, H. Zimmermann and Z. Luz, J. Chem. Phys. 69, 4126 (1978)

[45]   J.W. Emsley, G.R. Luckhurst and C.P. Stockley, Proc. R. Soc. London, Ser. A, <u>381</u>, 117 (1982)

[46]   E.T. Samulski, Israel J. Chem. <u>23</u>, 329 (1983)

[47]   A. Abe, Macromolecules <u>17</u>, 2280 (1984)

[48]   D.Y. Yoon and S. Bruckner, Macromolecules <u>28</u>, 65 (1985)

[49]   G. Ronca and D.Y. Yoon, J. Chem. Phys. <u>76</u>, 3295 (1982); <u>80</u>, 925 (1984)

[50]   K. Müller, K.-H. Wassmer, R.W. Lenz and G. Kothe, J. Polym. Sci. Polym. Lett. Ed. <u>21</u>, 785 (1983)

[51]   A.F. Martins, J.B. Ferreira, F. Volino, A. Blumstein and R.B. Blumstein, Macromolecules <u>16</u>, 279 (1983)

[52]   R.B. Blumstein, E.M. Stickless, M.M. Gauthier, A. Blumstein and F. Volino, Macromolcules <u>17</u>, 177 (1984)

[53]   R.B. Blumstein, E.M. Stickless and A. Blumstein, Mol. Cryst. Liq. Cryst. Lett. <u>82</u>, 205 (1982)

[54]   K.-H. Wassmer, E. Ohmes, M. Portugall, H. Ringsdorf and G. Kothe, J. Am. Chem. Soc. <u>107</u>, 1511 (1985)

[55]   H. Finkelmann, in Polymer Liquid Crystals, Eds. A. Ciferri, W.R. Krigbaum and R.B. Meyer, p. 35, Academic Press, New York 1982

[56]   V.P. Shibaev, S.G. Kostromin, N.A. Plate, S.A. Ivanov, V.Y. Vetrov and I.A. Yakovlev, Polym. Commun. <u>24</u>, 364 (1983)

[57]   H.J. Coles, Faraday Discuss. Chem. Soc. <u>79</u>, 201 (1985)

[58]   K.F. Wissbrun, J. Rheol. <u>25</u>, 619 (1981)

[59]   B. Wunderlich and J. Grebowicz, in Advances in Polymer Science, Eds. M. Gordon and N.A. Plate, Vol. 60/61, p. 1, Springer Verlag, Berlin 1984

[60]   U. Pschorn, H.W. Spiess, B. Hisgen and H. Ringsdorf, Makromol. Chem. <u>187</u>, 2711 (1986)

[61]   P. Meurisse, C. Friedrich, M. Dvolaitzky, F. Lauprêtre, C. Noel and L. Monnerie, Macromolecules <u>17</u>, 72 (1984)

[62]   V.J. McBrierty and D.C. Douglass, Macromol. Revs. <u>16</u>, 295 (1981)

[63]   C.H. Wade, Ann. Rev. Phys. Chem. <u>28</u>, 47 (1977)

[64]   R.Y. Dong, Israel J. Chem. <u>23</u>, 370 (1983)

[65]   T. Dippel, K.H. Schweikert, K. Müller, G. Kothe and F. Noack, to be published

[66]   D.P. Weitekamp, A. Bielecki, D. Zax, K. Zilm and A. Pines, Phys. Rev. Letters <u>50</u>, 1807 (1983)

[67]   R. Kreis, D. Suter and R.R. Ernst, Chem. Phys. Letters <u>123</u>, 154 (1986)

[68]   P. Meier, G. Kothe, P. Jonsen, M. Trecoske and A. Pines, J. Chem. Phys. 1987, in press

THE ELASTIC TRUMBBELL MODEL FOR DYNAMICS OF STIFF CHAINS

Daniel B. Roitman

Michigan Molecular Institute
1910 West St. Andrews Road
Midland, MI 48640 / USA

1. Introduction

The dynamical conformational behavior of macromolecules span over a very wide range of time scales, from the very fast localized side groups motions (10^{-9}s.) to the relatively slow rotations of the whole macromolecule in a viscous solvent (10^{-3}s.). In the present discussion we will be concerned mostly with the slower end of this spectrum. In particular we will present a model consisting of three beads connected by two rods and an elastic hinge (the "trumbbell" model) [1,4] to investigate the interactions between the rotations of the entire molecule and the slower bending modes of relatively stiff linear macromolecules in dilute solutions.

Several techniques can be used to investigate the conformational dynamics of macromolecules [5,6]. Viscoelasticity and transient electric birefringence (Kerr effect) are discussed in particular, but the calculations can, in principle, be applied to several other techniques (light scattering, dielectric relaxation, etc.).

The viscoelastic and transient Kerr effect behavior of dilute solutions containing stiff linear macromolecules cannot be explained in terms of models that assume rigid rod-like structures for the chain [6,7]. The storage modulus $G'(\omega)$ of these solutions is, for instance, one or two orders of magnitude larger than those of rod-like models. It has been proposed that bending modes of the backbone are responsible for this difference. Kerr effect studies of DNA solutions [8-10] have shown that as the fragment lengths are increased, the observed longest relaxation times (τ_1) become substantially smaller than those predicted on the basis of rigid rod behavior, suggesting that this difference is due to the internal flexibility of the DNA. In addition, multiple exponential behavior is observed for longer molecules and the relative contributions of the faster modes increase with contour length of the molecules L.

In this presentation the methods and results of the trumbbell model are reviewed in Section 2. In Sections 3 and 4, the results are compared with recent Kerr-effect relaxation measurements [8-10] and viscoelastic data [7]; and finally, suggestions for future developments are considered.

2. Theory and Methods

The most difficult problem regarding theoretical treatment of dynamics of semi-flexible polymers resides in the fact that the internal coordinates describing the shape of the chain change simultaneously with the rotations of the whole chain. In the trumbbell model there are four curvilinear non-orthogonal coordinates. The angle χ defines the bending (Fig. 1a) and the three Eulerian angles, α, β, γ (Fig. 1b), define the orientation of the trumbbell with respect to the laboratory coordinates. In addition,

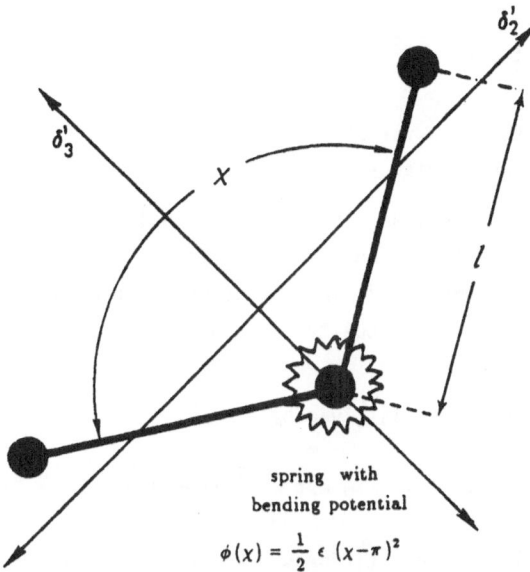

Fig. 1a: The elastic trumbbell and the "molecular" frame of reference.

spring with bending potential

$$\phi(\chi) = \frac{1}{2}\,\epsilon\,(\chi-\pi)^2$$

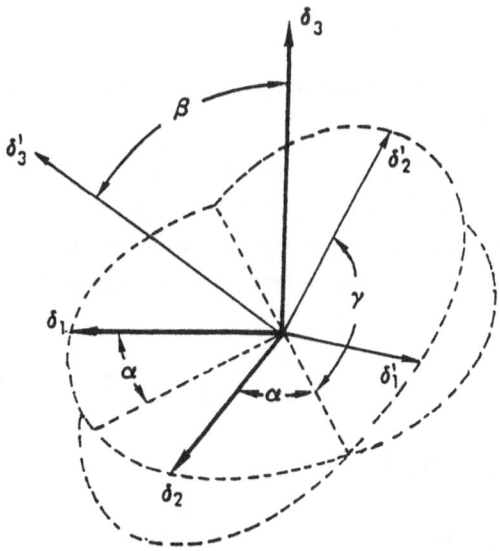

Fig. 1b: The laboratory frame of reference and the three Euler angles.

an elastic hinge at the middle is used to achieve varying degrees of chain stiffness. Although the model is a very drastic simplification of a semi-flexible macromolecule, it seems to be the most complex one for which the dynamic behavior has been completely calculated in the framework of the Kirkwood Diffusion Equation method.

The model consists of three beads having the same friction coefficient ζ without hydrodynamic interactions between them. The beads are connected by two frictionless rods of length ℓ with a joint at the middle bead. The bending potential at the joint is given by

$$\phi(\chi) = \frac{1}{2} \varepsilon (\chi - \pi)^2 \tag{2.1}$$

where ε is the spring stiffness, a variable parameter. This model is an extension of Hassager's free-hinge model, which has $\varepsilon = 0$ [11].

The dynamical behavior of the model is obtained by solving the linearized Kirkwood diffusion equation [12]. In following Kirkwood's formulation, all its underlying physical simplifications are assumed [13]; inertia effects are neglected, the solvent is considered a Newtonian fluid with viscosity η_s, and the drag forces on the beads are approximated with Stoke's law.

The Kirkwood diffusion equation is given by:

$$\frac{kT}{\zeta \ell^2} \, \underset{\sim}{L} \, \rho(\bar{q},t) - \frac{\partial \rho}{\partial t} = \underset{\sim}{Q} \exp\left\{-\frac{Z}{2} (\chi - \pi)^2\right\} \tag{2.2}$$

where $Z = \frac{\varepsilon}{2kT}$ is the relative spring stiffness, kT has the usual meaning, $\bar{q} = \alpha, \beta, \gamma, \chi$, and

$$\underset{\sim}{L} = \underset{\sim}{\nabla}^2 + \underset{\sim}{U}(Z,\bar{q}). \tag{2.3}$$

The operator $\underset{\sim}{Q}$ contains the external fields, while the operator $\underset{\sim}{U}$ arises from the internal bending potential ϕ.

In writing Eqs. (2.2) and (2.3) it is assumed that the distribution function $\Omega(\bar{q},t)$, which describes the temporal evolution of the model, can be expanded in powers of a parameter u which measures the departure from equilibrium (for instance the applied flow rate $\underset{\approx}{\kappa}_0$ of a viscoelastic experiment or the square of the applied electric field E_0^2 of the Kerr effect experiment):

$$\Omega(\bar{q},t) = \frac{f^{1/2}}{N} \left[f^{1/2} + u \rho(\bar{q},t) + \dots \right] \tag{2.4a}$$

where N is a normalization constant and f is the Boltzmann factor

$$f = \exp[-Z(\chi - \pi)^2] . \tag{2.4b}$$

Usually, only the lowest perturbation term in (2.4a) is retained. The distribution function then can be written in terms of functions Ψ_p (see Eq. (3.3)):

$$\rho(\bar{q},t) = \sum_{p=1}^{\infty} A_p \Psi_p(\bar{q}) \Gamma(\lambda_p,t) \tag{2.5}$$

where the coefficients A_p are given by the average

$$A_p = \int \underset{\sim}{Q} f^{1/2} \Psi_p \, d\bar{v} . \tag{2.6}$$

The functions Ψ_p are solutions of the eigenvalue problem

$$\underset{\sim}{L} \Psi_p + \lambda_p \Psi_p = 0 . \tag{2.7}$$

The eigenvalues λ_p represent the relaxation spectrum of the model and the eigenfunctions Ψ_p · represent the corresponding relaxation modes.

The eigenvalue problem Eq. (2.7) is solved using a numerical approximation called the "collocation method" [11]. This method consists of fitting trial functions F_p at a finite number of selected points (the collocation points) in the domain of the independent variables:

$$\Psi_p \approx F_p = \sum_{i=0}^{I-1} \sum_{i=0}^{J-1} \sum_{i=0}^{K-1} a_{ijk}^p \cos(2i\beta)\cos(2j\gamma)h_k(\chi,Z) \tag{2.8a}$$

where

$$h_k = f^{1/2} \sum_{s=0}^{k} b_{ks} \cos^s(\chi) . \tag{2.8b}$$

The coefficients b_{ks} are obtained from orthogonality relations between the functions h_k (χ,Z). The particular choice for the trial functions F_p, in References [1-3], is based on boundary conditions on the Eulerian angles and the asymptotic behavior for the model in the freely hinged case $(Z = 0)$ as well as the rigid-rod limit $(Z = \infty)$ [11], but other functions satisfying the proper boundary conditions could have been used. The sums of Eq. (2.8a) are carried over the number of collocation points ·K, J, K chosen for the β,γ,χ angles respectively. For symmetry reasons α does not enter. The coefficients a_{ijk}^p are determined by substitution of Eq. (2.8a) in Eq. (2.7) and by solving the corresponding algebraic eigenvalue problem.

Good convergence is quickly established when the collocation points of each variable I,J,K are varied from 1 to 5. The behavior of the relaxation spectrum λ_p as a function of $Z = \epsilon/2kT$ is shown in Fig. 2.

The calculations for the Kerr effect and visoelastic behavior (see Section 3) showed that the dominant relaxations are λ_1 and λ_2, while λ_3 and λ_4 are responsible for the remaining contributions (less than 10%). The modes associated with the

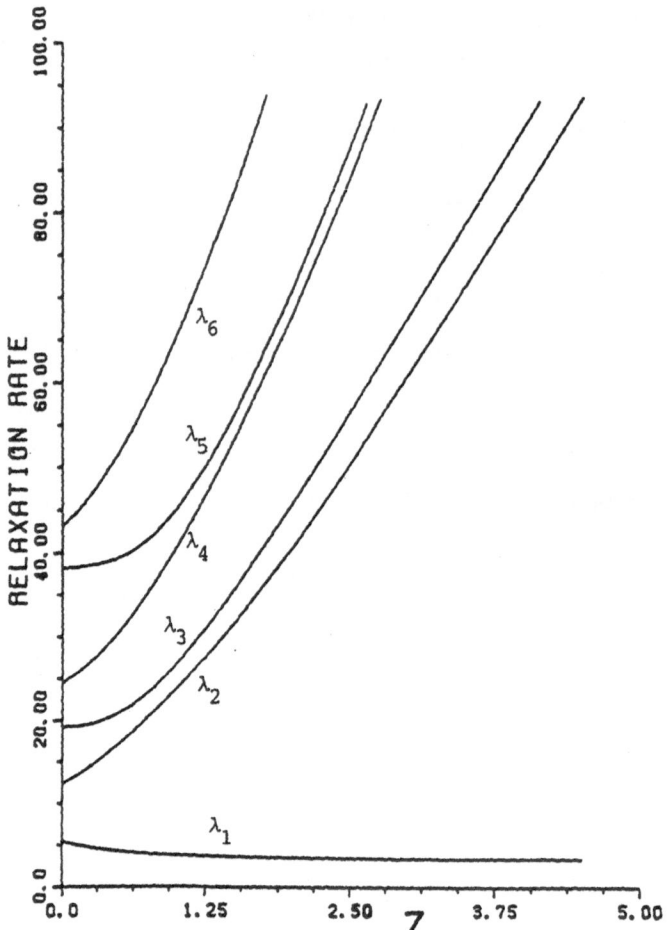

Fig. 2: Relaxation rates, λ_p, of the trumbbell as functions of the relative
spring siffness $Z = \varepsilon/(2kT)$.

dominant relaxations, Ψ_1 and Ψ_2, have complicated dependences on the bending angle
χ and orientation (β,γ) for small values of Z. This is an indication that bending
and rotations cannot be treated separately in this case. For larger Z, however, the
functions take the following asymptotic form:

$$\Psi_1 \approx h_0(\cos^2\theta_2 - \tfrac{1}{3}); \quad \cos\theta_2 = \sin(\beta)\sin(\gamma) \qquad (2.10)$$

$$\Psi_2 \approx \left[h_1(\chi,Z) + 0.18\, h_2(\chi,Z) + 0.03\, h_3(\chi,Z)\right] \times \left(\cos^2\theta_2 - \tfrac{1}{3}\right). \qquad (2.11)$$

The slowest mode Ψ_1 takes the angular dependence of a rigid rod, multiplied by a
factor proportional to $f^{1/2}$, which "modulates" the effective length of the rod. In
addition, Fig. 2 shows that λ_1 slows down towards $\lambda_R = 3$ as Z becomes larger ($\lambda_R = 3$

is the rigid-rod rotational relaxation limit). These two observations suggest that Ψ_1 is associated with the end-over-end rotation of the model when the hinge is "stiff" ($Z > 1$). When $Z > 1$, the next dominant mode, Ψ_2, approaches the form of a product of the angular dependence of the rigid rod times a bending function. Since the relaxation rate λ_2 increases almost proportionally to Z, it is proposed that Ψ_2 is associated with the bending of the model when the chain is stiff.

3. Kerr Effect and Viscoelasticity

In a transient Kerr effect experiment, an electric field $\bar{E} = E_0 \bar{\delta}_3 (t < 0)$ is applied to the solution. The macromolecules become oriented due to their permanent and induced dipole moments including birefringence to the solution. Then, the field is removed ($\bar{E} = 0$, $t \geq 0$). The birefringence decays as the macromolecules diffuse towards an isotropic distribution. The decay rates are determined by the shape, size and flexibility of the macromolecules. In the present distribution, we assume that the macromolecules have only induced dipole moments. In this case the electric potential energy V is given by:

$$V = -\frac{1}{2} (\bar{\bar{E}}\bar{\bar{E}} : \underset{=}{\pi}), \ t < 0 \tag{3.1}$$

where $\underset{=}{\pi}$ is the electrical polarizability of the model. This situation applies to DNA since there is evidence that DNA does not possess a permanent dipole moment [9].

In the following discussion it is assumed that the electric and optical polarizabilities differ only by a scalar factor. If the polarizability of each rod is anisotropic, with the principal axis along the direction of each rod, the polarizability of the trumbbell (in the molecular coordinate system) is given by:

$$\underset{=}{\pi}' = \pi_0 \ 2 \begin{pmatrix} 0 & 0 & 0 \\ 0 & \sin^2(\chi/2) & 0 \\ 0 & 0 & \cos^2(\chi/2) \end{pmatrix}. \tag{3.2a'}$$

The matrix of Eq. (3.2a) is then projected onto the laboratory coordinate system by a similarity transformation [2]:

$$\underset{=}{\pi} = \underset{=}{C}^{-1} \underset{=}{\pi}' \underset{=}{C}. \tag{3.2b}$$

Using the methods of the previous section, we find

$$\rho(\bar{q},t) = \frac{1}{2kT} \sum_{p=1}^{\infty} <\pi_{33}>_p \ \Psi_p(\bar{q})\Theta_p(t) \tag{3.3}$$

where

$$\Theta_p(t) = \exp\left\{ -\left(\frac{\lambda_p kT}{\ell^2 \zeta}\right) t \right\} \tag{3.4}$$

and

$$<\Pi_{33}>_p = \int \Pi_{33}\, f^{1/2}\, \Psi_p\, (\bar{q})d\bar{v}\ . \tag{3.5}$$

The difference between the principal refractive indices of the fluid in the $\bar{\delta}_3\bar{\delta}_2$ plane [12] is proportional to

$$B = n\left[(<\Pi_{33}> - <\Pi_{22}>)^2 + 4<\Pi_{32}>^2 \right]^{1/2} \tag{3.6}$$

where n is the number density of macromolecules and

$$<\Pi_{k\ell}> = \int \Pi_{k\ell}\, \Omega(\bar{q},t)d\bar{v}. \tag{3.7}$$

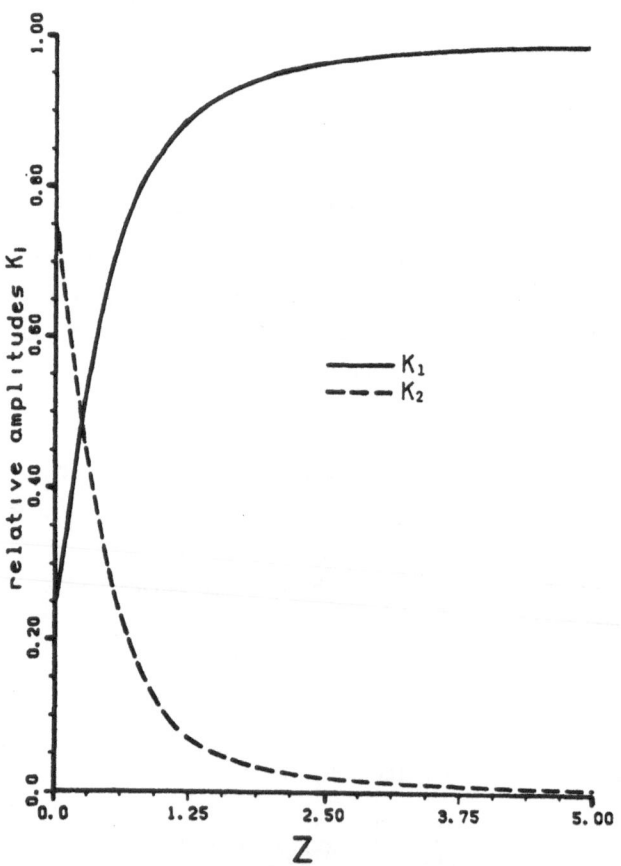

Fig. 3: Kerr effect amplitudes K_1 (continuous curve) and K_2 (broken line) against the relative spring stiffness.

Using the methods of the previous section, one finally obtains [2]

$$\frac{B(t)}{B(0)} = \sum_p K_p \, \Theta_p(t) \tag{3.8}$$

for the decay of the relative birefringence. In Eq. (3.8) the amplitudes K_p are

$$K_p = \frac{1}{D} \, (\langle \pi_{33} \rangle_p - \langle \pi_{22} \rangle_p) \langle \pi_{33} \rangle_p \tag{3.9a}$$

$$D = \sum_p \, (\langle \pi_{33} \rangle_p - \langle \pi_{22} \rangle_p) \langle \pi_{33} \rangle_p \tag{3.9b}$$

The first two terms (p = 1,2) dominate the sum (3.8). The contributions K_1 and K_2 are plotted in Fig. 3.

An important feature of the results is that the relative amplitudes of K_1 and K_2 are strong functions of the stiffness of the trumbbell. While the coefficient of the faster mode, K_2, is responsible for more than 3/4 of the relative birefringence at $Z = 0$ (free-hinge limit), its contribution becomes small when $Z > 0.75$. As a consequence, the birefringence decay curves appear bimodal for small Z, but become close to single exponentials for $Z \geq 0.75$ (Fig. 4).

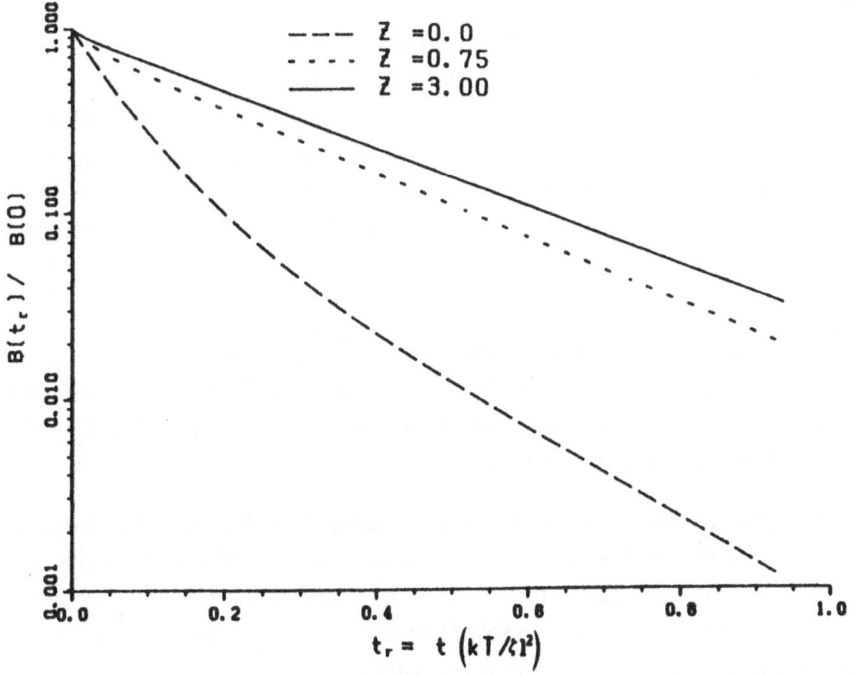

Fig. 4: Semi-log plot of the relative birefringence decay curves of·the model. The free-hinge case, $Z = 0$, shows multiple exponential behavior, while cases with $Z = 0.75$ and $Z = 3.0$ show nearly single exponential behavior. Notice that the slope of the curve for $Z = 0.75$ is steeper ($\lambda_1 = 3.87$) than the one for $Z = 3.0$ ($\lambda_1 = 3.22$), indicating substantially more bending of the trumbbell in the former case.

The viscoelastic behavior of the model was investigated as well [1]. The resulting expressions for the reduced storage and loss moduli $G_r'(\omega_F)$ and $G_\gamma''(\omega_F)$ respectively are [1,11]:

$$G_r'(\omega_F) = \omega_F^2 \sum_{p=1}^{\infty} \frac{b_p \lambda_p^{-2}}{1 + (\omega_F/\lambda_p)^2} \tag{3.10}$$

$$G_r''(\omega_F) = \omega_F \left[\frac{a}{2} + \sum_{p=1}^{\infty} \frac{b_p \lambda_p^{-1}}{1 + (\omega_F/\lambda_p)^2} \right] \tag{3.11}$$

a is given in Ref. [1], the reduced frequency is [6]

$$\omega_F = \omega \eta_s [\eta] \frac{Mw}{RT} \tag{3.12}$$

and the coefficients b_p are the integrals

$$b_p = \frac{\lambda_p^2}{8N\ell^4(\underline{\kappa}_0 : \underline{\kappa}_0)} \left[\int \int \left(\sum_{\nu=1}^{3} (\underline{\kappa}_0 : \bar{R}_\nu \bar{R}_\nu) \right) f^{1/2} \, \Psi_p d\bar{\nu} \right]^2 \tag{3.13}$$

where $\underline{\kappa}_0$ is the velocity gradient tensor (13) and $\bar{R}_\nu$ is the position of the ν-th bead with respect to the center of mass of the trumbbell.

The terms with coefficients b_1 and b_2 dominate the sums (3.10) and (3.11). Figure 5 shows the behavior of the model. G_r' and G_r'' are plotted in a log-log plot against ω_F for the cases when $Z = 0$ (free-hinge limit), $Z = 4$, and for the rigid rod limit ($Z = \infty$). The data, obtained by Hvidt et al. [7], correspond to solutions of light meromyosin (LMM), a very stiff muscular protein (persistence length $q = 155 \pm 20$ nm, contour length $L \approx 83$ nm).

The most striking feature of Fig. 5 can be seen in the storage curves G_r'. The elastic trumbell with $Z = 4$ shows a shallow gap between the first and second relaxation modes due to the large separation between these modes ($\lambda_2/\lambda_1 = 26$). G_r' of $Z = 4$ then climbs above the corresponding curves of the non-elastic models ($Z = 0$ and $Z = \infty$).

The data show good agreement with the elastic model ($Z = 4$). Unfortunately, data at higher frequencies is not available. Other stiff molecules show steady increases of their G_r' at higher frequencies [6]. This is due, presumably, to the contributions of their higher bending modes [6,7]. The trumbbell model, having a single hinge, cannot model the higher frequency behavior of these molecules.

The rise of G_r when $Z = 4$ above the non-elastic cases is explained as follows. Chains without internal potentials can only contribute to storage through the gradients

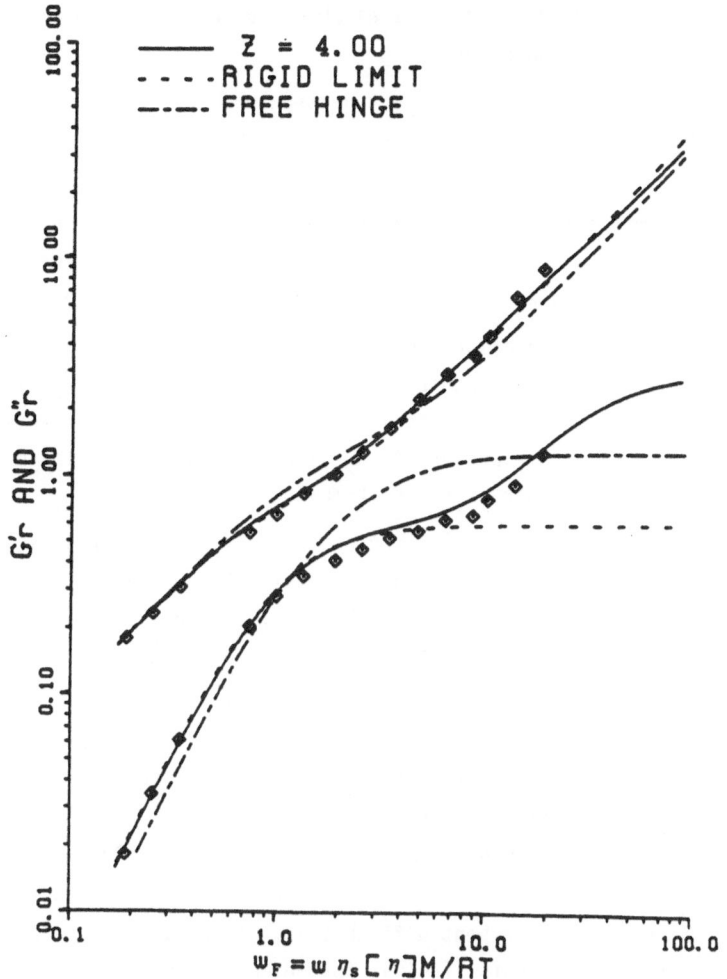

Fig. 5: Reduced viscoelastic moduli of light meromyosin (LMM) (obtained by Hvidt et al. [7]) against reduced frequency. The solid curves correspond to the elastic trumbbell with Z = 4, λ_2/λ_1 = 26. The dashed curves correspond to Hassager's free-hinged trumbbell (Z = 0) and rigid-rod limits (Z = ∞) as indicated. Notice the wide separation between the steeper parts of the solid curve for G_r^1, in agreement with the data.

in their distribution created by the flow (entropic storage), while chains with elastic potentials can contribute with mechanical as well as entropic storage.

A comparison between the viscoelastic results (Fig. 5) and the transient Kerr Effect curves (Fig. 4) suggests that the former appear to be more sensitive to internal bending dynamics of stiff chains (Z > 1) than the latter.

4. Kerr Effect. Comparison with Experiments

Studies of macromolecules by means of the transient Kerr effect have been performed for many years [5]. Polydispersity and aggregation have been serious problems in interpreting the results, however. Recent advances in preparation (recombinant DNA) and separation

techniques (gel electrophoresis) are now available, making it possible to obtain solutions that are practically monodisperse.

There is a broad qualitative agreement between results of the trumbbell model and
the findings of the studies of Stellwagen [9], Hagerman [8] and Lewis, Pecora and
Eden [10]:

(1) Short DNA molecules show single exponential decay. This behavior is expected
for the trumbbell with $Z > 1$. Unimodal decay is not necessarily indicative of rigid-rod
behavior, however (as it is clearly shown by the viscoelastic behavior, Fig. 5).

(2) Two exponential relaxation times (at least) τ_1 and τ_2, are observed for
longer molecules, and the relative contribution of the faster mode, τ_2, increases with
the molecular contour length L. The model shows mostly twofold exponential decay when
$Z < 0.75$, and the relative contribution of K_2 increases with the flexibility of the
hinge (this is analogous to increasing L keeping Z constant).

(3) The ratio of the measured relaxation times τ_1 to the relaxation τ_R (τ_R
is the relaxation time of a straight rigid-rod model with the same contour length L)
is always $\tau_1/\tau_R < 1$, i.e., the real molecules show faster rotational diffusivity than
their straight rigid counterparts. The slowest relaxation rate λ_1 of the model is always faster than the rotational relaxation rate of the rigid rod ($\lambda_2 = 3.00$). Note:
$\lambda_R/\lambda_1 \sim \tau_1/\tau_R$.

(4) Lewis et al. [10] found that the ratio of the measured relaxation time τ_1
to the observed faster relaxation τ_2 decreased as the contour length L of the molecules increased. This is in agreement with the behavior of the model (Fig. 2).

Quantitative comparisons present difficulties, since high electric fields
($E_0 \sim 1kV - 10kV$) were used in Refs. [8-10], for which the relative amplitudes of each
component on the decay curves showed dependence of the field strength. Lewis et al.,
however, found that the two dominant relaxation times, obtained using pulsed fields,
appear to be insensitive to the intensity of the fields or the duration of the pulses.

In order to attempt quantitative comparisons between the model and the experimental observations, we introduce the persistence length (q_T) of the trumbbell [4,14,15]

$$q_T = \frac{L/2}{1 - <\cos(\pi - \chi)>_{eq}} \quad ; \quad L = 2\ell \qquad (4.1)$$

where the average is taken over the equilibrium distribution function Eq. (2.4b). A plot
of q_T/L against Z (Fig. 6) shows that we can approximate

$$\frac{q_T}{L} \approx 0.23 + Z + \ldots \qquad (4.2)$$

within 10% for $Z > 0.75$ ($q_T/L \geq 1$). The above equation can be used to relate the relax-

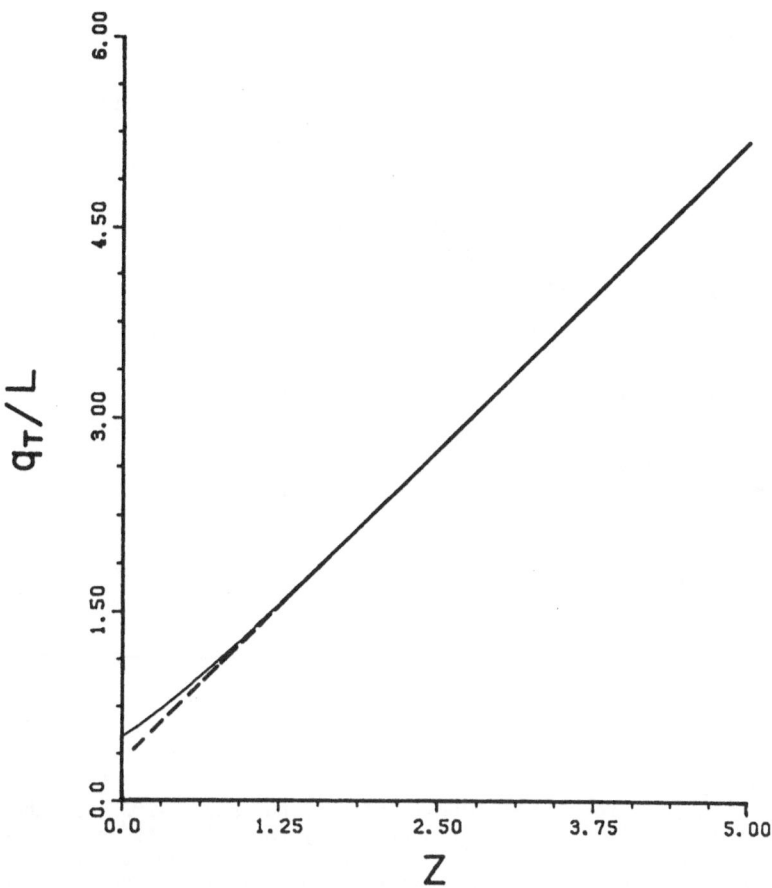

Fig. 6: Reduced persistence length of the trumbbell q_T/L against Z. The dashed line corresponds to the linear approximation Eq. (4.2).

ation rates λ_1 and λ_2 with q_T/L as follows:

1) - Plotting λ_R/λ_1 against q_T/L (or L/q_T).

2) - Relating λ_2/λ_1 with q_T/L .

The first method is illustrated in Fig. 7, where the solid curve corresponds to the trumbbell model and the data was obtained by Stellwagen [9] and Hagerman [8] for DNA solutions with different ionic strengths and contour lengths L. The relaxation times τ_R are those of rigid rods with the same L as the DNA fragments. They were calculated using Broersma's expression [16]:

$$\left(\frac{1}{\tau_R}\right) = \frac{18kT}{\pi\eta_s L^3}\left\{\ln\left(\frac{2d}{L}\right) - 1.57 + 7\left[\frac{1}{\ln(\frac{2d}{L})} - 0.28\right]^2\right\} \qquad (4.3)$$

where the diameter of the rod was assumed to be $d = 2.6$ nm and L was estimated assuming 0.34 nm for every base pair. The solvent is water at 20^0C. The following per-

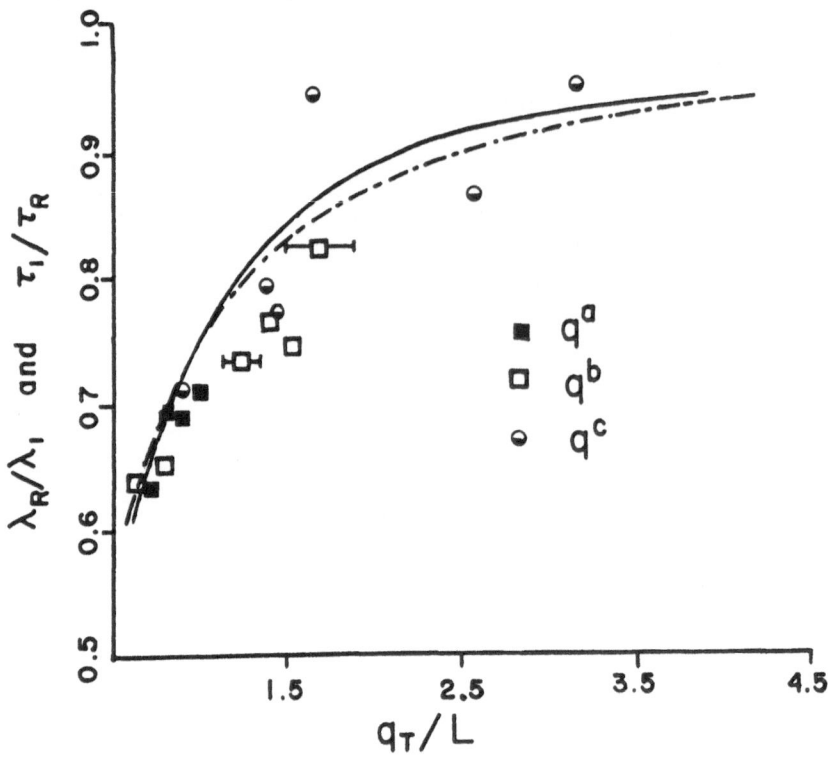

Fig. 7: Ratios λ_R/λ_1 for the trumbbell (solid line) and τ_1/τ_R for the Hagerman-Zimm model (dashed line) against the reduced persistence lengths q/L of each model. The data are the Kerr-effect relaxation time ratios determined by Stellwagen [9] and Hagerman [8] as indicated in the text. The error bars indicate the effect of 10% variations in the estimated persistence lengths.

sistence lengths were assumed:

q^a = 64 nm, DNA in 1.0 mM NaCl. L = 65.3 nm to 90.8 nm [8]

q^b = 110 nm, DNA in 0.2 mM NaCl. L = 65.3 nm to 176.8 nm [8]

q^c = 150 nm, DNA in 0.2 mM tris buff. L = 44.2 nm to 159 nm [9].

The decrease of the persistence length q with increasing ionic strength is attributed to the screening of the negative charges of the phosphate groups of the DNA by the Na^+ ions [8,17].

The broken line in Fig. 7 is the Hagerman-Zimm Monte Carlo simulation for the rotational relaxation of a worm-like chain [18]. Despite their differences, both models appear to show good agreement with the data and with each other. This agreement can be attributed to the insensitivity of the ratio λ_R/λ_1 (or λ_1/λ_R) to model details [14]. For the trumbbell, this ratio is bounded between 0.55 and 1.0 and it becomes rather insensisitive to changes of q_T when $q_T/L > 1.5$.

The second method (λ_2/λ_1 vs. q_T/L) is a more direct and sensitive test for the model since there is no need to resort to hypothetical rigid rods, and the ratio λ_2/λ_1 is quite sensitive to changes in q_T/L. Using Fig. 2 and Eq. (4.2) one obtains the approximate expression:

$$\frac{\lambda_2}{\lambda_1} = -3.6 + 7.1 \left(\frac{q_T}{L}\right) \tag{4.4}$$

which is within 10% of the tabulated results of the trumbbell model [1,14] when $Z \geq 1$ ($\lambda_2/\lambda_1 \geq 6.5$ and $q_T/L \geq 1.3$).

Since $\lambda_1 \approx \lambda_R$ ($Z > 2$) and $\lambda_R \sim L^{-3}$ for a rigid rod (Eq. (4.3)), it is found that $\lambda_2 \sim L^{-4}$ (large Z). This behavior is in agreement with previous results [19,20].

Table 1: Persistence Lengths of Stiff Molecules Estimated from λ_2/λ_1

Molecule	Method	λ_2/λ_1	L (nm)	q_T (nm)	q(nm) (reported)	Ref.
LMM[7]	Viscoelastic	26	83	352	155 ± 20	7,14[*]
MR[21]	Viscoelastic	7.8 ± 0.2	144	244 ± 10	120 ± 30	7,14[*]
DNA[10] (367 bp)	Kerr Effect	6.26	125	158[**]	50	10
DNA[10] (762 bp)	Kerr Effect	4.73	259	253[**]	70	10
DNA[10] (1010 bp)	Kerr Effect	4.62	343	324[**]	100	10

[*]The authors of Ref. [7] analyzed the results of several laboratories to conclude that both LMM and MR have persistence lengths q = 130 ± 40 nm. The entries in this column are an alternative interpretation of those results [14].

[**]Obtained by interpolation from Tables 4.3 and 4.7 of Ref. [14]

Table 1 shows the estimated persistence lengths from λ_2/λ_1 for several molecules. The values of λ_2/λ_1 for Myosin Rod (MR), Light Meromyosin (LMM) (LMM is a subfragment of MR) were obtained by fitting the model [15] to the viscoelastic measurements of Hvidt et al. [7,10]. The values of λ_2/λ_1 for DNA were obtained by Lewis et al. [10].

The persistence lengths q_T predicted by the model using the ratio λ_2/λ_1 are systematically longer than those obtained by other methods (including the first method discussed above), as it is indicated in Table 1. The trends, however, are consistent with those results: MR appears to be more flexible (shorter q) than LMM, and the persistence lengths of the DNA fragments of Reference 10 appear to increase as the contour length L increases. This latter trend is not well understood [10].

5. Conclusion

The dynamical behavior of the elastic trumbbell model was discussed. Two relaxation modes, the slowest one (λ_1) associated to rotation and the faster one (λ_2) to bending, dominate the relaxation spectrum of viscoelasticity and Kerr Effect. As the stiffness of the spring increases, the slowest relaxation mode Ψ_1 becomes dominant in the Kerr Effect case $(K_1 \gg K_2)$, and the decay curves fit nearly single exponentials if $Z \geq 0.75$. This, however, does not mean that substantial bending is not present. The effect of bending is apparent in the fact that the end-over-end relaxation rate λ_1 is faster than the rigid rod limit λ_R (30% faster for $Z = 0,75$). Viscoelasticity, on the other hand, shows a large contribution from the internal bending mode Ψ_2 to the storage modulus G_r'. The bimodal character of the curve is more pronounced the stiffer the chain. Viscoelastic measurements of very stiff macromolecules (q/L > 2) are, however, difficult to perform. In most cases [6,7,21,22], however, the elastic trumbbell model has failed to reproduce the high frequency behavior of these semistiff molecules be-cause higher bending modes appear to contribute [6] to the relaxation spectrum.

Previous theoretical work related to the dynamics of semi-stiff molecules used models that assumed small departures from rigid-body behavior [4,20,23]. Since most stiff molecules that have been investigated, on the other hand, possess $4 < \lambda_2/\lambda_1 < 15$ (LMM is exceptionally rigid with $\lambda_2/\lambda_1 \approx 26$), these models appear to be inadequate. It is clear that models with more degrees of freedom than the trumbbell must be tackled. The Kirkwood method that we have used here provides an exact treatment, but this method is too difficult for any but a few simple cases. A five beads model [24] (Pentabbell) is probably tractable with methods similar to the ones presented here. A different approach that it is being developed consists of performing dynamic simulations of multibead models [25,26]. Finally, the most challenging approach to this problem would be to solve the dynamics of the worm-like chain in the presence of external fields. Fixman [27] has recently proposed new ideas towards a solution of this problem.

Acknowledgments

This work was mostly done as part of the Doctorate Thesis of this author under the di-rection of Professor B.H. Zimm. The kind invitation of Professor R. Pecora for partici-pating in the Symposium and the hospitality of the Zentrum für interdisziplinäre For-schung der Universität Bielefeld and Professor Th. Dorfmüller are greatly appreciated. This manuscript was written with the support of MMI.

References

[1] D.B. Roitman and B.H. Zimm, J. Chem. Phys. 81, 6333 (1984)
[2] D.B. Roitman and B.H. Zimm, J. Chem. Phys. 81, 6348 (1984)
[3] D.B. Roitman, J. Chem. Phys. 81, 6356 (1984)

[4] N. Nagasaka and H. Yamakawa, J. Chem. Phys. 83, 6480 (1985)

[5] C. O'Konski, Ed., "Molecular Electro-optics", Vols. I and II. (M. Dekker Inc., New York 1976 and 1978)

[6] J.D. Ferry, "Viscoelastic Properties of Polymers", 3rd Ed. (Wiley, New York 1980)

[7] S. Hvidt, J.D. Ferry, D.L. Roelke and M.L. Greaser, Macromolecules 16, 740 (1983)

[8] P. Hagerman, Biopolymers 20, 1503 (1981)

[9] N. Stellwagen, Biopolymers 20, 399 (1981)

[10] R.J. Lewis, R. Pecora and D. Eden, Macromolecules 19, 134 (1986)

[11] O. Hassager, J. Chem. Phys. 60, 4001 (1974)

[12] J.G. Kirkwood, J. Polym. Sci. 12, 1 (1954)

[13] R.B. Bird, O. Hassager, R. Armstrong and C.F. Curtiss, "Dynamics of Polymeric Liquids", Vol. 2, (Wiley, New York 1977)

[14] D.B. Roitman, Ph.D. thesis, University of California, San Diego, 1984

[15] C. Post, Biopolymers 22. 1087 (1983)

[16] S. Broersma, J. Chem. Phys. 32, 1626 (1960)

[17] M. Troll, D.B. Roitman, J. Conrad and B.H. Zimm, Macromolecules 19 1186 (1986)

[18] P. Hagerman and B.H. Zimm, Biopolymers 20, 1481 (1981)

[19] N. Ookubo, M. Komatsubara, H. Nakajima and Y. Wada, Biopolymers 15, 929 (1976)

[20] S.R. Aragón S. and R. Pecora, Macromolecules 18, 1868 (1985)

[21] S. Hvidt, F.H.M. Nestler, M.L. Greaser and J.d. Ferry, Biochemistry 21, 4064 (1982)

[22] C.J. Carriere, E.J. Amis, J.L. Schrag and J.D. Ferry, Macromolecules 18, 2019 (1985)

[23] K. Iwata, J. Chem. Phys. 71, 931 (1979)

[24] C.F. Curtiss and R.B. Bird, J. Non Newt. Fluid Mech., 392 (1977)

[25] S.A. Allison, Macromolecules 19, 118 (1986)

[26] R.J. Lewis and R. Pecora, personal communication

[27] M. Fixman, "Brownian Dynamics of Chain Polymers", to be published

ROTATION OF LARGE MOLECULAR IONS AND TRANSIENT
DIELECTRIC RELAXATION EFFECTS

Kenneth G. Spears
Department of Chemistry
Northwestern University
Evanston, Illinois, 60201, USA

Introduction

In this manuscript we discuss the rotational motions of charged molecules and briefly comment on transient dielectric response in the chemical dynamics of ions. The molecular volumes range from 100-400 $\mathring{A}^3$ and the molecular shapes and charges vary according to structure. We discuss the observed rotational rates in terms of hydrodynamic models, local interactions of solvent with molecular charge and the long range dielectric friction effects of a rotating dipole. The following sections on rotational motions include a brief history, a review of basic theory and experimental techniques, recent data from our laboratory, a solvent coordination model, a discussion of dielectric friction theory and experiment, and a summary of some other work on viscosity effects. A short discussion of transient dielectric effects includes a brief discussion of ionic photodissociation and the role of dielectric relaxation in transient processes.

History

Molecules of large size have sufficiently slow rotational motions that the time resolved techniques of picosecond spectroscopy could be used to monitor the rotational correlation time. The early work of Chuang and Eisenthal [1] was followed by Fleming et al. [2], Spears and Cramer [3], and a number of other workers [4-7]. We will not attempt to review all of the work, but emphasize those results which relate to the main themes of this paper.

Several different themes have motivated the published work on large molecule rotations. The dominant framework for discussion has always used the Debye-Stokes-Einstein model for rotational diffusion, with modifications for molecular shape and hydrodynamic boundary conditions. Additional considerations of molecular charge and solvent interactions have included short range solvent coordination [3] and long range effects of dielectric friction [8].

The rotational motions of ionic solutes is an important probe of solvent coordination and solvent dielectric response. For large ionic molecules these solvent interactions with the ionic charge are averaged over the "long" time scale of molecular rotation. Consequently, the structural information about the solvent micro-environment or the dielectric response of the environment is sufficiently time averaged to allow statistical models. The motivation for studying rotations of charged molecules includes our desires to understand the special micro-environment of a charged molecule in a variety of solvents and solvent conditions. Furthermore, the transition state dynamics of charged molecules has exaggerated dependences on solvent properties. The experimental and theoretical probe of rotational motions of ionic molecules can contribute towards our understanding of dynamical processes involving ions, especially as technological improvements allow measurements of smaller ionic solutes and rotational time scales on the order of solvent dynamic response.

Basic Theory and Experiment

The dominant framework for understanding the rotation of large molecules in solution has been the Debye-Stokes-Einstein (DSE) model modified by frictional [9] and shape factors [1-4]. The predicted orientational relaxation time, T_{or}, is given by

$$T_{or} = \frac{S}{\lambda} \frac{\eta V}{kT} \tag{1}$$

where

η : viscosity

V : molecular volume

kT : Boltzmann's constant (k), temperature (T)

S : shape factor for ellipsoid (S = 1, sphere)

λ : slip factor .

The various predictions for molecular rotation include estimates of S and V with a computation of λ for the two limits of hydrodynamic boundary conditions. The shape factor S controls the value of λ in the slip limit. The shape factor equations[2] for a prolate ellipsoid with an axial ratio, ρ, of the long/short axes are given by

$$S = \frac{3\rho[(2\rho^2 - 1)S' - \rho]}{2(\rho^4 - 1)} \tag{2}$$

$$S' = \frac{\ln[\rho + (\rho^2 - 1)^{1/2}]}{(\rho^2 - 1)^{1/2}} \tag{3}$$

For example, an axial ratio of 1.99 yields a λ-value of 4.20 in the slip limit.

Experimentally, the most important methods for measuring rotational motion of large molecules are by polarized excitation and probe techniques. Excited state rotation times are measured by pumping a population of molecules with a polarized laser pulse and then observing the time response of fluorescence for several polarizations. Transient absorption techniques use a polarized absorption to some long lived state to deplete the ground state population, and then use polarized absorption of a weak probe beam to observe the rotational randomization of the ground state population.

The prediction of experimental results depends on the molecular absorption axes relative to the emission axes and moments of inertia. The general case has up to 5 exponentials, although many simpler cases exist. For example, if one is studying fluorescence emission with detection at 90^0 to excitation, the plane of the input laser direction and fluorescence detection axis can be used to reference the polarization. If we define the input polarization perpendicular to this plane, then detection polarization either parallel or perpendicular to the input will contain rotational information, while the "magic angle" of 54.7^0 only retains the intrinsic fluorescence decay, T_F. In general, the observed intensities, I(t), of fluorescence decays can be used to compute the function

$$I_{||}(t) - I_{\perp}(t) = R(t) \exp(-t/T_f) \ . \tag{4}$$

The anisotropy function R(t) is a function of the diffusion moments D_x, D_y, D_z of the molecule about the inertial axes x, y, z. For prolate (long and narrow) ellipsoids with the excitation dipole on the long axis, one predicts that R(t) is a single exponential. Oblate (flat, large area) ellipsoids are more theoretically complicated. Experimentally, the data on oblate ellipsoids with charges has been dominated by a single rotational time constant.

The experimental technique of time correlated photon counting [10] has been used in our group to measure rotational correlation times by fluorescence emission. This method has been extensively described in the literature, and is well-suited for rotational times greater than 50 picoseconds. Faster times can be studied with streak camera detection or fluorescence upconversion methods. The newer photomultiplier tubes based on microchannel plate technology can reduce the lower limit of time correlated photon counting.

Data for Large Molecule Rotation

The published studies of molecular rotation have established a range of behaviors [1-7, 11,12]. Large uncharged molecules nominally follow the slip limit with molecular volumes and shapes defined by van der Waals sizes. Positively charged molecules (usually of the substituted amine structures) show little evidence for strong hydrogen bond effects and often exhibit rotations similar to the stick limit for oblate molecules, or between stick and slip for prolate molecules. The deviations from pure ellipsoid shapes, often described as molecular "roughness", make it difficult to assess the relative importance of charge or shape changes. The negative ions (usually oxygen anions) rotate much slower than stick predictions in the hydrogen bonding solvents such as alcohols, but they show behavior in dipolar aprotic solvents between slip and stick limits for oblate, "rough", ellipsoids. The case of prolate, negatively charged molecules of high symmetry present a wider range of solvent dependent behavior, and this data will be discussed in more detail. Solvent temperature variations appear to yield activation energies for rotation which are similar to macroscopic viscosity [5,12], which is expected from Eq. (1) and large molecules. Other parametric changes such as solvent density and solvent mixtures will be discussed in a later section.

Recent work in our group with resorufin anion [1,12] has demonstrated a very significant change in rotation rate with hydrogen bonding environment. This type of molecular structure, shown in Figure 1, is a symmetric, delocalized anion structure which has

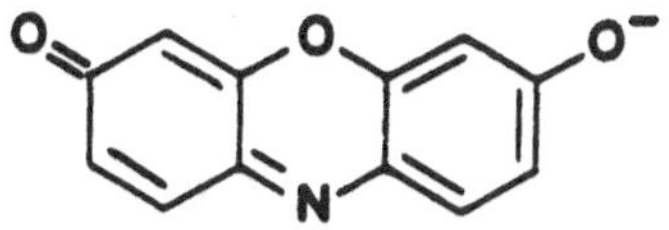

Fig. 1: Structure of resorufin anion represented as a localized anion, as might be present in a salt. The symmetric oxazine structure probably has charge delocalization to oxygen atoms at both ends of resorufin in a solution environment.

little "roughness". The rotation times in its ground state [13] are similar to the excited state [12]. A plot of rotation time versus viscosity for a single temperature would show that three classes of solvent (the alcohols, protic amides and dipolar aprotics) all have a different dependence on viscosity. In Table 1 we tabulate the effective volume of rotation in these solvents as TT_{or}/η. The tabular representation clearly shows a 5-6 fold slowing of rotation in the alcohol solvents compared with the dipolar aprotic solvents. As we will see, the slowest rotation is far slower than the stick limit. These data show dramatic effects of solvent on rotation, and we will pursue an interpretation in the next section. In the following section we will examine the importance of dielectric friction on the general rotation problem.

Model for Resorufin Rotation

We assume that the resorufin molecule is a resonance structure with a partial negative charge at each end of the molecule. The alternative, an asymmetric ionic structure, requires that solvent stabilization greater than the resonance energy be localized at one end of the molecule. On a statistical basis this does not seem likely, compared with a symmetric solvent stabilization of reduced charge. In addition, the absorption spectra and fluorescence spectra in dipolar aprotic and alcohol solvents show modest shifts which can be explained by conventional interpretations based on dispersion and electrostatic effects. Additional comments on the charge symmetry will be made in the next section.

The data in Table 1 show that the dipolar aprotic solvents such as dimethyl sulfoxide (DMSO) and dimethylacetamide (DMA) have rotation times similar to a predicted slip limit of $0.76 \times 10^4 ps^o K/cp$ rather than the stick limit of $2.05 \times 10^4 ps^o K/cp$.

Table 1: Summary of Resorufin Excited State Rotation Times

Solvent[a]	T, °C	$TT/\eta \times 10^{-4}$	T, ps[b]
MeOH	20	4.89	100
EtOH	20	4.46	183
TFE	24.1	6.30	380
1-PrOH	20	4.77	367
2-PrOH	23.1	5.83	427
1-BuOH	20	5.26	529
i-BuOH	24	6.03	671
2-BuOH	21.2	5.40	660
t-BuOH	21.5	6.13	1074
H_2O	20	2.56	88
D_2O	17.6	2.32	106
NMF	24.4	2.59	150
NMA	30.9	2.39	296
EG	20	1.67	1134
DMF	22.7	1.83	50
DMA	24.4	0.80	53
DMSO	29.8	1.14	69

a. The solvent symbols are methanol (MeOH), ethanol (EtOH), 2, 2, 2 trifluor-
   ethanol (TFE), N-methylformamide (NMF), N-methylacetamide (NMA), ethylene
   glycol (EG), dimethylformamide (DMF), dimethylacetamide (DMA), dimethyl-
   sulfoxide (DMSO).

b. The rotations T are computed at the specified T and corresponding
   viscosity from data sets characterized by TT/η .

In these solvents there is little tendency for strong hydrogen bonding to the anion
sites of resorufin. Consequently, we expect that the slip limit should be a good prediction for a smooth ellipsoid in these solvents, even with a molecular negative charge.
The reasonable agreement of the data with the slip limit suggests that we can use the
slip limit as a measure of "normal" rotational behavior when considering the other
solvents.

The rotation of the resorufin molecule in hydrogen bonding solvents can be considered as a case of strong solvent interaction at exposed anion sites. The particular
anion site of the resorufin molecule is very similar to the rose bengal molecule pre-

viously studied in our group [3,14]. For rose bengal, the alcohol solvents interacted in a site specific manner at the oxygen anions, which supported the concept of hydrogen bonding to the oxygen anion. Those studies demonstrated that the singlet-triplet electronic coupling in rose bengal was dramatically affected by hydrogen bonding to the oxygen anion. The changes were very large, and they correlated with the expected changes in hydrogen bond strength with alcohol structure [14]. Significant alcohol residence times, comparable to the fluorescence lifetimes, are required to show such structure specific shifts. Average solvation structures on the order of one nanosecond are reasonable inferences, although any single solvent molecule might have a coordination time less than one nanosecond.

The interpretation of the data in Table 1 requires a model for solute-solvent interactions. Theoretically some progress has been made in this direction [15], although there are no results in a form suitable for experimental comparison. The physical concept of a "solvent torque" was used by the author in earlier work on the rotation of rose bengal [3]. A quantitative description of solvent torque may be capable of bridging the gap between the slip limit and a complete solvent coordination limit. Solvent interactions provide a range of retarding torques according to the solvent-solvent forces and solute-solvent interactions. Unfortunately, there is no quantitative theory of solvent torque so that only qualitative discussions are possible. The data on resorufin rotation is qualitatively much clearer than rose bengal because of the higher symmetry of resorufin. In particular, the solvent size effects in a coordination model are much more amenable to study for resorufin.

The preceding discussion supports testing a model of solvation which leads to an average increased size of resorufin, with rotation in the slip limit. This approach assumes that solvent coordination leads to an average increase in dynamical size without requiring rigid coordination of the same solvent molecules over the time scale of rotation. We summarize the assumptions of the model calculation:

1. Use the slip limit of an "enlarged" resorufin molecule.
2. Use 2 sites for solvent coordination.
3. Define the new axis ratio by using the solvent length in Angstroms, L_e, to obtain a major axis $(12.2 + 2 L_e)$ and minor axis $(2 L_e)$.
4. Define the ellipsoid model volume, V_m, by computing the volume of an ellipsoid with the new axes.

Comparison with the data is made by converting the data to an effective volume, V_e. This conversion uses the proposed axial ratio (Assumption 3) to compute the friction factor and shape correction. The ratio of V_e/V_m should be unity for those cases where average solvent coordination is a reasonable model. The results are shown in Table 2.

The results for all of the alcohols cluster around a V_e/V_m ratio of unity, especially if one uses the less extended geometry for primary propanol and butanol. The deviations from unity for the alcohols have no correlation with molecular volume.

Table 2: Model Calculation for Resorufin with Solvent Coordination[a,b,c]

	L_e(A)	V_s($Å^3$)	ρ	S	F	V_e($Å^3$)	V_e/V_m (σ)
MeOH	4.37	36.1	2.396	0.554	0.338	1110	1.32(0.23)
EtOH	5.69	53.1	2.072	0.642	0.261	1520	0.95(0.06)
TFE	6.46	65.7	1.944	0.682	0.225	2640	1.20(0.07)
1-PrOH	7.00	70.1	1.871	0.706	0.205	2270	0.84
1-PrOH	6.35	70.1	1.96	0.677	0.230	1940	0.92
2-PrOH	6.16	70.1	1.99	0.668	0.238	2260	1.16
1-BuOH	8.63	87.1	1.707	0.762	0.157	3530	0.77
1-BuOH	7.00	87.1	1.87	0.706	0.204	2510	0.93
i-BuOH	6.12	87.1	1.825	0.721	0.191	3140	1.02
2-BuOH	6.12	87.1	1.997	0.666	0.239	2070	1.08(0.08)
t-BuOH	6.13	87.1	1.995	0.666	0.239	2360	1.23
H_2O	3.68	20.6	2.658	0.493	0.394	443	0.80(0.09)
NMF	5.54	60.2	2.101	0.634	0.268	843	0.56
NMA	5.54	77.2	2.101	0.634	0.268	775	0.52(0.04)
EG	7.26	60.8	1.84	0.716	0.196	842	0.29(0.01)
EG	4.3	60.8	2.419	0.548	0.343	355	0.44(0.02)

a. The headings are L_e, effective solvent length; V_S, solvent volume; ρ, major axis/ minor axis; S, the shape correction;, F, the friction correction using "slip"; V_e, the effective volume from experiment; V_e/V_m, the ratio of the experimental effective volume to the model volume from the ellipsoid (standard deviations based on data).

b. Solvent abbreviations are the same as in Table 1.

c. Model uses $(12.2 + 2L_e)$ for the major axis and $2L_e$ for the minor axis.

The various solvent classes show some interesting deviations. In particular, the mono-substituted amides and ethylene glycol exhibit much faster rotation than expected by the model. Water has a value of 0.8, which suggests a slightly faster rotation. The size of H_2O is very small, much smaller than any other solvent, so that this deviation from unity may be dependent on the sensitivity of the model to size assumptions.

The model calculation has clearly identified the monosubstituted amides and ethylene glycol as solvents where the assumption of a symmetric increase in effective size is not appropriate. However, the alcohols and water appear to be consistent with the model. The unusual behavior in monosubstituted amides and ethylene glycol is not from solute-solvent differences in hydrogen bonding, because these solvents have similar hydrogen bond strengths to oxygen anions. For NMA, NMF and ethylene glycol we have unusual dual site solvent-solvent interactions which allow linear chains and networks. Unlike water, these sites bridge a molecular length of substantial size and create macroscopic properties of solvent boiling point and viscosity which are quite different from the alcohols. Qualitatively, the faster rotation of resorufin in monosubstituted amides and ethylene glycol probably relates to a greatly reduced average coordination volume because of solvent-solvent forces overwhelming the solute-solvent interactions and reducing the average coordination. In ethylene glycol the rotation is faster than the stick limit so that a modest solvent torque may result from solute-solvent interactions, without requiring significant solvent coordination around both ends of the molecule.

In summary, the model of symmetric solvent coordination of resorufin, with rotation of the coordinated molecule in a slip limit, is consistent with the data for alcohol solvents and water. For ethylene glycol and monosubstituted amide solvents the observed rotation is near stick limit values, which suggests that solvent coordination is not a major factor in these solvents. The simple molecular model of solvent coordination really doesn't focus on the molecular level questions. For example, single solvent molecule exchange rates versus solute rotation rates control the viability of the "average" coordination model. When fast exchange with bulk solvent occurs, the models should analyze solvent torque from selective interactions. Additional experiments with other charged molecules and molecular dynamics models may be required in order to fully understand the rotational motions of charged molecules.

Dielectric Friction

The rotation of a dipole moment in a dielectric medium requires that the medium readjust its polarization to the probe motion. In fluids which have a time lag in dielectric relaxation, there is a retardation of the rotational motion which is called dielectric friction. If a molecule has a localized charge distribution which is not symmetric about the center of rotation, then there is a net dipole moment, μ, defined by

$$\mu = Zex$$

where Ze is the magnitude of the charge and x is the displacement from the center of mass. The complete theory of dielectric friction [16] has been simplified to estimate this effect [8]. For example, a charge of one unit which is displaced by 6 Ångstroms in a 200 $Å^3$ molecule will increase the rotation time by 60% in a fluid of 5ps dielectric relaxation (T_D) and a static dielectric constant (ϵ) of 30. While the full equation for this increase depends on the exact assumptions, for larger dielectric constant fluids one finds that the percent increase in rotation is roughly proportional to $T_D(\mu)^2/\epsilon$ where ϵ is the permanent dipole of the solute and T_D and ϵ are properties of the fluid. One expects that larger dielectric relaxation times and lower static dielectric constants will enhance this effect.

The experimental study of the dielectric friction effect requires a separation of local solvent interactions from dielectric friction. Comparative experiments are needed which can define these two effects, but several practical problems are present in interpreting the published data in these terms. The molecules for which data are available usually have an electronic resonance sharing of ionic charge, which greatly reduces the dipole moment of typical dye molecules. A molecule such as cresyl violet may be sufficiently asymmetric to localize the charge, and cresyl violet was analyzed as a possible example of dielectric friction [8]. The resorufin molecule, with a dramatic rotational slowdown in alcohols, would seem to be another candidate for dielectric friction effects except that all spectroscopic evidence suggests that the charge is symmetric. Furthermore, the rotation rate is quite fast in ethylene glycol and very slow in alcohols, even though the dielectric friction component, if it existed, should be in same direction. Another theoretical treatment would be useful, which explicitly considers a separated charge distribution as two coulombic polarizing units, to examine the effects of dielectric saturation in high dielectric fluids. For a large charge separation distance between two ends of a prolate ellipsoid, one might expect "local friction" contributions from each end of the molecule. Of course, such a model should be very dependent on dielectric constant and charge separation, but will eventually reduce to the case of negligible dielectric friction for small charge separations.

Since the meeting, two papers have appeared [17] which use the concept of dielectric friction to explain the slowing of rotation rates in alcohols and different rates at isoviscosity points of two component solvent mixtures. This work has applied dielectric friction to explain resorufin data in alcohols. We disagree with this application to resorufin, and obviously did not use this as an example in our earlier work on dielectric friction. The contribution of dielectric friction should be examined with our concept of "local friction", but these molecular structures do not allow a good isolation of solvent coordination and friction effects.

Rotation and Macroscopic Viscosity

The study of rotational motion as a function of pressure has been done for the ionic dye molecule rhodamine 6G [18]. This study used hydrostatic pressure on ethanol, from 1 to 6 kbar, to compare with the similar viscosity range achieved by different alcohol solvents. The ratio of T/η is about 1.8 times smaller for the pressure study than the alcohol study. The pressure control of viscosity changes a different class of solvation parameters than the change of solvent. Clearly, the simple hydrodynamic interpretation of viscosity is overwhelmed by other, more subtle changes in the interactions responsible for rotation of this large, somewhat "rough", cationic, oblate molecule.

Viscosity can be changed by using solvent mixtures, in particular the alcohols and water create conditions of constant viscosity with two different compositions.

The word of Beddard et al. [19] has demonstrated that cresyl violet has quite different rotation times in ethanol/water mixtures which have constant viscosity. Such effects are difficult to interpret because of the variations in molecular interactions which can occur. As we previously noted, cresyl violet is a candidate for dielectric friction effects, so that this contribution may be a factor which modulates the unknown effects of solvent interactions. Despite the problems of interpretation, this is an intriguing example of how macroscopic viscosity does not always correlate with rotation times of large, ionic molecules. Recent workers have further explored the effects of solvent mixtures [17].

Transient Dielectric Effects

The role of transient dielectric response in ionic kinetic processes is related to solvent effects on rotational motion of ionic molecules. For example, transition states and reaction rates can be heavily modulated by solvent response. We will briefly describe the processes which occur during an ionic photodissociation which we have studied in our laboratory [20].

The reaction of a substituted triphenyl carbonium ion (malachite green) and a cyanide ion yield a colorless compound, which can be re-ionized in a dielectric solvent by an ultraviolet light pulse. From the lowest lying excited state this process proceeds through an ionic transition state, whose rate of activation depends on the solvent dielectric constant according to simple dipole solvation energetics, rather than hydrogen bonding or other energetics. From a second excited state the ionization kinetics is sub-picosecond, so that transient effects are measurable with 0.5 ps resolution. There are two areas for which the data provides new insights.

The dependence of initial ion pair recombination on dielectric constant illustrates the importance of dielectric longitudinal relaxation times, T_L, in defining a time scale for solvent stabilization. The longitudinal relaxation time is given by $\frac{\varepsilon_\infty}{\varepsilon} T_D$, where ε_∞ is the infinite frequency dielectric response. The experiment varied this time by varying ε, and we observed a dramatic effect on ion recombination yields which suggests that the vibrationally excited ions define the main recombination path unless the solvent response time for stabilization is significantly faster than vibrational lifetime.

A second insight into solvent relaxation effects is that this initially formed ion pair has a tetrahedral geometry of the carbonium ion which relaxes to a planar carbonium ion. The rates of this process in polar, non-hydrogen bonding solvents behave as expected, for a solvent where motional and dielectric relaxation times are faster than the times of conformer change. However, the conformer change does involve a charge delocalization onto a dimethylaniline group so that when ethyl alcohol, but not methyl alcohol, is used as a solvent one finds a significant reduction of rate despite the probable similarity of energetics. This is the first demonstration of the theoretically predicted case of solvent "dynamic" control of a transition state crossing [21] In this case, we have created a reduction of rate because the solvent cannot respond to the transition state charge configuration as fast as was observed for other solvents. The need for charge displacements suggests that one origin of the effect can be thought of as dielectric friction. A molecular model of the effect would use the individual solvent reorientation times as dominating the local solvation environment. The complete explanation of such processes requires more experimental examples.

Summary

The rotation rates of charged molecules show dramatic changes as a function of solvent environment. The hydrodynamic models of rotational diffusion in the slip limit are useful estimates of the rotation rate for large, symmetrically charged molecules, if there are no strong solvent interactions. For cases of strong interactions, the rotation rate reduces by a factor of 5-6. For the symmetric, anionic dye molecule resorufin, we successfully modeled the rotation times in alcohols and water as a larger, solvent coordinated resorufin molecule in the slip limit. This model of average coordination was not successful in solvents of ethylene glycol or monosubstituted amides. These solvents have strong solvent-solvent interactions which probably reduced the average coordination to a point where the rotational rate is more like the hydrodynamic stick limit. Dielectric friction reduction of the rotation rate is possible for molecules which have an asymmetric charge distribution. Experimental evidence for this contribution exists in limited form; although more experiments are needed to separate dielectric friction from local coordination or solvent torques due to selective solvent interactions.

Several avenues for additional work were examined in this manuscript. We encourage developing a quantitative model of solvent torque to encompass the intermediate solvent interaction case, which is between coordination behavior and the slip limit. Dielectric friction theory should examine the conditions under which a symmetric placement of two charges can be treated in a "local friction" model. Models of pressure effects and solvent size effects, along with more data, will be helpful in experimental interpretations. Unusual rotational behaviour in mixtures of solvents represent a new frontier, although liquid volume and solvent interactions undoubtedly play a large role in such solvents.

The measurement of rotational motions can be useful in characterizing solvent interactions in a long term and transient sense. Interactions of solvent with ionic transition states include dielectric relaxation and motional effects. We briefly demonstrated these effects in the ionic photodissociation of malachite green leucocyanide to cyanide ion and the malachite green carbonium ion.

References

[1] T.J. Chuang and K.B. Eisenthal, Chem. Phys. Lett. $\underline{11}$, 368 (1971)

[2] G.R. Fleming, J.M. Morris and G.W. Robinson, Chem. Phys. $\underline{17}$, 91 (1976);
     G.R. Fleming, A.E.W. Knight, J.M. Morris, R.J. Robbins and G.W. Robinsons,
     Chem. Phys. Lett. $\underline{51}$, 399 (1977)

[3] K.G. Spears and L.E. Cramer, Chem. Phys. $\underline{30}$, 1 (1978)

[4] A. von Jena and H.E. Lessing, Chem. Phys. $\underline{40}$, 245 (1979); Chem. Phys. Lett. $\underline{78}$,
     187 (1981)

[5] D.H. Waldeck and G.R. Fleming, J. Phys. Chem. $\underline{85}$, 2614 (1981)

[6] D.W. Phillion, D.K. Kuizenga and A.E. Siegman, Appl. Phys. Lett. $\underline{27}$, 85 (1975)

[7] W.W. Mantulin and G. Weber, J. Chem. Phys. $\underline{66}$, 4092 (1977)

[8] D. Kivelson and K.G. Spears, J. Phys. Chem. $\underline{89}$, 1999 (1985)

[9] C.M. Hu and R. Zwanzig, J. Chem. Phys. $\underline{60}$, 4354 (1974)

[10] K.G. Spears, L.E. Cramer and L. Hoffland, Rev. Sci. Instr. $\underline{49}$, 255 (1978)

[11] K.G. Spears, K.M. Steinmetz-Bauer and T.H. Gray in Picosecond Phenomena, Vol. II,
     R. Hochstrasser, W. Kaiser and G.V. Shank (eds.), Springer, New York 1980,
     pp. 106-110

[12] K.G. Spears and K.M. Steinmetz, J. Phys. Chem. $\underline{89}$, 3623 (1985)

[13] E.F.G. Templeton, E.L. Quitevis and G.A. Kenney-Wallace, J. Phys. Chem. $\underline{89}$,
     3238 (1985)

[14] L.E. Cramer and K.G. Spears, J. Amer. Chem. Soc. $\underline{100}$, 221 (1978)

[15] R. Perolta and R. Zwanzig, J. Chem. Phys. $\underline{70}$, 504 (1979)

[16] P.A. Madden and D. Kivelson, J. Phys. Chem. $\underline{86}$, 4244 (1982);
     Adv. Chem. Phys. $\underline{56}$, 467 (1984)

[17] E.F.G. Templeton and G.A. Kenney-Wallace, J. Phys. Chem. $\underline{90}$, 2896 (1986);
     J. Phys. Chem. $\underline{90}$, 5441 (1986)

[18] L.A. Philips, S.P. Webb, S.W. Yeh and J.H. Clark, J. Phys. Chem. $\underline{89}$, 17 (1985)

[19] G.S. Beddard, T. Doust and J. Hudales, Nature $\underline{294}$, 145 (1981)

[20] K.G. Spears, T.H. Gray and D. Huang, J. Phys. Chem. $\underline{90}$, 779 (1986)

[21] G. van der Zwan and J.T. Hynes, J. Chem. Phys. $\underline{76}$, 2993 (1982)

DYNAMICS OF SEMIRIGID MACROMOLECULES IN DILUTE SOLUTION: STUDIES OF DNA RESTRICTION FRAGMENTS

Roger J. Lewis and R. Pecora
Department of Chemistry
Stanford University
Stanford, California 94305 / USA

I. Introduction

Macromolecules in solution exhibit a variety of complex translational, rotational, and bending motions. Measurement of time constants associated with these motions is important in characterizing macromolecules. For instance, measurements of translational and rotational diffusion coefficients are often performed since these quantities may be used to obtain information about macromolecular shape, size, and size distribution. In addition, the dynamics itself is often of importance in the function or application of a given system. This is especially true for biological systems where the internal flexing motions are often of central importance in the biological function of the molecule. Most work has centered on the extreme cases of rigid rod and Gaussian coil macromolecules in both dilute and semidilute solutions, although many important macromolecular systems are composed of semistiff rodlike macromolecules.

In this article we give a summary of some of our work studying the dynamics of semistiff chains in dilute solution, centering on a series of experiments on a model system of monodisperse DNA restriction fragments. In the next section a very brief review of some of the essential features of the dynamics of rigid rods, Gaussian coils, and semistiff chains is given. Section III contains a description of the DNA restriction fragments that we are studying, a description of the transient electric birefringence apparatus which we have used to study rotational and internal motions of these fragments, and a discussion of the data analysis methods which are necessary to study systems with multiple relaxation modes. The experimental results are discussed in Section IV and a comparison of the results with theory, including Brownian dynamics simulations, is given in Section V, VI, and VII.

II. Dynamics of Rigid Rods, Gaussian Coils, and Semistiff Chains

Many theories have been presented for the rotational and translational diffusion constants of rigid rod macromolecules in dilute solution. These theories are hydrodynamic in origin, treating the rod as a particle immersed in a continuum. They may be used to calculate the diffusion coefficients in terms of the rod length, axial ratio, and solution viscosity [1-6]. In addition, since translational motion parallel to the long

molecular axis is faster than translational motion perpendicular to it, the rate of
a given translational displacement depends on the orientation of the molecule, giving
rise to translational-rotational coupling. The theory of this effect has been formulated by many authors, including calculation of the time correlation functions that
are measured in polarized dynamic light scattering experiments [7-10]. Tobacco mosaic
virus has been an especially useful system for testing hydrodynamic theories of stiff
rodlike particles. Many electric birefringence decay experiments and depolarized light
scattering experiments have measured its rotational decay time. Polarized dynamic light
scattering experiments have been performed on tobacco mocaic virus over the past 18
years and, to highlight the experimental difficulties, only very recently has work been
reported in which the effects of translational-rotational coupling may have been detected [11,12].

At the other extreme of very high macromolecular flexibility, the dynamic Gaussian coil model [13,14] -- now often called the Rouse-Zimm model -- has been very successful in providing a description of the translational and the long wavelength internal modes of motion of a polymer. In the Rouse-Zimm formulation the polymer chain is
treated as a linear collection of beads connected by a set of Hookean springs. The distances between the beads are Gaussian. When these distances stray from the root-mean-
square value, an entropic restoring force (obeying Hooke's law) is exerted to return
them to this equilibrium separation. Other forces exerted on the beads are frictional
forces proportional to the bead velocity, fluctuating stochastic Brownian forces, and
interbead hydrodynamic interactions. When the frictional forces on the beads are small
and the beads are far apart, hydrodynamic interactions are unimportant (free-draining
limit). At the opposite extreme, when frictional forces are high and the beads are
placed close together hydrodynamic interactions become very important (non-free-draining limit). In the free-draining and the non-free-draining limits a normal coordinate
analysis may be readily performed. The relaxation times for the resulting normal mode
coordinates have been calculated in terms of the radius of gyration of the coil, the
solution viscosity and, in addition, in the free-draining limit, the macromolecular
translational diffusion coefficient.

Most of the dynamic experiments on flexible coils have been performed on very
high molecular weight synthetic polymers in organic solvents (see for instance the review article by Han and Schaefer [15]) and appear to give results in accord with this
model in the non-free-draining limit. These experiments have for the most part not been
able to resolve the individual normal modes but have measured average quantities connected with the set of mode relaxation times. For instance, a major source of information has been the average relaxation time (first cumulant) measured by polarized dynamic
light scattering experiments [15].

Although it is difficult to experimentally measure such quantities as the anisotropy of the translational diffusion coefficient and resultant translational-rotational coupling, it is believed that for the relatively simple long rigid rod system, the

theory is probably essentially correct with perhaps some quantitative refinements to be made in the future. The flexible coil case is more difficult to assess but, under theta conditions at least, the Rouse-Zimm approach appears to provide a good semiquantitative description of the translation and long wavelength internal modes of motion of a very flexible coil.

Macromolecules with stiffness intermediate between those of the rigid rod and the Gaussian coil are of great importance in both biology and technology. Stiff chain polymers such as the aromatic polyamides are used to produce strong, lightweight, corrosion resistant, thermally stable materials that are replacing metals in many applications. Biologically important macromolecules such as collagen, myosin, actin and DNA are all semistiff. In most cases the molecular stiffness and dynamics are important in the molecular function. In spite of their importance, however, little is known of the dynamics of these systems. Both the theoretical and experimental difficulties are formidable. There are few experimental techniques that can be applied to study the modes of motion of these molecules because, in most techniques, all modes contribute and it is difficult to separate them experimentally. In addition, most of these macromolecules, especially the synthetic ones, are difficult to prepare in a monodisperse state. Since the properties that we are interested in are strongly dependent on molecular weight, the measured quantities are often an average over all the species present as well as an average over the different modes of motion, rendering an analysis of the data extremely difficult.

The theoretical situation, in contrast to the rigid rod and Gaussian coil cases, is hardly any better. The theories used are usually some version of the Kratky-Perod wormlike chain model [16] in which the polymer chain is treated as a space curve with a "persistence length", defined as the distance along the chain in which the correlation function of the dot product of two vectors tangent to the curve decays to $1/e$ of the value when the distance is zero. Various authors have derived equations for the rotational and translational diffusion constants of the coil as a function of the contour length, persistence length, and the solvent viscosity. Some authors use the average configuration of the chain supplemented with hydrodynamic equations to compute these quantities [17,18]. Zimm and Hagerman [19,20] calculate the rotational diffusion coefficient of a semistiff chain by generating a distribution of chain shapes by Monte-Carlo techniques, computing the rotational diffusion coefficient for each shape, treating it as a rigid structure, and then averaging the result.

The internal dynamics, as well as the coupling between the internal modes and the translation and rotation, are much more difficult to describe in a quantitative manner. The theory of Harris and Hearst [21] treats the chain as a weakly bending rod with a force constant for bending that is inversely proportional to the persistence length. These authors also include a force constant for stretching in their model. As has been later pointed out by Harris and Hearst [21] and also by Soda [22], in the limit of stiff chains this theory is known to give incorrect results. Aragon and

Pecora [23] have recently modified the Harris-Hearst theory by omitting the force constant for stretching and supplementing the equation of motion with a constraint that the contour length of the chain be constant. Thus the Aragon and Pecora equation for the internal dynamics of the chain is the equation for a weakly bending rod with the constraint that only the solutions with a given contour length may be used. Aragon has used this model to calculate the depolarized light scattering time correlation function for a semistiff chain [24]. As noted below, this correlation function should be similar to that expected for a simple model of dynamic birefringence decay. It should be emphasized that this theory neglects hydrodynamic interactions entirely, and that the overall translation and rotation of the molecule must be treated separately -- no interactions of the translation and the rotation with the internal bending modes are included in the theory.

Given the complexity of these theories, there have been two other approaches to modeling these systems. One approach is to construct a simpler model with fewer degrees of freedom, in which one can include coupling between some of the modes of motion (for instance, rotational and bending), and which can be solved analytically (or nearly so). In that case one can obtain a qualitative understanding of how the various modes of motion contribute in a given type of experiment. The other approach is to simulate the Brownian motion of a model of the macromolecule on a computer. In this case the Langevin equation of motion is written for the simplified chain dynamics and then numerically integrated using a high speed digital computer. The data generated by this procedure for positions of the elements (usually "beads") making up the polymer chain, as a function of time, are then used to compute time correlation functions that are measured in experiments such as transient electric birefringence decay and depolarized and polarized dynamic light scattering.

The first approach has been taken by Roitman and Zimm [25-27] who have constructed what they call a "trumbell model" for semistiff chains. This model replaces the polymer chain by three beads, one placed at the center of the molecule and the other two at each end. The ends of the molecule can bend about the center with a Hooke's law restoring force proportional to the angular deviation of the beads from a linear configuration. Friction is exerted on the beads as they move through the solvent. Hydrodynamic interaction is ignored in the Roitman-Zimm version of this model, but was subsequently added by Yamakawa [28].

The Brownian dynamics approach has been taken by Allison [29,30] who has modeled DNA chains as a linear collection of beads with force constants for bending at each bead. In addition to the bending force, Allison adds a very strong stretching restoring force to each bond. Hydrodynamic interactions are included in the Langevin equation as well as stochastic Brownian forces.

Our approach has been to do experiments on well-defined monodisperse model systems and to use sophisticated, well-tested data analysis techniques to obtain information about the individual modes of motion contributing to the experiment. The results of these experiments are then compared with present theories of the dynamics (including Brownian dynamics simulations) of semistiff chains. Because of the relative ease of preparing homologous series of monodisperse samples, we have started our experiments on monodisperse DNA restriction fragments. These systems have the complication of being polyelectrolytes. It is, however, probable that the polyelectrolyte effects are subsumed in the constants that appear in a given dynamical model and need not be treated explicitly, with the exception perhaps of experiments such as transient electric birefringence in which the molecule is oriented by an external electric field.

III. Methods

One of the principal advantages of DNA as a model semistiff macromolecular system is that, because of its biological significance, techniques for its production, purification, and characterization are extremely well worked out [31]. Genetic engineering techniques make it possible to produce useful quantities (i.e. hundreds of micrograms) of monodisperse DNA samples in a variety of molecular weights. We have constructed three DNA plasmids (circular pieces of DNA that replicate within bacteria) that allow us to produce samples of DNA of length 367, 762, 1010, and 2311 base pairs (Fig. 1) [32]. One base pair (bp) corresponds to a contour length of 3.36 Å and a molecular weight of approximately 660 Daltons [33]. The persistence length of DNA is probably between 500 and 1000 Angstroms (150-300 base pairs) in low ionic strength solutions [34-38] and thus our DNAs range from 1-2 to 8-15 persistence lengths. This size range should show a range of behavior from that of a flexible rod to that of a nearly random coil.

The plasmids are propagated in the common laboratory bacterium *B. coli*, strain HB101 [31], and purified according to the procedure of Marko, Chipperfield, and Birnboim [39] with a minor modification [32]. This purification procedure relies on multiple selective precipitation steps to purify the superhelical plasmid DNA from contaminating proteins, lipids, chromosomal DNA, and RNAs. After purification, the plasmids are digested with the appropriate restriction enzyme(s) in order to cut the DNA framgents of interest from the rest of the plasmid (Figure 2). The fragments are purified from the other sections of the plasmid by electrophoretic separation on an agarose gel and then isolated by removing the appropriate band from the gel. The purity of the resulting DNA fragment is checked with ultraviolet absorption measurements at 260 and 280 nm [31] and analytic agarose gel electrophoresis. Typically we estimate our samples contain a few percent of contaminating RNA, mostly transfer RNA.

It should be noted that the 367 bp sample, which is obtained by restriction enzyme digestion of the 762 bp fragment, actually consists of an equimolar mixture of 367 bp, 368 bp, and 27 bp fragments. The 27 bp fragment constitutes only 3.5% of the mixture by weight. It is not expected to measurably alter the experimental results because of its small optical and electrical polarizability and because its rotational decay time will be faster than the behaviors that we wish to study.

Plasmid Restriction Maps

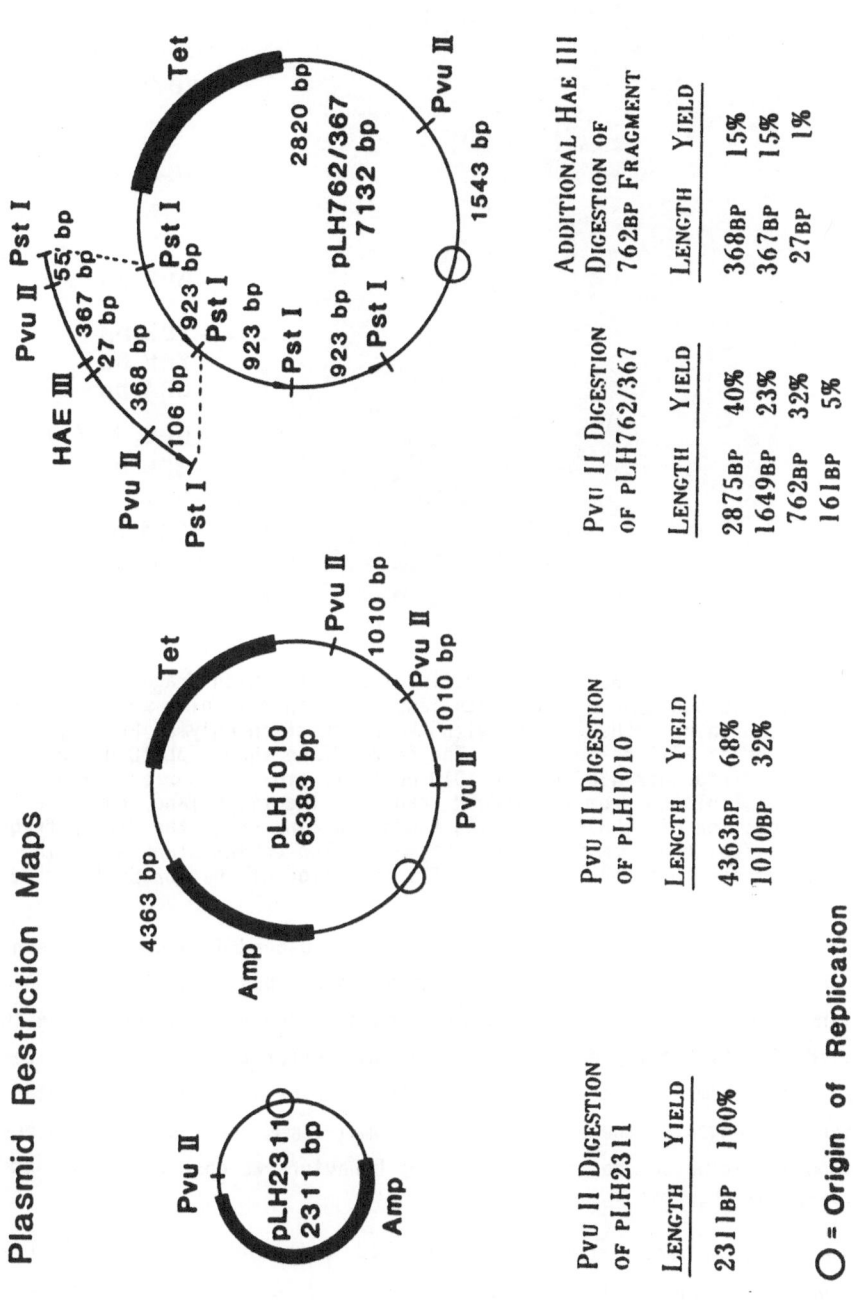

PvII Digestion of pLH2311

Length	Yield
2311bp	100%

Pvu II Digestion of pLH1010

Length	Yield
4363bp	68%
1010bp	32%

Pvu II Digestion of pLH762/367

Length	Yield
2875bp	40%
1649bp	23%
762bp	32%
161bp	5%

Additional Hae III Digestion of 762bp Fragment

Length	Yield
368bp	15%
367bp	15%
27bp	1%

○ = Origin of Replication

Figure 1: Restriction maps of the three plasmids used to produce monodisperse DNA samples. In each case the theoretical yield by weight of the fragments of interest is at least 30%. The plasmid pLH762/367 yields a fragment 762 bp in length when digested with the restriction enzyme Pvu II. When this fragment is isolated and digested with the restriction enzyme Hae III we obtain a sample with greater than 96% by weight 367 bp and 368 bp fragments and the rest a 27 bp fragment.

Preparative Restriction Digests

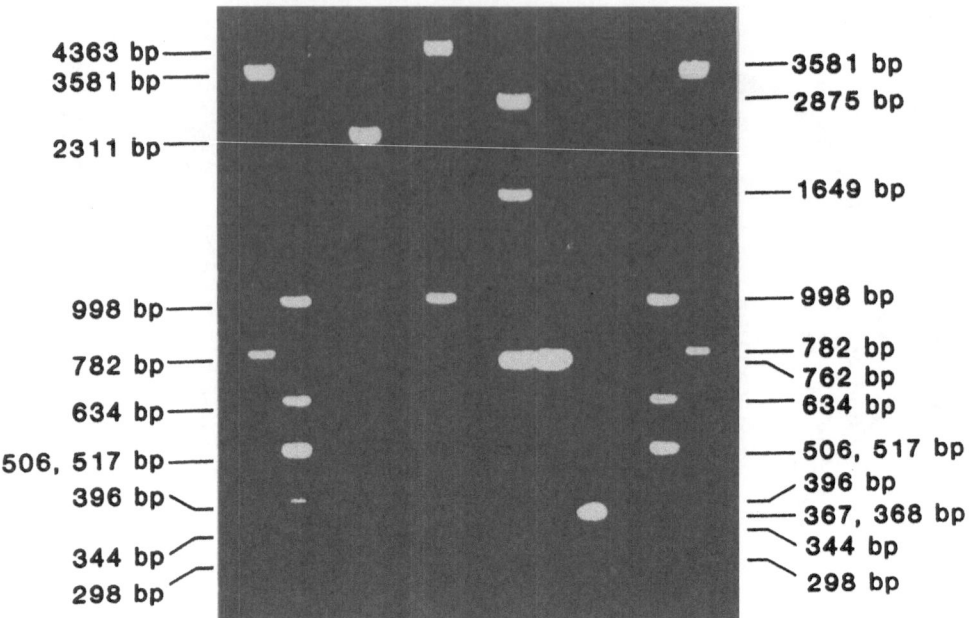

Figure 2: Agarose gel electrophoresis demonstrating the restriction fragments shown in Figure 1. The first, second, eighth, and ninth lanes contain molecular weight standards. The third lane contains pLH2311 cut with the restriction enzyme Pvu II, producing a single fragment 2311 bp in length. The fourth lane shows pLH1010 cut with Pvu II, producing fragments 4363 bp and 1010 bp in length. For our experiments the 1010 bp fragment is isolated from the larger fragment. The fifth lane shows the result of Pvu II digestion of pLH762/367 and the sixth lane shows the 762 bp fragment after it has been isolated from the other fragments. The seventh lane shows the 367 bp and 368 bp fragments that result from Hae III digestion of the 762 bp fragment.

Purified and ethanol-precipitated DNA fragments are resuspended in sodium phosphate buffer (1 mM in phosphate) at pH = 7.0 and a concentration of approximately 15-20 µg/ml. These solutions are then dialyzed against several changes of the same buffer at 0-4 °C over 8-24 hours. The DNA is finally diluted with the same buffer to a final concentration of 5-10 µg/ml. Measurement of the impedence of the samples at 1 kHz demonstrates that the final ionic strength of the samples ranges from 1-3 mM. The final DNA concentration is low enough that we can be assured that the behaviors we observe correspond to those of a single DNA chain.

The transient electric birefringence apparatus used is that of Don Eden of San Francisco State University and has been previously described [40,41]. This apparatus has a number of advantages; high sensitivity allowing the use of relatively small electric field strengths, excellent temperature control minimizing the effects of convection, and a compensation cell which removes the transient electric birefringence contribution of the solvent. The use of the smallest field strengths possible is important, as it minimizes electrophoresis and convection artifacts. Presumably the simplest possible dynamic behavior will be observed in the low-field linear-response regime as well.

The transient electric birefringence data is taken at 20 °C and averaged over 5000 pulses. Often the pulses are of reversing polarity to minimize net electrophoresis of the DNA. Details of the pulse duration and field strengths used may be obtained from our previous publications [42,43], although they are discussed as needed here. The birefringence is measured as a transient electric intensity variation that is then converted to true birefringence using the known geometry of the optical polarizing components. The zero-field decay of the data is then normalized and multiplied by negative one. The negative birefringence of DNA is thus made positive so that we may use nonnegativity constraints during data analysis, as is discussed below.

Data analysis is performed on a Universe 68 computer, a M68000-based system manufactured by Charles River Data Systems. Two programs DISCRETE and CONTIN, both written by Stephen Provencher [44-48], are used to analyze the data and thus determine the number of separate decay processes, their amplitudes, and their decay times.

DISCRETE fits the observed decay data to a sum of discrete decaying exponentials with nonnegative amplitudes, i.e. the data is fit to the form

$$g(t) = \sum_{i=0}^{n} A_i \exp(-\Gamma_i t) \tag{1}$$

where the A_i are the amplitudes and the Γ_i are the decay constants. Γ_0 may be set to zero to allow for a constant additive term. DISCRETE automatically determines the minimum number of terms required in the solution in order to adequately fit the experimental data. No initial guesses are made as to the number of decay processes, their amplitudes, or their decay times. Thus DISCRETE is able to give us an unbiased estimate of the number of separate decay processes occurring in the experiment, as long as a sum of discrete decaying exponentials is an appropriate model for the experimental system. In many cases, such as a monodisperse solution of rigid rods or a monodisperse solution of DNA, it is reasonable to assume that a number of discrete decay processes contribute to the electric birefringence decay. Consider, however, a polydisperse solution of rigid rods with a monomodal but wide distribution of lengths. The birefringence decay would consist of a monomodal but wide distribution of decay times and, with any reasonable signal-to-noise ratio, many different exponential decays would have to be added together to adequately represent the data. In this case DISCRETE would overestimate the number of decay processes occurring. In fact there would be only one distinct decay process, end-over-end rotation.

In cases such as the hypothetical polydisperse solution of rigid rods, the appropriate general theoretical form for the data is

$$g(t) = \int_{\Gamma_0}^{\Gamma_1} G(\Gamma) \exp(-\Gamma t) \, d\Gamma \tag{2}$$

where $G(\Gamma)$ is a continuous weighting function that determines the contribution of each decay time, equal to $1/\Gamma$, to the decay. As $G(\Gamma)$ is a continuous amplitude or weighting function, analogous to the amplitudes A_i above, we constrain it to be non-negative as well. CONTIN fits the experimental data to the theoretical form of Equation (2), subject to the nonnegativity constraint on $G(\Gamma)$. In addition, it finds the smoothest or least complex (in a figurative sense) form of $G(\Gamma)$ that adequately fits the experimental data. In this way CONTIN avoids overestimating the number of decay processes that occur in the data. Given the electric birefringence data from a poly-disperse solution of rigid rods, the solution from CONTIN would simply be a monomodal but wide distribution of decay times, the correct result.

IV. Results of Transient Electric Birefringence Experiments

In the vast majority of analyses of transient electric birefringence data from our monodisperse DNA samples, CONTIN and DISCRETE agree reasonably well as to the number of decay processes, their weights, and their decay times. This agreement strongly sug-gests, first of all, that we are correctly determining the number of decay processes and, second of all, that the separate decay processes have single decay times, or at least a very narrow distribution of decay times.

One might expect several different relaxation processes to contribute to the zero-field birefringence decay of a semiflexible macromolecule like DNA. The slowest relax-ation would correspond to a process resembling end-over-end rotation. The faster relax-ations might stem from rotations around other temporary axes, or from internal bending motions. The characterization of particular decay processes as "rotation" or "bending" is probably artificial as the true dynamical modes presumably each contain contributions from rigid rotation around several axes and from internal bending. Nonetheless we refer to the slowest observed decay mode as the "rotational" decay mode. We refer to the faster observed modes as "internal bending" modes for reasons that will become clearer below.

Figure 3 shows the CONTIN analysis of the birefringence decay from a solution of 367 bp DNA after orienting pulses of 50 μs and 2 μs duration. After a 50 μs pulse we observe two decays, the rotational decay around 14 μs and the first internal bending decay around 2.3 μs. When we use a 2 μs orienting pulse, the rotational decay mode does not have time to become fully excited. Thus the ratio of the amplitude of the rotational mode to that of the first internal mode is decreased. In addition, we now observe an even faster decay mode, with a decay time around 0.5 μs.

The CONTIN analyses of a similar experiment performed on a solution of the 762 bp DNA is shown in Figure 4. With a 300 μs orienting pulse we clearly see the rotational and first internal modes at approximately 75 and 18 μs, respectively. The use of a faster 20 μs pulse allows the additional detection of a second internal bending mode, with a decay time of 2-5 μs. With a very short pulse of 2 μs duration the data obtained was not accurate enough to clearly resolve the decay modes, although it is clear that the very short orienting pulse selectively excites the fastest decaying modes.

AMPLITUDE (arbitrary units)

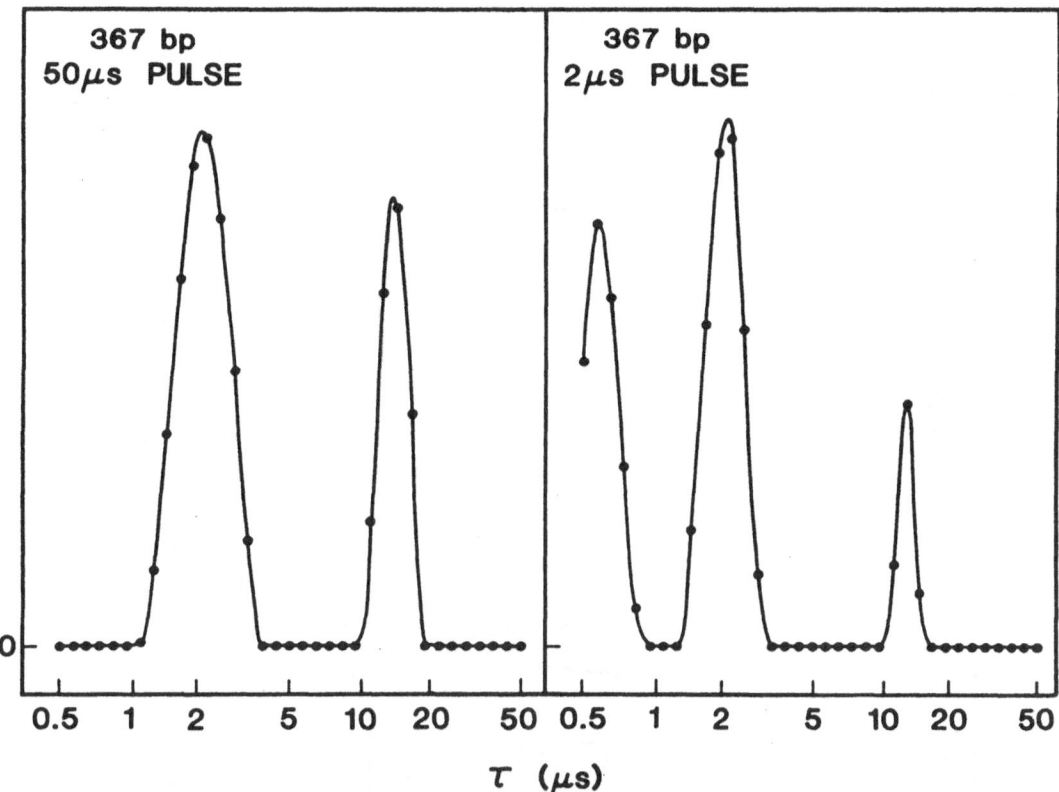

τ **(μs)**

Figure 3: CONTIN analyses of the zero-field birefringence decays of a solution of the 367 bp DNA. The first panel shows the result after an orienting pulse of 50 μs duration, during which almost steady-state birefringence is obtained. Two decay processes are detected, the slowest with a decay time around 14 μs corresponds to approximately 75% of the amplitude and the faster, with a decay time around 2.3 μs, corresponds to approximately 25% of the amplitude. The second panel shows the result with a short orienting pulse of only 2 μs duration. This short pulse results in a different weighting of the various decay modes, although their decay times are essentially unchanged. With the shorter orienting pulse the slowest decay mode corresponds to approximately 50% of the amplitude, the mode around 2 μs contributes 45% to the amplitude, and an even faster mode around 0.5 μs contributes the final 5%.

Figure 5 shows the analyses of data from a solution of the 1010 bp DNA. In this case it is only with the longest pulse, 400 μs, that we are able to resolve the decay modes from each other. The rotational mode is located at about 180 μs, the first internal mode at approximately 40 μs, and the second internal mode around 5 μs. With the shorter pulses of 50 and 5 μs we can see a shift in the decay towards the faster decay processes but cannot resolve the modes.

Figure 6 shows the analyses of data from a solution of the 2311 bp DNA. These CONTIN analyses are undersmoothed, meaning that although they represent the best fits to the experimental data, they may contain more information than is justified by the

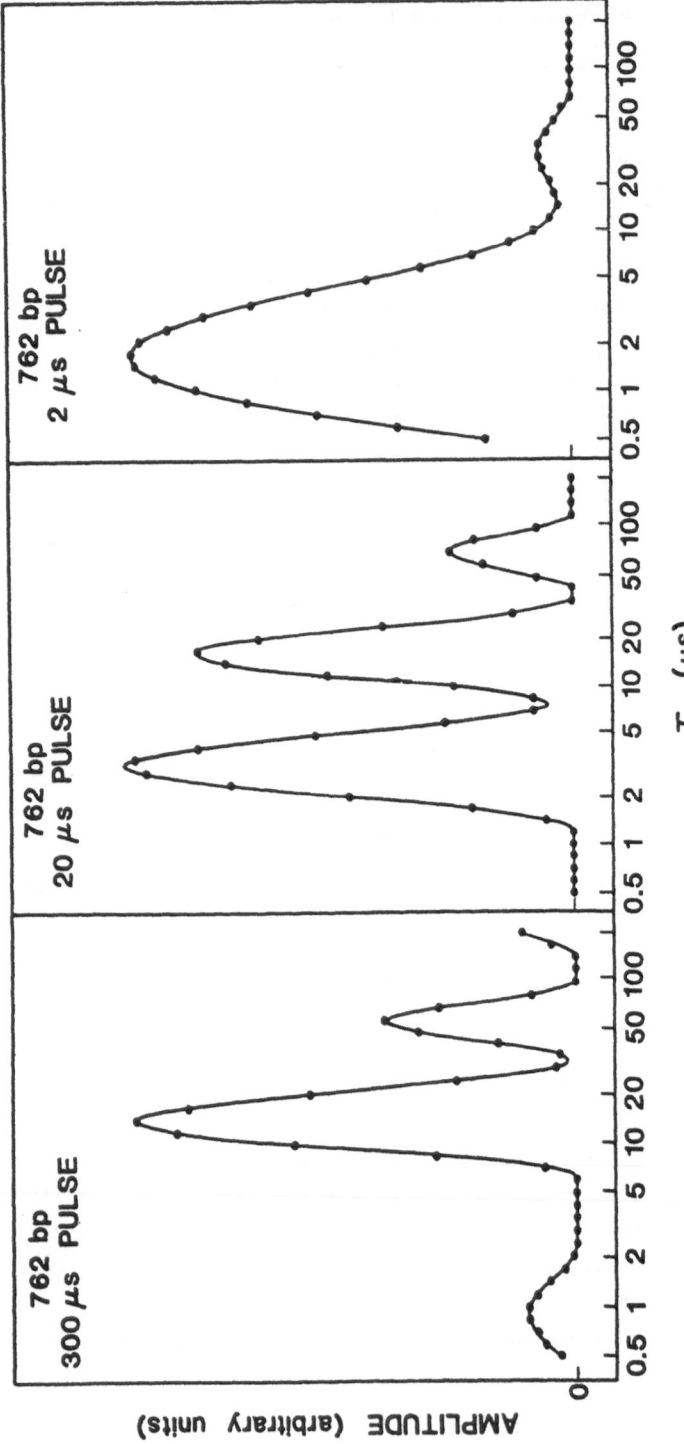

Figure 4: CONTIN analyses of data from a solution of the 762 bp DNA. The first panel shows the results after an orienting pulse of 300 μs. The rotational decay occurs at 75 μs and constitutes 51% of the decay while the first internal mode occurs at 18 μs and is 45% of the amplitude. The small peak at 0.5-2 μs is probably an artifact. The second panel shows the results after a 20 μs pulse. The first two peaks are essentially the same although the weight has shifted towards the first internal decay mode. In addition, we now see a peak around 2-5 μs, although its position varied from run to run. With the 2 μs pulse the data was of insufficient accuracy to resolve the decay processes from each other, these results are shown in the third panel.

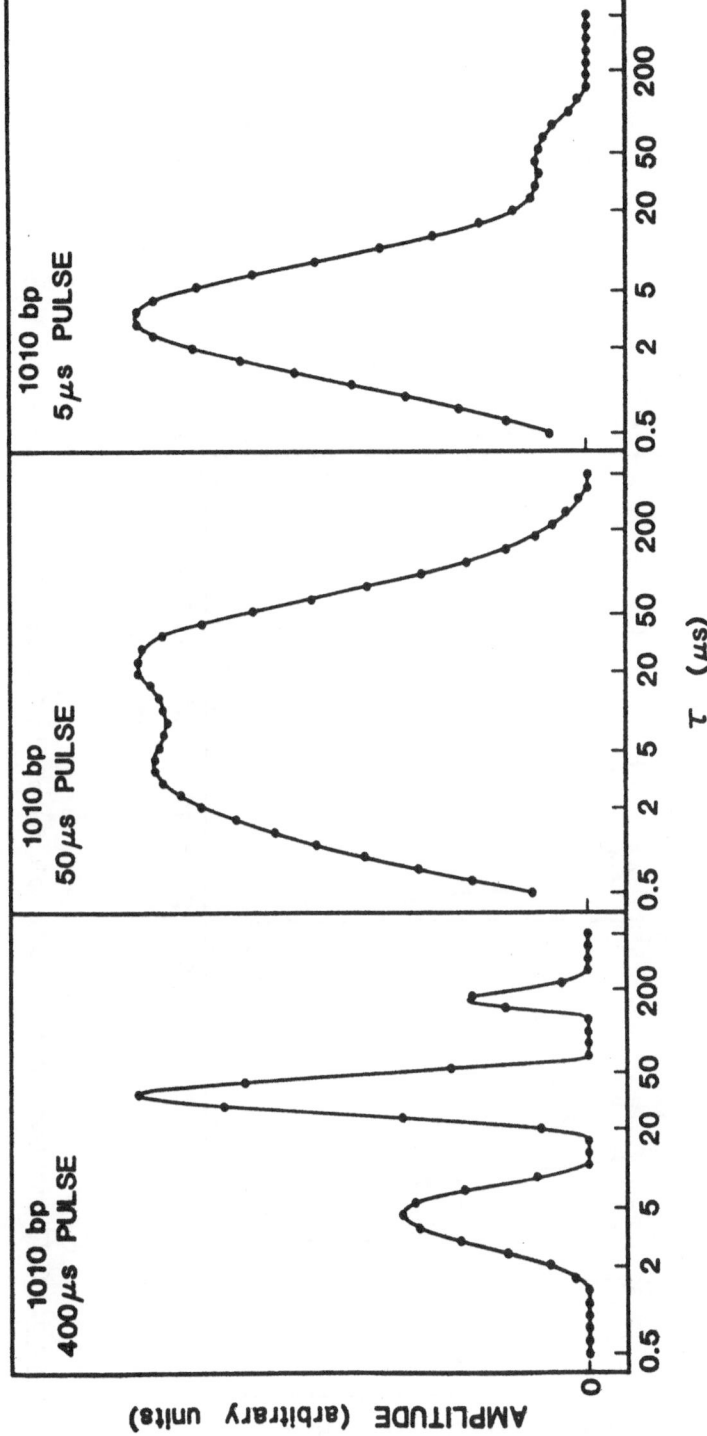

Figure 5: CONTIN analyses of data from a solution of the 1010 bp DNA. With the longest pulse of 400 µs the rotational mode made up 41% of the decay and was located at 179 µs. The first internal mode was at 40 µs and corresponded to 54% of the decay. The fastest mode at approximately 5 µs made up the other 5% of the decay.

232

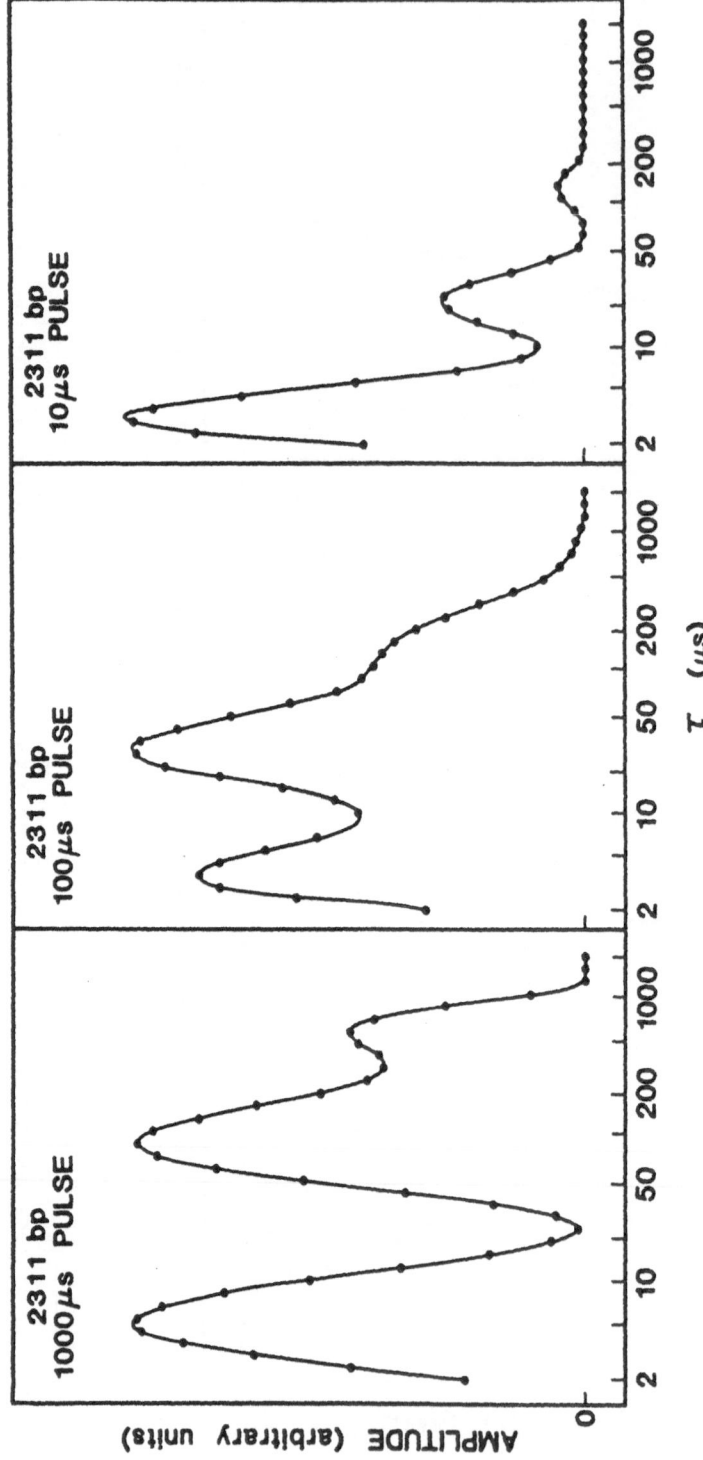

Figure 6: CONTIN analyses of data from a solution of the 2311 bp DNA. See text for discussion.

precision of the data [42]. Similar to the results from the other DNAs, we observe multiple decay modes, the slowest around 700 μs. The data obtained after a 10 μs pulse probably shows the first, second, and possibly the third internal modes.

The experimental data whose CONTIN analyses are shown above were also analyzed by DISCRETE. We believe the CONTIN analyses give strong evidence that the birefringence decays consist of multiple discrete exponential decays. If that is the case then DIS-CRETE, which assumes a sum of discrete exponentials to be the form of the decay, should give the most accurate and consistent analysis of the data. The results of the DISCRETE analyses are given in Table I.

Table I: Rotational and Internal Decay Times as Analyzed by DISCRETE

DNA Length	Rotational Decay Time, μs	First Internal Decay Time, μs	Second Internal Decay Time, μs
367 bp	14.4	2.3	< 1.0
762 bp	78.	16.5	2-4
1010 bp	171.	37.	5-6
2311 bp	688.	204.	40-60

In order to interpret the meaning of the various decay times that we have observed in the 2311 bp DNA, it is useful to have a good estimate of that DNA's radius of gyration under our experimental conditions. In order to obtain such an estimate we performed total intensity light scattering experiments on solutions of the 2311 bp DNA [49]. Results from such a total intensity measurement are shown in Figure 7. The angular dependence of the scattered intensity is fit with the theoretical form factor for a Gaussian coil [49] in order to get a measure of the radius of gyration. These values for the radius of gyration, obtained at a variety of DNA concentrations, are then extrapolated back to zero DNA concentration. This extrapolation is shown in Figure 8. These experiments, performed at an ionic strength of 100 mM, give a measured radius of gyration of 1044 Å. This is in good agreement with the 1030 Å radius of gyration that one would calculate assuming a wormlike chain of contour length 7760 Å and a persistence length of 500 Å [50]. If we assume that the persistence length under the low-ionic-strength conditions of the birefringence experiments to be 1000 Å, then the radius of gyration of the 2311 bp DNA would increase to 1350 Å [50].

Using these electric birefringence measurements of the decay times of the DNAs and the value of the radius of gyration of the 2311 bp DNA, we are now in a position to compare our results with various models of the motion of semistiff chains, as well as previous experimental work. We review these theoretical models and experimental results below.

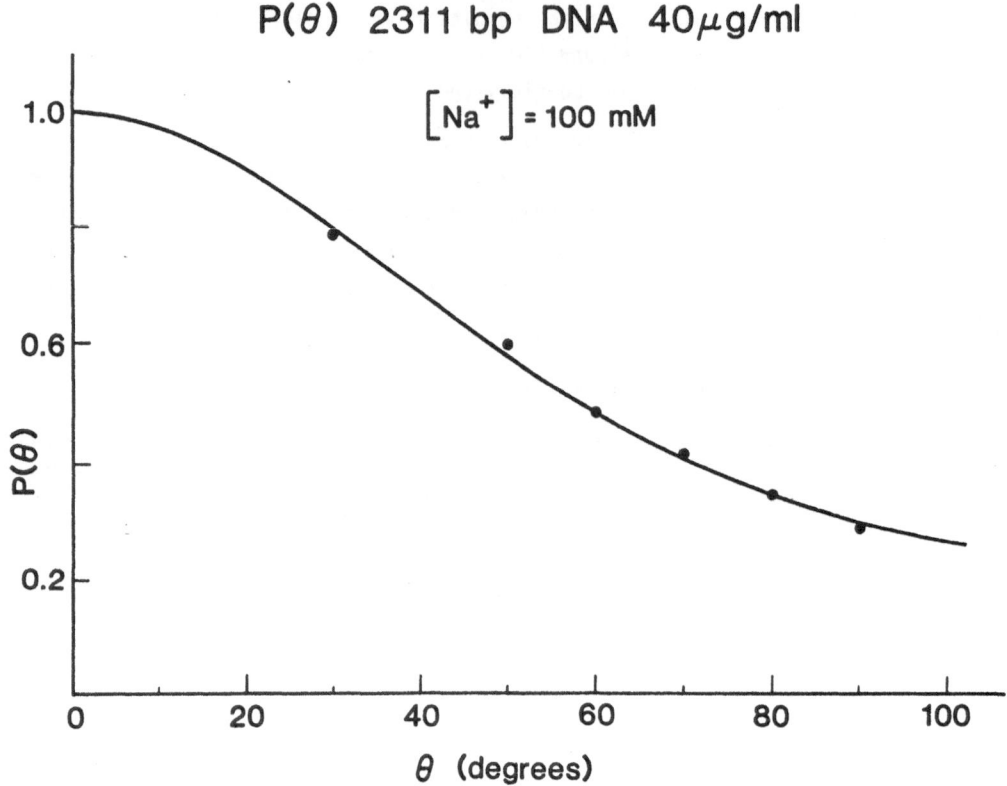

Figure 7: Integrated intensity light scattering determination of the apparent radius of gyration of the 2311 bp DNA at a concentration of 40 µg/ml. The smooth curve shows the best fit of the theoretical form factor for a Gaussian coil to the experimental points.

V. Discussion of Transient Electric Birefringence Results

Roitman and Zimm have published three papers detailing theoretical calculations of the hydrodynamic behavior of the trumbell [25-27]. A trumbell consists of three beads held together by two frictionless rigid bonds of constant length. The trumbell may bend at its center bead, subject to a restoring force proportional to the angular displacement and a bending constant Z. Roitman and Zimm's work was based on calculations on the freely bending trumbell published in 1974 by Hassager [51,52]. The model assumes that there is no hydrodynamic interaction between the beads and that the interaction of the trumbell with an external electric field is due to an anisotropic polarizability of the two bonds. The behavior of the trumbell in the transient electric birefringence experiment is calculated in their second paper [26]. The birefringence decay is predicted to consist of a sum of discrete exponentials. The slowest decaying exponential is due to a dynamic mode that resembles rigid rotation, although there is

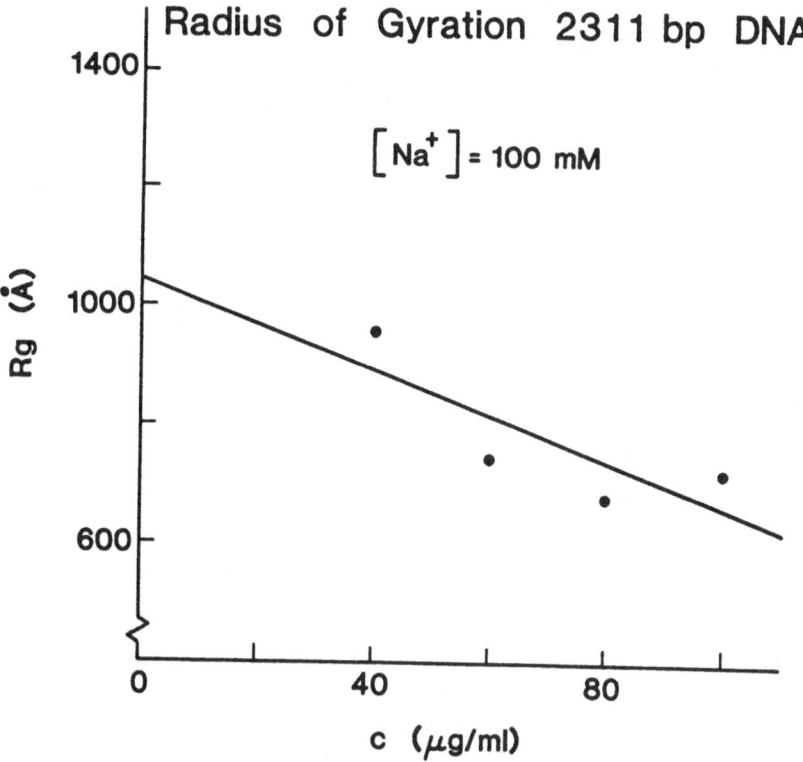

Figure 8: Extrapolation of the apparent radius of gyration of the 2311 bp DNA to zero concentration. The line represents the best least-squares fit to the four data points. The zero-concentration intercept is at 1044 Å, in good agreement with the theoretical prediction of 1030 Å. See text.

some flexing as well. We denote the decay time of this mode as T_1. The next mode, whose decay time we call T_2, consists of a coupled bending rotational motion. There are additional modes as well, but their contribution to the birefringence decay is less significant.

It is important to note the dependence of the decay times T_1 and T_2 on the stiffness parameter Z. As Z increases, meaning the trumbell is becoming stiffer, the slowest decay mode slows down until the decay time is simply that of the rotation of a rigid linear array of three beads. As Z increases, however, the decay time of the first bending mode decreases. Thus the ratio of the rotational decay time to that of the first internal bending mode is a sensitive function of the stiffness parameter Z, i.e. the larger Z the larger this ratio. This is shown explicitly in Table II. As is discussed below, we expect the shorter DNA to behave more like a rigid trumbell (a large value for Z) and the longer DNA to behave like a more flexible trumbell (a small value of Z).

Table II: Ratio of the first decay time to the second decay time for the model
of Roitman and Zimm [25]

Stiffness Parameter (Z)	Ratio of T_1 to T_2
1.25	7.8
1.00	6.5
0.75	5.3
0.50	4.1
0.25	3.1

The Nagasaka and Yamakawa alternate formulation of the trumbell theory includes
the effects of hydrodynamic interactions between the beads [28]. For the cases of interest here there is little quantitative and no qualitative difference between the theories. For the purposes of discussion we limit ourselves to the Roitman and Zimm theory.

Recently Aragon and Pecora [23] have formulated a new theory of the dynamics of
semiflexible macromolecules. Aragon [24] has given a comparison of the electric birefringence data presented here with the Aragon and Pecora theory. As previously noted,
this theory explicitly assumes that the rotational and translational motions of the
semiflexible rod are independent of the internal dynamics. The internal bending motions
are assumed to be of the same form as those of an elastic rod. Hydrodynamic interactions
are neglected. The interested reader is referred to the original articles for more details of this theory and its applicability to analysis of our birefringence data.

The Rouse-Zimm model briefly described in Section II predicts a set of discrete
decay times [13,14]. In the free-draining limit, where the effects of hydrodynamic interactions between different sections of the molecule are ignored, the decay times may
be expressed as

$$T_K = R_g^2/\pi^2 D K^2 \tag{3}$$

where R_g is the radius of gyration, D is the translational diffusion constant,
and K is the mode number. In the non-free draining limit, where the hydrodynamic
interactions essentially limit the dynamics of the chain, the expression for the decay
times is

$$T_K = 11.8\, \eta R_g^3/kT\, \lambda'_K \tag{4}$$

where η is the solvent viscosity, k is Boltzmann's constant, T is the absolute temperature, and λ'_K is an eigenvalue given in Reference [14].

In addition to the theories mentioned above, a number of pertinent experimental results have appeared in the recent literature. Hagerman has published theoretical work with Zimm [19] and experimental work [20] aimed at determining the rotational diffusion constant of semistiff chains. Using Monte Carlo techniques, Hagerman and Zimm [19] created a rigid ensemble of chains and directly calculated the rotational diffusion constant around the center of mass. After including a small correction for the difference in location of the center of mass and the center of rotational resistance, they empirically fit this data to a simple formula. Their result gives the ratio of the observed rotational diffusion constant to that of a rigid rod of the same length, as a function of the ratio of the contour length to the persistence length. In the second article [20], Hagerman reports transient electric birefringence experiments on DNA in low ionic strength solution. By measuring the slowest decay time, presumably related to rotational motion of the DNA, Hagerman was able to compare it to the theoretical results of Hagerman and Zimm [19].

Stellwagen has also presented transient electric birefringence results on DNA [53]. Here as well, she analyzed the longest decay time of the DNA in order to characterize the rotational motion of the molecule. A comparison of Hagerman's and Stellwagen's results with ours is presented below.

Diekmann et al. [54] have also published significant results on DNA restriction fragments. Using a multiexponential analysis of the linear dichroism decay from solutions of DNA, they were able to demonstrate and characterize two decay times for lengths of 95 bp to 258 bp. For DNA lengths of 603 bp and greater it took greater than two decay processes to account for the observed decays. The relative weights of the decay processes were dependent on the magnitude of the electric field, but the decay times were not. This is analogous to our observation that the weights of the various decay processes are dependent on the duration of the orienting pulses but the decay times are not.

As a simple check of our experimental methods and data analysis techniques, it is useful to compare our measured rotational decay times with those obtained by Stellwagen [53] and Hagerman [20]. In order to directly compare the decay times we can correct Stellwagen's and Hagerman's results to 20 ^{o}C by allowing for the difference in thermal energy, kT, and water viscosity at different experimental temperatures. Figure 9 shows our results compared with those of Stellwagen and Hagerman. Within the experimental variation our measurements of the rotational decay times agree with the previous measurements. This strongly suggests that CONTIN and DISCRETE are correctly separating the various decay modes from each other.

As mentioned above, a prediction of the Roitman-Zimm model for the trumbell is that the ratio of the rotational decay time to the first internal decay time decreases

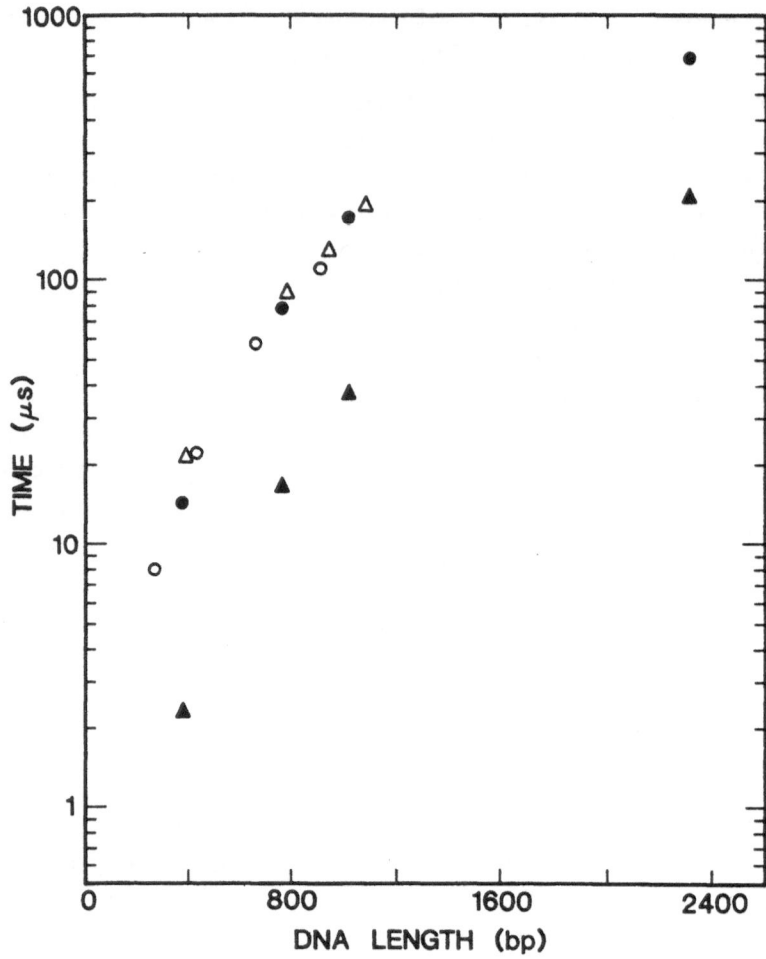

Figure 9: A comparison of the rotational decay times determined by our work and those determined by Stellwagen [53] and Hagerman [20]. The open circles show the rotational times determined by Hagerman and the open triangles show the results of Stellwagen. In both cases their results have been corrected to 20 °C for comparison with our results. The solid circles show our measured rotational decay times and the solid triangles show the decay times for the first internal mode.

as the stiffness parameter Z decreases, or as the semistiff molecules that we are modeling increase in length. Figure 10 shows the measured ratio of the first two decay times to each other. As predicted by the Roitman-Zimm theory the ratio decreases with increasing DNA length. It is interesting to note that the ratio approaches a value between 3 and 4 for the largest DNA. This is the range of values predicted by the Rouse-Zimm model of a Gaussian coil.

How well does the Rouse-Zimm model account for the decay times observed for the largest DNA? Using our value for the radius of gyration of the 2311 bp DNA, the measured translational diffusion constant [50], and Equations (3) and (4) we may predict the decay times from the Rouse-Zimm model. In order to match the radius of gyration

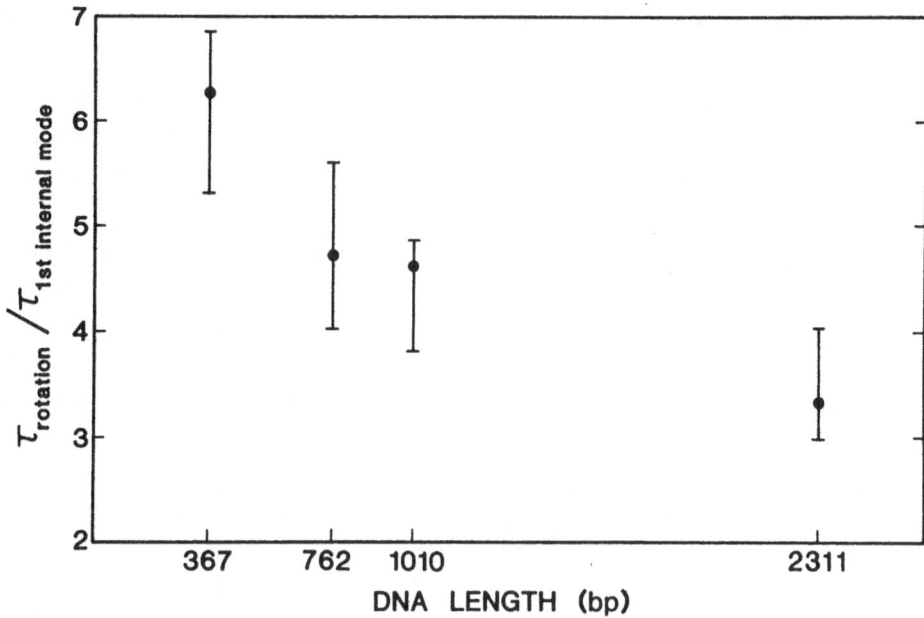

Figure 10: Ratio of the rotational to the first internal decay times as a function of DNA length. Decay times were determined by analysis of several data sets with DIS-CRETE. The plotted points show the ratio of the mean of the observed rotational times to the mean of the observed first internal decay times. The error bars show the limits of the worst-case errors; i.e. the upper end of the bar shows the ratio of the longest measured rotational decay time in any data set to the shortest observed first internal mode decay time in any data set. Even with this conservative error estimate it is clear that the ratio of the decay times decreases as the DNAs increase in length.

and the translational diffusion constant we may use as an explicit model a 10 bead chain, with an 86 nm average bond length, and a bead diameter of 6.8 nm [50]. This set of parameters gives a draining parameter [13] of 0.23, suggesting that our DNA should show behavior closer to the free-draining limit than to the non-free-draining limit. Table III shows the predicted Rouse-Zimm decay times (calculated for a radius of gyration of 135 nm and a translational diffusion constant of 3.16×10^{-8} cm^2/s) in the free-draining and the non-free-draining limits. As can be seen from the data, the free-draining Rouse-Zimm model comes very close to accounting for all three observed decay

Table III: Predicted Rouse-Zimm decay times and experimental decay times for the 2311 bp DNA (in μs) [50]

Mode Number	Free-draining limit	Non-free-draining limit	Observed
1	584	1780	688
2	146	561	204
3	65	297	40-60

times. As predicted, the DNA decay times are between the two theoretical limits, but much closer to the free-draining limit.

At this point it may seem that we have arrived at a reasonably self-consistent picture of DNA dynamics. The trumbell model explains at least some of the observations on the smaller DNAs and the Rouse-Zimm model accounts for the results from the largest DNA. In order to fully understand the electric birefringence decay, however, it is necessary to understand the orientation process the DNA undergoes in the electric field. Unfortunately the present evidence suggests that this process is quite complicated. Only two points will be made here in this context. First of all, we have performed analyses of electric birefringence decays from solutions of the 1010 bp DNA holding the pulse length constant but varying the electric field strength. In agreement with the

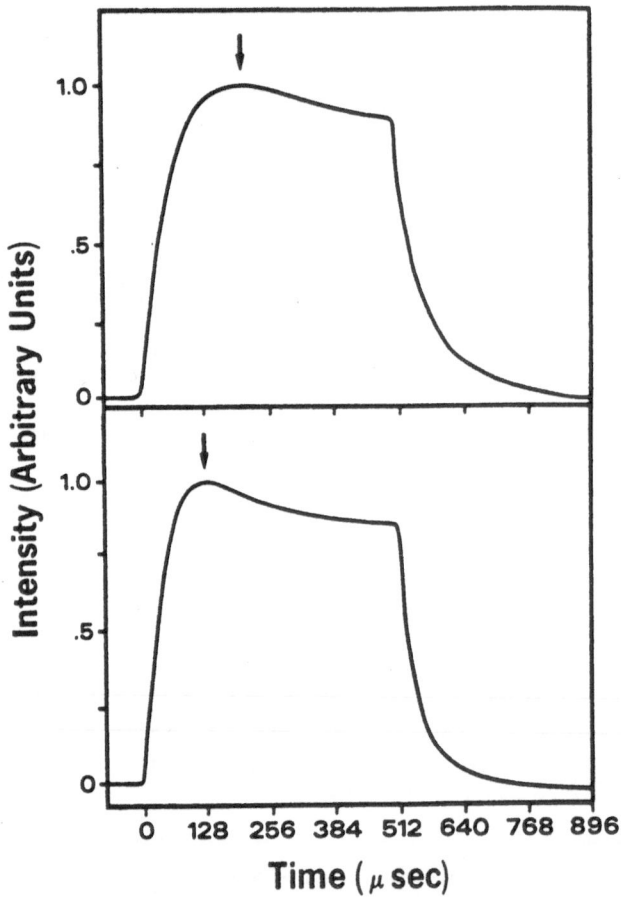

Figure 11: The birefringence excitation and decay curve for the 1010 bp DNA as a function of orienting field strength. The electric field is on from 0 to 512 μs. The top panel shows the birefringence both during and after an orienting pulse of 375 V/cm. The arrow points to the maximum in the birefringence, approximately 200 μs after the beginning of the pulse. The bottom panel shows the results with an orienting pulse of 1300 V/cm. With the higher field strength the maximum in the birefringence occurs at approximately 130 μs. The vertical scales are different for the two panels.

findings of Diekmann et al. [54] we have found the weights of the zero-field decay processes are dependent on the strength of the electric field [43]. This suggests that the dynamics that we are studying are easily distorted by our perturbation, i.e. the linear regime is quite limited. The second point is that the birefringence itself goes through a maximum even while the electric field remains constant. This is shown in Figure 11. In order to explain this observation one must postulate additional processes, such as motion of counterions or exchange of energy between various dynamic modes [43], that have not been included in the discussion up to this point. Much work remains to be done in this area. Until we have a better understanding of the orientation process, however, interpretation of electric birefringence data will remain somewhat tentative.

VI. Methods of Brownian Dynamics Simulations of the Trumbell and the 367 bp DNA

Brownian dynamics simulations of various models of macromolecular motion may be used to assess the degree to which a particular model can account for experimental observations. Building upon earlier work [29] Allison has recently performed Brownian dynamics simulations of a depolarized light scattering experiment on a short DNA fragment [30]. His results show excellent agreement with the bending theory of Barkley and Zimm [55]. This work has been extended [56] in order to verify some of the predictions of the Roitman-Zimm trumbell model and to model the 367 bp DNA that we have studied. The autocorrelation function from a simulated dynamic depolarized light scattering experiment should be directly comparable with the transient electric birefringence decay [56] if the electric-field orientation mechanism is simply due to the interaction of the electric field with a constant anisotropic polarizability of the DNA chain segments.

The trumbell that we have modeled consists of three beads of radius 43.25 Å, bond lengths of 544.27 Å, and a stiffness parameter Z, equal to 1.0. The solvent viscosity was assumed to be 1.0 centipoise and the temperature 20 °C. The effects of hydrodynamic interactions were ignored in the simulation. The simulated depolarized light scattering autocorrelation function was calculated and analyzed by CONTIN as above.

The simulated dynamic depolarized light scattering autocorrelation function was also calculated for a 10-bead model of the 367 bp DNA. The size and spacing of the beads and the stiffness of each of the bonds was chosen to give the appropriate persistence length, the correct average of the squared end-to-end distance, and a particular ensemble-averaged rotational diffusion tensor [56]. The values for the elements of the ensemble-averaged rotational diffusion tensor were determined by creating a Hagerman-Zimm type chain [19] with many more subunits. Again the calculated autocorrelation function was analyzed with CONTIN as above.

VII. Results and Discussion of Brownian Dynamics Simulations

The Roitman-Zimm trumbell model predicts that three separate decay processes contribute significantly to the birefringence decay of the trumbell that we have simulated.

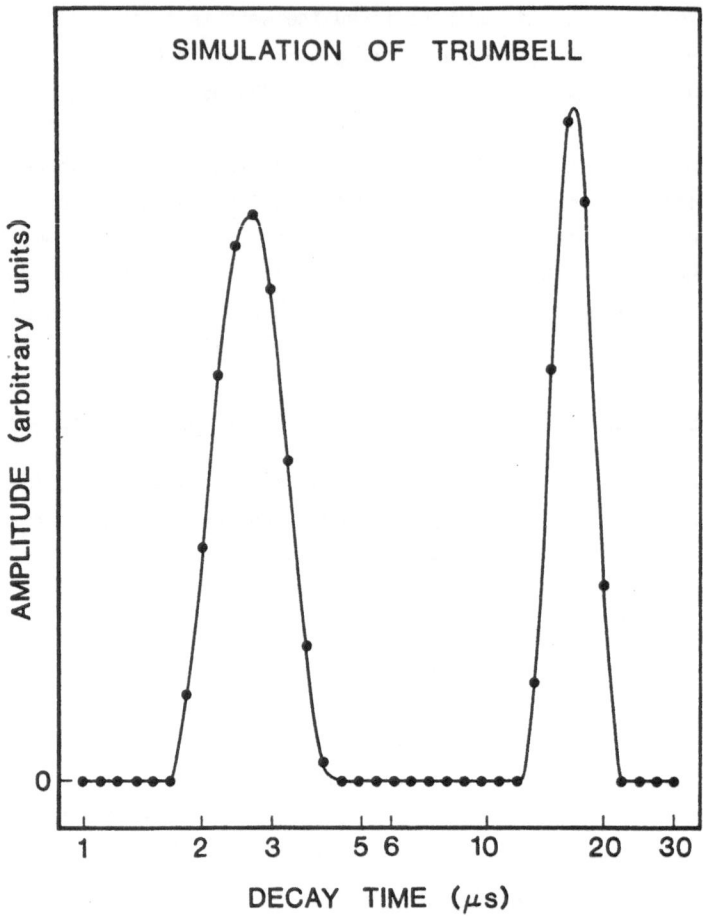

Figure 12: CONTIN analysis of a simulation of a trumbell in solution. The dynamic depolarized light scattering autocorrelation function was calculated from a dynamic simulation of a trumbell. The effects of hydrodynamic interactions between the three beads
were neglected. Two decay peaks are obtained. The slower, at 17.0 μs, constitutes 81%
of the decay while the faster, at 2.77 μs constitutes the other 19% of the decay. As
discussed in the text, this agrees well with the predictions of the Roitman-Zimm theory.

The slowest decay process is predicted to occur at 16.25 μs and accounts for 85% of the
decay, the next process should occur at 2.51 μs and constitutes 11% of the decay, and
the fastest significant process should occur at 2.25 μs and constitutes 2% of the decay.
The rest of the decay amplitude is contained in faster decay processes. The two decay
processes at 2.51 μs and 2.23 μs are probably too close together for us to resolve with
CONTIN and we would expect them to be detected as a single peak at 2.47 μs (the weighted average of the times) with a total amplitude of 13% of the decay. The results of the
CONTIN analysis of the simulated trumbell decay is shown in Figure 12. The slowest process occurs at 17.0 μs with an amplitude of 81% and the faster process at 2.77 μs with
an amplitude of 19%. This is in excellent agreement with the predictions of the Roitman-Zimm model.

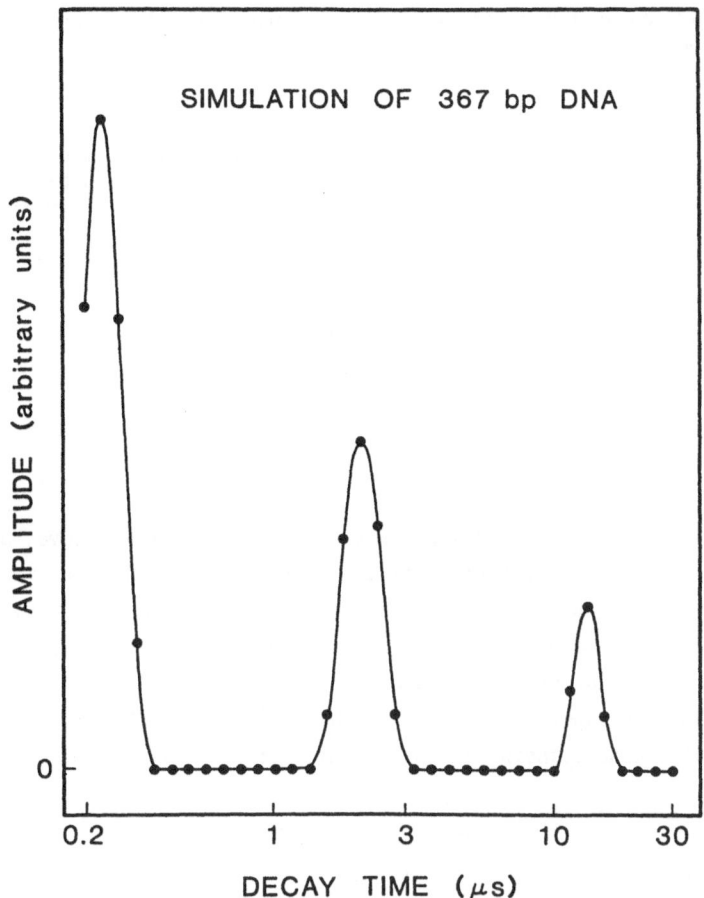

Figure 13: CONTIN analysis of a simulation of the 367 bp DNA in solution. The dynamic depolarized light scattering autocorrelation function was calculated from a dynamic simulation of a ten bead model of the 367 bp DNA. The effect of hydrodynamic interactions between the beads was included, although the interaction was preaveraged, and the persistence length was assumed to be 600 Å. Three decay processes are detected. The slowest decay occurs at 14.4 μs and presumably represents primarily rotational motion. The next faster decay occurs at 2.19 μs and the fastest decay occurs at 0.25 μs. Comparison with Figure 3 shows good agreement between experiment and simulation. Results of other simulations are given in Table IV.

Dynamic depolarized light scattering experiments were simulated for 10-bead models of the 367 bp DNA, assuming three different persistence lengths, 400 Å, 600 Å, and 900 Å. A CONTIN analysis of the 600 Å persistence length data is shown in Figure 13. Three peaks are resolved. The slowest occurs at 14.4 μs, the next at 2.19 μs, and the fastest at 0.25 μs. This is in excellent agreement with the experimental results shown in Figure 3. Table IV shows how the results of the simulations vary with the assigned persistence length of the chain. As expected, the slowest, presumably rotational, decay time decreases with decreasing persistence length. When the persistence length is increased to 900 Å the dynamics approach more closely that of a rigid rod and we detect only two decay times. The first internal mode is relatively fast, as would be expected for a quite rigid molecule.

Table IV: Comparison of TEB Decay Times and Computer Simulation for the 367 bp DNA [56]

Method	Rotational Decay Time, μs	First Internal Decay Time, μs	Second Internal Decay Time, μs
Experiment	14.4	2.3	< 1.0
Simulations:			
PL = 400	9.5	2.15	0.24
PL = 600	14.4	2.19	0.25
PL = 900	15.5	0.53	----

The Brownian Dynamics simulations, whose results are shown above, demonstrate two points. First of all, by direct verification of the trumbell theory we have checked the trumbell theory, the dynamics simulation software, and the data analysis method. Secondly, the fact that we are able to simulate the dynamics of the 367 bp DNA and duplicate with reasonable accuracy the decay processes demonstrated in the transient electric birefringence experiment, suggests that we may have the basic elements necessary for a detailed understanding of the dynamics of semistiff chains at hand.

VIII. Conclusions

We have reviewed our work using monodisperse DNA restriction fragments as model systems for the dynamics of semistiff polymer chains. This work is only a beginning. Much work remains to be done. For instance, one of the difficulties in the transient birefringence experiments arises from a lack of knowledge of the orientation mechanism of DNA in an electric field. Thus, since it is not known which modes are excited, it is difficult to quantitatively interpret the birefringence decay curve. Depolarized and polarized dynamic light scattering experiments, which are currently in progress in our laboratory, should provide data which is free of this uncertainty. In addition, it is clear that Brownian dynamics simulations of increasingly realistic models of macromolecules, performed in close conjunction with the experiments, are invaluable tools in elucidating the dynamics of these complex systems.

IX. Acknowledgments

The use of the electric birefringence apparatus in the laboratory of Professor Don Eden of San Francisco State University is gratefully acknowledged. All of the dynamics simulations were performed by Professor Stuart Allison at Georgia State University. Work in our laboratory was supported by NSF grant CHE-8511178, a grant to the Center for

Materials Research at Stanford University, and, in its early stage, by NIH grant 2R01 GM 22517 to R.P. R.J.L. was supported by the NIH Medical Scientist Training Program at the Stanford University School of Medicine.

References

[1] Riseman, J. and Kirkwood, J.G.: J. Chem. Phys. 18, 512 (1950)

[2] Broersma, S.; J. Chem. Phys. 32, 1626 (1960)

[3] Broersma, S.; J. Chem. Phys. 32, 1632 (1960)

[4] Broersma, S.; J. Chem. Phys. 74, 6989 (1981)

[5] Tirado, M.M. and Garcia de la Torre, J.; J. Chem. Phys. 71, 2581 (1979)

[6] Tirado, M.M. and Garcia de la Torre, J.; J. Chem. Phys. 73, 1986 (1980)

[7] Maeda, H. and Saito, N.; J. Phys. Soc. Jpn. 27, 984 (1969); Polymer J. 4, 309 (1973)

[8] Gierke, T.D., Ph.D. Thesis, Department of Chemistry, University of Illinois, 1974

[9] Rallison, J.M. and Leal, L.G.; J. Chem. Phys. 74, 4819 (1981)

[10] Aragon, S.R. and Pecora, R.; J. Chem. Phys. 82 5346 (1985)

[11] Wilcoxon, J. and Schurr, J.M.; Biopolymers 22, 849 (1983)

[12] Kubota, K., Urabe, H., Tominaga, Y., and Fujime, S.; Macromolecules 17, 2096 (1984)

[13] Zimm, B.H.; J. Chem. Phys. 24, 269 (1956)

[14] Zimm, B.H., Roe, G.M. and Epstein, L.F.; J. Chem. Phys. 24, 279 (1956)

[15] Schaefer, D.W. and Han, C.C. in Dynamic Light Scattering: Applications of Photon Correlation Spectroscopy, R. Pecora (Ed.), Plenum, New York, 1985

[16] Kratkey, O., and Porod, G.; Rec. Trav. Chim. 68, 1106 (1949)

[17] Hearst, J.E. and Stockmayer, W.H.; J. Chem. Phys. 37, 1425 (1962)

[18] Yamakawa, H. and Fujii, M.; Macromolecules 6, 407 (1973)

[19] Hagerman, P.J. and Zimm, B.H.; Biopolymers 20, 1481 (1981)

[20] Hagerman, P.J.; Biopolymers 20, 1503 (1981)

[21] Harris, R.A. and Hearst, J.E.; J. Chem. Phys. 44, 2595 (1966)

[22] Soda, K.; J. Phys. Soc. Jpn. 35, 866 (1973)

[23] Aragon, S.R. and Pecora, R.; Macromolecules 18, 1868 (1985)

[24] Aragon, S.R.; Macromolecules, 20, 370 (1987)

[25] Roitman, D.B. and Zimm, B.H.; J. Chem. Phys. 81, 6333 (1984)

[26] Roitman, D.B. and Zimm, B.H.; J. Chem. Phys. 81, 6348 (1984)

[27] Roitman, D.B.; J. Chem. Phys. 81, 6356 (1984)

[28] Nagasaka, K. and Yamakawa, H.; J. Chem. Phys. 83, 6480 (1985)

[29] Allison, S.A. and McCammon, J.A.; Biopolymers 23, 363 (1984)

[30] Allison, S.A.; Macromolecules 19, 118 (1986)

[31] Maniatis, T., Fritsch, E.F. and Sambrook, J.; Molecular Cloning, A Laboratory Manual; Cold Spring Harbor Laboratory; Cold Spring Harbor, New York, 1982

[32] Lewis, R.J., Huang, J.H. and Pecora, R.; Macromolecules 18, 1530 (1985)

[33] Dickerson, R.E.; Sci. Am. 249, 94 (1983)

[34] Harrington, R.E.; Biopolymers 17, 919 (1978)

[35] Cairney, K.L. and Harrington, R.E.; Biopolymers 21, 923 (1982)

[36] Borochov, N. Eisenberg, H. and Kam, Z.; Biopolymers 20, 231 (1981)

[37] Kam, Z., Borochov, N. and Eisenberg, H.; Biopolymers 20, 2671 (1981)

[38] Rizzo, V. and Schellman, J.; Biopolymers 20, 2143 (1981)

[39] Marko, M.A., Chipperfield, R. and Birnboim, H.C.; Anal. Biochem. 121, 382 (1982)

[40] Eden, D. and Elias, J.G.; in Measurement of Suspended Particles by Quasi-Elastic Light Scattering, B.E. Dahneke (Ed.), Wiley: New York, 1983

[41] Highsmith, S. and Eden, D.; Biochemistry 24, 4917 (1985)

[42] Lewis, R.J., Pecora, R. and Eden, D.; Macromolecules 19, 134 (1986)

[43] Lewis, R.J., Pecora, R., and Eden, D.; Macromolecules, in press

[44] Provencher, S.W.; Biophysics Journal 16, 27 (1976)

[45] Provencher, S.W.; J. Chem. Phys. 64, 2772 (1976)

[46] Provencher, S.W., Hendrix, J., De Maeyer, L. and Paulussen, N.; J. Chem. Phys. 69, 4273 (1978)

[47] Provencher, S.W.; Makromol. Chem. 180, 201 (1979)

[48] Provencher, S.W.; "CONTIN User's Manual"; European Molecular Biology Laboratory Technical Report EMBL-DA02, Heidelberg, 1980

[49] Lewis, R.J.; Ph.D. Thesis, Stanford University, Stanford, CA (1985)

[50] Lewis, R.J. and Pecora, R.; Macromolecules 19, 2074 (1986)

[51] Hassager, O.; J. Chem. Phys. 60, 2111 (1974)

[52] Hassager, O.; J. Chem. Phys. 60, 4001 (1974)

[53] Stellwagen, N.C.; Biopolymers 20, 399 (1981)

[54] Diekmann, S. Hillen, W., Morgeneyer, B., Wells, R.D. and Porschke, D.; Biophys. Chem. 15, 263 (1982)

[55] Barkley, M.D. and Zimm, B.H.; J. Chem. Phys. 70, 2991 (1979)

[56] Lewis, R.J., Allison, S.A., Pecora, R. and Eden, D.; in preparation

SLOW AND ULTRASLOW ROTATIONAL MOTIONS OF MACROMOLECULES IN THE
VICINITY OF THE GLASS TRANSITION AND IN LIQUID CRYSTALLINE
POLYMERS AS REVEALED BY PULSED DEUTERON NMR

H.W. Spiess
Max-Planck-Institut für Polymerforschung
Postfach 3148, D-6500 MAINZ, FRG

Introduction

The detailed characterization of molecular mobility should provide the basis for the
understanding of the macroscopic behaviour of materials on a molecular level. Besides
scattering methods nuclear magnetic resonance (NMR) has recently proven to yield especially clear-cut information on slow and ultraslow molecular motions in polymers
and related materials. The reason is that once a highly resolved solid-state NMR
spectrum is generated, the orientation of specific bond directions can accurately be
monitored via the corresponding frequencies in the spectrum. By correlating the NMR
signals following pulsed irradiation with time delays between the rf-pulses molecular
motions can be monitored over an extraordinary wide range of correlation times.
Pulsed deuteron (^{2}H) NMR techniques as well as their applications to polymers have
been reviewed already [1-3], therefore the main features only will be discussed here.

Pulsed Deuteron NMR

The main advantages of pulsed ^{2}H-NMR can be summarized as follows [1-3]: Deuterons
represent well-defined nuclear spin lables since they monitor the orientation of individual C-H bond directions by virtue of their quadrupole coupling. Different motional
mechanisms can clearly be distinguished because they lead to often vastly different
NMR spectra. In particular, highly restricted motions lead to broad ^{2}H-NMR spectra,
narrow spectra on the other hand are expected for highly flexible chains.

The dynamic range over which polymer mobility can be studied is extremely high
(10^{10} Hz - 10^{-2} Hz). This is achieved by exploiting the NMR signals following different pulse sequences and in addition analyzing angular dependent spin-lattice relaxation.

Motional heterogeneity can be detected not only in semi-crystalline systems but also
in amorphous polymers. From a line shape analysis the distribution function of correlation times can be obtained and by combination with spin-lattice relaxation *homogeneous* and *heterogeneous* distributions can clearly be distinguished [4].

Examples

Ultraslow motions in the Vicinity of the Glass Transition

Deuteron spin alignment [5] is particularly suited to characterize the ultraslow motions associated with the glass transition of amorphous polymers. For the polystyrene backbone a diffusive type of reorientation by small angle motions has been detected [6]. Slightly above the glass transition this motion corresponds to correlation time τ between 1h and 1s and thus shows the typical WLF behaviour of macroscopic dynamic properties [7].

Thermotropic Liquid Crystalline Polymers

Due to its selectivity deuteron NMR is particularly useful for studying the molecular mobility of the different building blocks of liquid crystalline polymers [8-9]. In a particular side group system involving phenylbenzoate as a mesogen the molecular dynamics of the mesogenic side groups is largely frozen at the glass transition, a 180° flip-motion of phenylene rings, however, is observed in the glassy state [8-9]. Remarkably the molecular motion cannot be described by a single correlation time. Heterogeneous mobility, characteristic of amorphous polymers is observed also in these highly orientationally ordered glassy systems [9].

Polymer Model Membranes

Deuteron NMR has been used to characterize in detail the restrictions on the molecular mobility of a model membrane which is stabilized by polymerization in its hydrophylic part [10-11]. Although the polymer model membrane exhibits a phase transition into a liquid crystalline phase as revealed by DSC, the ^{2}H-NMR study shows that the mobility at the polymer chain, the spacer which should decouple the head group from the polymer chain, and the head group itself is largely hindered. In particular the motional freedom measured via the number of conformations accessible for a specific segment of the monomer unit is drastically reduced by the polymerization. On the other hand, the mobility of the lipid chains in the hydrophobic part is much less affected, which is important for biological applications.

New Developments

The developments of solid state NMR techniques for studying molecular structure and mobility in polymers has by no means come to an end yet. Therefore we would like to mention some specific developments presently persued in our laboratory:

Analogies Between Quasielastic neutron Scattering and Deuteron Spin Alignment [12]

We have recently shown that quasielastic neutron scattering and deuteron spin alignment yield largely equivalent information about single particle reorientations, although on different time scales. Both techniques can discriminate between different motional mechanisms by systematically varying a phase factor. The two methods nicely supplement each other, where deuteron spin alignment is used to monitor slow motions with correlation times $\tau_c < 10^{-4}$ s, whereas neutron scattering can detect fast motions with $\tau_c < 10^{-8}$ s.

Two-Dimensional NMR Techniques for Studying Molecular Motions [13]

Although different motional mechanisms can be distinguished through different NMR observables, e.g. line shapes as a function of pulse spacings, etc., specific motional models have to be used for data analysis. In unfavorable cases ambiguities may exist. If the pulse spacings are varied systematically, however, a two-dimensional exchange spectrum [14] can be generated. We have recently shown that the jump angles resulting from rotational motions can directly be projected into such 2D-exchange spectra of ^{2}H-spin alignment or ^{13}C-stimulated echo spectra. Thus the jump angles are detected without the need for interfacing a model. This technique will be able, e.g., to uniquely specify conformational transitions in solid polymers and the complex molecular mobility in the vicinity of the glass transition.

References

[1]   H.W. Spiess, Colloid & Polymer Sci. 261, 193-209 (1984)

[2]   H.W. Spiess, Advan. Polym. Sci. 66, 23-58 (1985)

[3]   H.W. Spiess, Pure & Appl. Chem. 57, 1617-1626 (1985)

[4]   C. Schmidt, K.J. Kuhn and H.W. Spiess, Progr. Colloid & Polymer Sci. 71, 71-76 (1985)

[5]   H.W. Spiess, J. Chem. Phys. 72, 6755-6762 (1980)

[6]   S. Wefing, F. Fujara, H. Sillescu and H.W. Spiess, to be publised

[7]   J.D. Ferry, Viscoelastic Properties of Solid Polymers, J. Wiley, London (1980)

[8]   C. Boeffel, B. Hisgen, U. Pschorn, H. Ringsdorf and H.W. Spiess, Israel Journal of Chemistry 23, 388-394 (1983)

[9]   U. Pschorn, H.W. Spiess, B. Hisgen and H. Ringsdorf, Makromol. Chem. 187, 2711-2723 (1986)

[10]   R. Ebelhäuser and H.W. Spiess, Makromol. Chem. Rapid Comm. 5, 403-411 (1984)

[11]   R. Ebelhäuser and H.W. Spiess, Ber. Bunsenges. Phys. Chem. 89, 1208-1214 (1985)

[12]   F. Fujara, S. Wefing and H.W. Spiess, J. Chem. Phys. 84, 4579-4584 (1986)

[13]   C. Schmidt, S. Wefing, B. Blümich and H.W. Spiess, Chem. Phys. Letters 130, 84-90 (1986)

[14]   R.R. Ernst, G. Bodenhausen and A. Wokaun, Principles of Nuclear Magnetic Resonance in One and Two Dimensions, Oxford Univ. Press (1987)

Lecture Notes in Mathematics

Lecture Notes in Physics